Advanced Processing, Properties, and Applications of Starch and Other Bio-Based Polymers

Advanced Processing, Properties, and Applications of Starch and Other Bio-Based Polymers

Edited by

FARIS M. AL-OQLA
Department of Mechanical Engineering
Faculty of Engineering
The Hashemite University
Zarqa, Jordan

S.M. SAPUAN
Advanced Engineering Materials and Composites
Research Centre (AEMC)
Department of Mechanical and Manufacturing Engineering
Universiti Putra Malaysia
Serdang, Selangor, Malaysia

Laboratory of Biocomposite Technology
Institute of Tropical Forestry and Forest Products
Universiti Putra Malaysia
Serdang, Selangor, Malaysia

ELSEVIER

Elsevier
Radarweg 29, PO Box 211, 1000 AE Amsterdam, Netherlands
The Boulevard, Langford Lane, Kidlington, Oxford OX5 1GB, United Kingdom
50 Hampshire Street, 5th Floor, Cambridge, MA 02139, United States

Library of Congress Cataloging-in-Publication Data
A catalog record for this book is available from the Library of Congress

British Library Cataloguing-in-Publication Data
A catalogue record for this book is available from the British Library

ISBN: 978-0-12-819661-8

For information on all Elsevier publications visit our website at
https://www.elsevier.com/books-and-journals

Publisher: Matthew Deans
Acquisitions Editor: Edward Payne
Editorial Project Manager: Charlotte Rowley
Production Project Manager: Kiruthika Govindaraju
Cover Designer: Alan Studholme

Typeset by TNQ Technologies

List of Contributors

Hairul Abral
Department of Mechanical Engineering
Andalas University
Padang, Sumatera Barat, Indonesia

H.A. Aisyah
Laboratory of Biocomposite Technology
Institute of Tropical Forestry and Forest Products
Universiti Putra Malaysia
Serdang, Selangor, Malaysia

Amani M. Al-Ghraibah
Al-Ahliyya Amman University
Amman, Jordan

Faris M. AL-Oqla
Department of Mechanical Engineering
Faculty of Engineering
The Hashemite University
Zarqa, Jordan

Maha Al-Qudah
Primary Health Care Corporation (PHCC)
Doha, Qatar

Mochamad Asrofi
Laboratory of Material Testing
Department of Mechanical Engineering
University of Jember
Jember, East Java, Indonesia

M.R.M. Asyraf
Department of Aerospace Engineering
Universiti Putra Malaysia
Serdang, Selangor, Malaysia

M.S.N. Atikah
Department of Chemical and Environmental Engineering
Universiti Putra Malaysia
Serdang, Selangor, Malaysia

M.N.M. Azlin
Institute of Tropical Forestry and Forest Products
Universiti Putra Malaysia
Serdang, Selangor, Malaysia

School of Industrial Technology
Department of Textile Technology
Universiti Teknologi
MARA Negeri Sembilan
Kuala Pilah Campus
Kuala Pilah, Negeri Sembilan, Malaysia

Manik Chandra Biswas
Doctoral Fellow
Fiber and Polymer Science
Textile Engineering
Chemistry and Science
NC State University
Raleigh, NC, United States

Ahmed Edhirej
Advanced Engineering Materials and Composites Research Centre
Department of Mechanical and Manufacturing Engineering
Universiti Putra Malaysia
Serdang, Selangor, Malaysia

Department of Mechanical and Manufacturing Engineering
Sabha University
Sabha, Libya

Mohd Nor Faiz Norrrahim
Research Centre for Chemical Defence (CHEMDEF)
Universiti Pertahanan Nasional Malaysia
Kuala Lumpur, Malaysia

Osama O. Fares
Electrical Engineering Department
Isra University
Amman, Jordan

M.D. Hazrol
Advanced Engineering Materials and Composites Research Centre (AEMC)
Department of Mechanical and Manufacturing Engineering,
Universiti Putra Malaysia
Serdang, Selangor, Malaysia

Md Enamul Hoque
Department of Biomedical Engineering
Military Institute of Science and Technology (MIST), Mirpur Cantonment, Dhaka, Bangladesh

M.R.M. Huzaifah
Laboratory of Biocomposite Technology
Institute of Tropical Forestry and Forest Products
Universiti Putra Malaysia
Serdang, Selangor, Malaysia

M.I.J. Ibrahim
Advanced Engineering Materials and Composites Research Centre
Department of Mechanical and Manufacturing Engineering
Universiti Putra Malaysia
Serdang, Selangor, Malaysia

Department of Mechanical and Manufacturing Engineering
Sabha University
Sabha, Libya

Laboratory of Biocomposite Technology
Institute of Tropical Forestry and Forest Products
Universiti Putra Malaysia
Serdang, Selangor, Malaysia

Rushdan Ibrahim
Pulp and Paper Branch
Forest Research Institute Malaysia
Kepong, Selangor, Malaysia

R.A. Ilyas
Biocomposite Technology & Design
Advanced Engineering Materials and Composites Research Centre (AEMC)
Department of Mechanical and Manufacturing Engineering
Universiti Putra Malaysia (UPM)
Serdang, Selangor, Malaysia

Laboratory of Biocomposite Technology
Institute of Tropical Forestry and Forest Products
Universiti Putra Malaysia
Serdang, Selangor, Malaysia

Latifah Jasmani
Pulp and Paper Branch
Forest Research Institute Malaysia
Kepong, Selangor, Malaysia

Ridhwan Jumaidin
Fakulti Teknologi Kejuruteraan Mekanikal dan Pembuatan
Universiti Teknikal Malaysia Melaka
Durian Tunggal, Melaka, Malaysia

Abudukeremu Kadier
Department of Chemical and Process Engineering
Faculty of Engineering and Built Environment
National University of Malaysia (UKM)
Bangi, Selangor, Malaysia

Research Centre for Sustainable Process Technology (CESPRO)
Faculty of Engineering and Built Environment
National University of Malaysia (UKM)
Bangi, Selangor, Malaysia

Mohd Sahaid Kalil
Department of Chemical and Process Engineering
Faculty of Engineering and Built Environment
National University of Malaysia (UKM)
Bangi, Selangor, Malaysia

Research Centre for Sustainable Process Technology (CESPRO)
Faculty of Engineering and Built Environment
National University of Malaysia (UKM)
Bangi, Selangor, Malaysia

A. Khalina
Institute of Tropical Forestry and Forest Products
Universiti Putra Malaysia
Serdang, Selangor, Malaysia

Santhana Krishnan
Centre of Environmental Sustainability and Water Security (IPASA)
Research Institute of Sustainable Environment (RISE)
School of Civil Engineering
Faculty of Engineering
Universiti Teknologi Malaysia (UTM)
Johor Bahru, Johor, Malaysia

C.H. Lee
Institute of Tropical Forestry and Forest Products
Universiti Putra Malaysia
Serdang, Selangor, Malaysia
INTROP
Universiti Putra Malaysia
Serdang, Selangor, Malaysia

S.H. Lee
Institute of Tropical Forestry and Forest Products
Universiti Putra Malaysia
Serdang, Selangor, Malaysia

Tariq Mahbub
Department of Mechanical Engineering
Military Institute of Science and Technology (MIST), Mirpur Cantonment, Dhaka, Bangladesh

Sushmita Majumder
Department of Materials and Metallurgical Engineering
Bangladesh University of Engineering and Technology (BUET), Mirpur Cantonment, Dhaka, Bangladesh

S. Misri
Laboratory of Biocomposite Technology
Institute of Tropical Forestry and Forest Products
Universiti Putra Malaysia
Serdang, Selangor, Malaysia

N. Mohd Nurazzi
Board of Technologists (MBOT)
Futurise
Persiaran APEC
Cyberjaya, Selangor, Malaysia

Siti Nuraishah Mohd Zainel
Fakulti Teknologi Kejuruteraan Mekanikal dan Pembuatan
Universiti Teknikal Malaysia Melaka
Durian Tunggal, Melaka, Malaysia

A. Nazrin
Laboratory of Biocomposite Technology
Institute of Tropical Forestry and Forest Products
Universiti Putra Malaysia
Serdang, Selangor, Malaysia

M.N.F. Norrahim
Research Centre for Chemical Defence (CHEMDEF)
Universiti Pertahanan Nasional Malaysia
Sungai Besi, Kuala Lumpur, Malaysia

Mahmoud M. Rababah
Department of Mechanical Engineering
The Hashemite University
Zarqa, Jordan

Tilottoma Saha
Department of Materials and Metallurgical Engineering
Bangladesh University of Engineering and Technology (BUET)
Dhaka, Bangladesh

S.M. Sapuan
Advanced Engineering Materials and Composites Research Centre (AEMC)
Department of Mechanical and Manufacturing Engineering
Universiti Putra Malaysia
Serdang, Selangor, Malaysia
Laboratory of Biocomposite Technology
Institute of Tropical Forestry and Forest Products
Universiti Putra Malaysia
Serdang, Selangor, Malaysia

Nasmi Herlina Sari
Department of Mechanical Engineering
Mataram University
Mataram, West Nusa Tenggara, Indonesia

Ahmed Sharif
Department of Materials and Metallurgical Engineering
Bangladesh University of Engineering and Technology (BUET)
Dhaka, Bangladesh

R. Syafiq
Laboratory of Biocomposite Technology
Institute of Tropical Forestry and Forest Products
Universiti Putra Malaysia
Serdang, Selangor, Malaysia

Edi Syafri
Department of Agricultural Technology
Politeknik Pertanian
Payakumbuh, Sumatra, Indonesia

Jarin Tusnim
Department of Materials and Metallurgical Engineering
Bangladesh University of Engineering and Technology (BUET)
Dhaka, Bangladesh

Tengku Arisyah Tengku Yasim-Anuar
Department of Bioprocess Technology
Faculty of Biotechnology and Biomolecular Sciences
Universiti Putra Malaysia
Serdang, Selangor, Malaysia

E.S. Zainudin
Advanced Engineering Materials and Composites Research Centre (AEMC)
Department of Mechanical and Manufacturing Engineering
Faculty of Engineering
Universiti Putra Malaysia
Serdang, Selangor, Malaysia

M.Y.M. Zuhri
Advanced Engineering Materials and Composites Research Centre (AEMC)
Department of Mechanical and Manufacturing Engineering
Faculty of Engineering
Universiti Putra Malaysia
Serdang, Selangor, Malaysia

Contents

CHAPTER 1

Biopolymer Composites and Sustainability

MAHMOUD M. RABABAH • FARIS M. AL-OQLA

1 INTRODUCTION

It is very obvious that plastic pollution has negative impact on the environment as well as the climate change. Unfortunately, the pollution occurred along the whole production cycle of the plastic from its production using the fossil fuels until its disposal by burning. Beside the plastic pollution, deforestation, greenhouse effect, industrial pollution, and many other factors are responsible in causing the negative impact on the environment by blowing more gases to air as carbon dioxide, methane, SO_2, nitrous oxide, and many others (AL-Oqla et al., 2016; Alaaeddin et al., 2019b).

Greener plastic composites can be obtained of renewable resources in a more ecological responsible manner. This is achieved using the biotechnology and was improved using the nanotechnology, which is a promising approach that would greatly affect the value chains of the plastic industry worldwide. Some steps are already achieved in developing sustainable plastics. In fact, photodegradable plastics with a balance amount of both antioxidants and catalysts are developed. The catalysts initiate a controlled degradation while maintaining the performance properties of the plastics. These photodegradable plastics are possessing similar performance properties to conventional plastics at close costs. However, at the moment, they still use fossil fuels and they are not able to fully degrade to H_2O and CO_2 in the soil (Khabbaz et al., 1999). Besides, photofragmentation may occur if no control is performed causing a litter increase. Degradable polymers are developed without using antioxidant or with prooxidants that help in a slow degradation. Comparing to photodegradable polymers, degradable polymers possess similar performance properties, cost structure, and production of other degradation products than H_2O and CO_2, such as alcohols, alkenes, esters, and ketones (Jakubowicz, 2003). Therefore, developing sustainable plastics from biodegradable and renewable resources is a demanding goal.

An amount of 260 billion bounds of plastic were annually produced in the world at the end of the last century, with an industry value of 1 trillion dollars (Halley and Dorgan, 2011). This amount is subjected to a massive increase because of the high demanding due to the population increase and the new developed consumers' habits. Great amount of petroleum is consumed for plastic production. However, as it is finite supply, its prices will increase more and more. In addition, the environmental pollution caused from producing, using, and disposing of plastic materials is of a great concern due to greenhouse gases and the global warming effects. The decaying of world reserve from petroleum and the increasing demands from developing countries such as China and India are both cause the prices of oil to reach unprecedented levels. These high prices drive a similar increase in petroleum-based plastics. This leads for mining of lower-grade crude oil such as the Canadian heavy oil (Deffeyes, 2008). The heavy oil is less economical and more environmentally harmful than the light oil. However, plastics can be of a great assist to humanity by increasing the agricultural production, decreasing the food loses, reducing the fuel consumption, offering lighter and cheaper alternatives for many products, improving the healthcare, etc. In other words, plastic materials are essential in our modern societies. Unfortunately, the energy issues directly impact the plastics industries. What will be the impact on our daily lives, our health, our environment, and on the plastic industry itself (more than 1 million employees in the United States alone) if the sustainable technologies do not reach maturity so soon or if they are not widely adopted? Developing appropriate methods and approaches for producing green composites has been a demanding priority for some time. However, the evolving economical and technical

Advanced Processing, Properties, and Applications of Starch and Other Bio-Based Polymers. https://doi.org/10.1016/B978-0-12-819661-8.00001-9

problem limits pursuing such approaches on large scales (AL-Oqla and Sapuan, 2014a,c). Even though the need of developing bioplastic and biocomposite materials is demanding, such materials must first be cost competitive.

2 PLASTIC

2.1 Origin of Plastics

The polyethylene polymer used in plastic bags production is derived from petroleum. Petroleum is a complex mixture of carbon and hydrogen compounds with heavy metals as nickel and vanadium and other components as sulfur. As petroleum contains high concentration of chlorinated hydrocarbons as well as heavy metals, it is toxic to animals and plants.

The process of extraction petroleum is composed of four stages: first, crude oil is obtained by deep drilling from natural reservoirs below the sea or offshore. This crude oil is shipped to the refineries. In the second stage, the crude oil is separated by evaporation/condensation process at different temperatures. In the third stage, the compounds are yielded to conversion process in the presence of heat, pressure, and catalyst (for instance, platinum). In this process, the shape of the compounds and its molecular weight are changed. The compounds obtained from the process serves as fuel to automobiles, factories, etc. Some of these compounds can be delivered to factories for upgrading (last stage) to produce fertilizers and different plastic products. For instance, ethylene is upgraded from these refined compounds and used for plastic products. Ethylene is explosive, inflammable, toxic, and carcinogenic.

Extraction petroleum stages: Drilling → Separation → Conversion → Upgrading

During the refining stages of the crude oil to be converted to fuel, plastics, and other petro-based products, many types of gases are emitted to air. These gases are carrying harmful components like carbon monoxide, hydrocarbons, sulfur dioxide, and nitrous oxide. Unfortunately, these components remain in the final petroleum compounds after the separation process. Their effects on our environment and on the ecosystems are catastrophic and can lead to acid rains, and unfortunately, these effects are irreversible.

2.2 Applications of Plastics

Plastics become more and more very essential in our modern societies. It is used in a wide variety of applications, such as in packaging, automobile industry, aerospace, agriculture, and household products, etc. Its availability, flexibility, durability, lightweight, and most important its cheap prices help plastics to dominate a great portion of the current production markets. Plastics are mainly categorized into two main groups: thermoplastics and thermosets. Thermoplastics can in general be melted and recycled, some examples of thermoplastic materials are PE, PP, PS, polyethylene terephthalate (PET), and polyamide. On the other side, thermosets have can neither be melted nor be recycled. This is because the polymer chains for these plastics are connected in strong cross-link bonds as the case in epoxy resin, polyurethane, and unsaturated polyester. As the petro-based plastics do not degrade, they cause pollution. The solution to this is to develop and use biodegradable bioplastics as alternatives to the conventional plastics. These bioplastics will require shorter time to decompose after been disposed. Also, they can fertilize the soil in the composting process, where they can be mix with soil in order to degrade by the help of bacteria. The life of the biodegradable bioplastics begins from renewable resource, such as cellulose and starch, and ends eco-friendly comparing with petro-based plastics.

2.3 Chemical Pollution from Plastics

Great portion of the chemical pollution occurs during energy generation; such energy is used for generating electricity, mining industry, transportation, etc. The pollution strikes our environment through global warming, acid rains, and through producing carcinogenic substances in air and all around. Tons of harmful gases are blown in the atmosphere in most of the industrial sectors as coal mining, uranium processing, petrochemical industry, etc. The pollution is expected to increase with the increase population, the developed consumption habits of people, and the growing and the spreading of technology. Direct burning of plastics in air spreads very harmful and toxic gases into the atmosphere. Some of these gases are alkanes, alkenes, and chlorinated and aromatic hydrocarbons (PAHs, PCDDs, and PCDFs). In fact, huge amount of gases are accumulated in ecosystem. Gases like carbon monoxide, nitrogen oxide, and volatile particulates when accumulated in atmosphere form dark smog. In the extrusion processes where the melting temperature is reached (between 150 and 300°), many gases are leaked to the ecosystem due to the attraction between polymers and the additives, long thermal exposure, or aging.

2.4 Initiatives Against Plastics Pollution

Some people who are aware about the harmful impact of using conventional plastics in our daily life are trying to raise the awareness of their societies by distributing eco-friendly bags fabricated from natural fibers (AL-Oqla and Salit, 2017; AL-Oqla and Sapuan, 2018a; Alaaeddin et al., 2019c). United Kingdom took a pioneer step and prevented using microplastics in personal care products as well as in cosmetics. This step is taken, according to the Department of Environment, Food and Rural Affairs, since the microplastics are more harmful than the large-size plastics. As the microplastics are infinitesimal, they cannot be removed or separated from the ecosystem easily, and thus, they can easily reach the food chain. A student named Jonsson from Iceland introduced new bottles by mixing powdered agar with water. These bottles help in developing sustainable environment and replace the conventional bottles made normally from plastics. Some countries such as Sri Lanka started to enforce the plastic manufacturers to produce their products with certain standards. Some cities from all over the world are happily declaring that their cities are plastic-free zones. Most of these cities started by banning the use of plastic bags and announced a deadline for their use.

3 RENEWABLE-BASED PLASTICS

In the past, renewable-based plastics were lacking the sufficient properties for many applications beside their high costs. Nowadays, great successes are being achieved in new bioplastic martials commercially. These achievements are directly reflected from the successful applications of the industrial biotechnology. Examples include Mirel (polyhydroxyalkanoates polymers) for injection-molded products, polylactides for flexible films, and compost bags (AL-Oqla et al., 2018a; AL-Oqla et al., 2015c; AL-Oqla and Sapuan, 2014b). Other examples are the soy oil-based materials that are produced on commercial scale by number of companies in order to be used in urethanes industries (AL-Oqla et al., 2018b; Fares et al., 2019).

Moreover, many global companies are working on commercializing further renewable-based plastics as PBS (polybutylene succinate) for use as flexible films in agricultural and packaging applications. Besides, many of the soft drink manufacturers have converted to utilize PET from renewable resources. All these cumulative successes strongly encourage in replacing the petroleum plastics produced in billions of pounds annually. Fortunately, generating renewable-based plastics are becoming more and more possible. Better understanding of the impacts of plastics on the environmental is now possible through the life cycle life-cycle analysis (LCA) of these plastics. In other words, the LCA studies the overall impact of the material along its life (Black et al., 2011). Improvement strategies for the processes are also introduced and analyzed for their impacts. For example, planting crops with low water and fertilizer needs in marginal regions can reduce deforestation as well as the pressures on food supplies. Similarly, the fast and wide developing steps in the biotechnological industry are facilitating the conversion of the biomass into fuel or useful chemicals and are making it very feasible. Finally, nanotechnology is contributing in enhancing the materials performance properties.

Bioplastics are obtained by many stages as described in the scheme shown in Fig. 1.1. The number of the chemical transformations required to convert from the raw biomass to the final polymer is called a stage. For example, polylactides is considered as three-stage bioplastics: plant is converted into sugar, then the sugar is fermented to lactic acid, and finally the lactic acid is polymerized. In a two-stage process, the plant is directly fermented and polymerized as the case with polyhydroxybutyrates/polyhydroxyalkanoates. Another example of two-stage bioplastics is the polyaminoundecanoic acid known as Nylon-11: castor oil is first extracted and then chemically converted to polymers. Finally, in one-stage bioplastics, the biomass itself contains the targeted polymer. Most of the common biopolymers, such as the natural rubber, starch, and cellulose, are obtained in one-stage.

The genetic engineering contributes to the advances in biotechnology by moving genes over species, for example, moving some genes from the bacteria that produce PHAs into sugarcane. Direct production of desired polymers from sunlight and CO_2 has many attractive benefits. However, the LCA should first be performed. Besides, such genetic engineering faces many ethical, social, environmental, and regulatory issues.

4 STARCH-BASED BIOPLASTICS

Starch-based plastics have many industrial applications in food packaging, injection molding, and as flexible films. They are composed of starch, plasticizer agents, and additives. Starch-based plastics are considered very attractive choices in terms of economical and sustainable aspects due to the low cost, the inherent biodegradability, and the large content of the renewable resources in its composition. However, they are water sensitive, and they reveal poor performance properties in severe environmental conditions (Iannotti et al., 2018). Recently developed researches improved the water resistivity of the starch-based plastics while maintaining their biodegradability, and hence, their

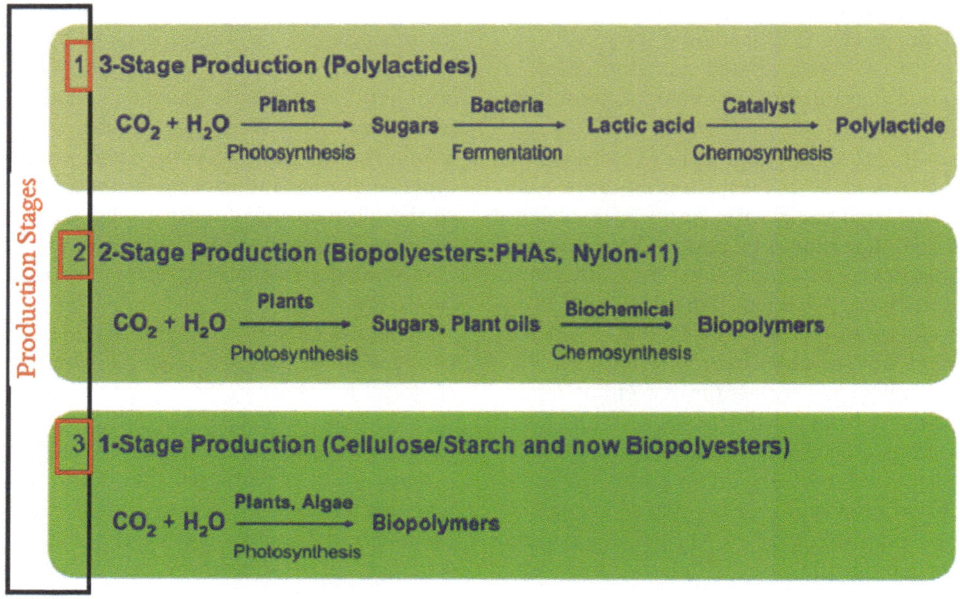

FIG. 1.1 Bioplastics features based upon the number of biochemical transformations required to achieve the final polymer.

applications were extended to new aspects. Some grades of these plastics are commercially produced as shown in Fig. 1.2.

Blending starch with synthetic polymers such as polyethylene or ethylene vinyl acetate has been extensively investigated. The advantages of the obtained plastics are the low cost, the good mechanical properties, the good packaging properties, and the ability to manufacture using conventional machines. The disadvantages of these plastics are the nonrenewable synthetic components and the partial degradability (Alaaeddin et al., 2019a,d).

5 BIOPOLYESTERS

Regardless of whether the polyesters are composed of renewable resources or from fossil resources, they are often degradable as the ester bonds can be easily hydrolyzed. Polyesters from renewable sources as PLAs and PHBs are now commercially available. On the other side, polyesters from fossil resources are also commercially available as PBS. It is noteworthy that significant efforts are currently pushing to produce commercial PBS from renewable resources. There are wide and extensive investigations in the literature on incorporating biopolyesters, petroplastics, and other bioplastics in polymer blends (AL-Oqla and El-Shekeil, 2019; Alaaeddin et al., 2019c; Valerio et al., 2016).

6 BIOCOMPOSITES AND BIO-NANOCOMPOSITES

Significant attention is also raised for natural fiber reinforced composites lately. The fibers are obtained from abaca, flax, jute, hemp, palm, kenaf, and many more plants (AL-Oqla, 2017; AL-Oqla and Sapuan, 2018b; AL-Oqla et al., 2015b). These fibers are used to produce biocomposites with matrices of bioplastic or petroplastic materials. Among these fibers, kenaf is considered one of the most promising natural fibers for many reasons, including the low emission of odor. Until now, biocomposites are mainly devoted for sheet applications, more specifically as interior parts in automobiles (AL-Oqla et al., 2015a). On the other side, nanocomposites are the promising key to overcome many of the drawbacks of the biocomposites. However, there are still many challenges in their development (AL-Oqla and Omari, 2017; AL-Oqla and Salit, 2017; AL-Oqla et al., 2014; Sadrmanesh et al., 2019). Massive research has been conducted on nanoclay reinforced bioplastics. Adding nanocellulose or carbon nanotubes to biopolymers can improve a set of thermal properties. An increasing attention is paid to using nanocellulose in bio-based materials as its cost is less expensive than many conventional petroplastics (\$0.20– \$0.25/lb). Renewable-based polymers are preferred over petroplastics. However, for many applications such as in automobile industry, and in building constructions,

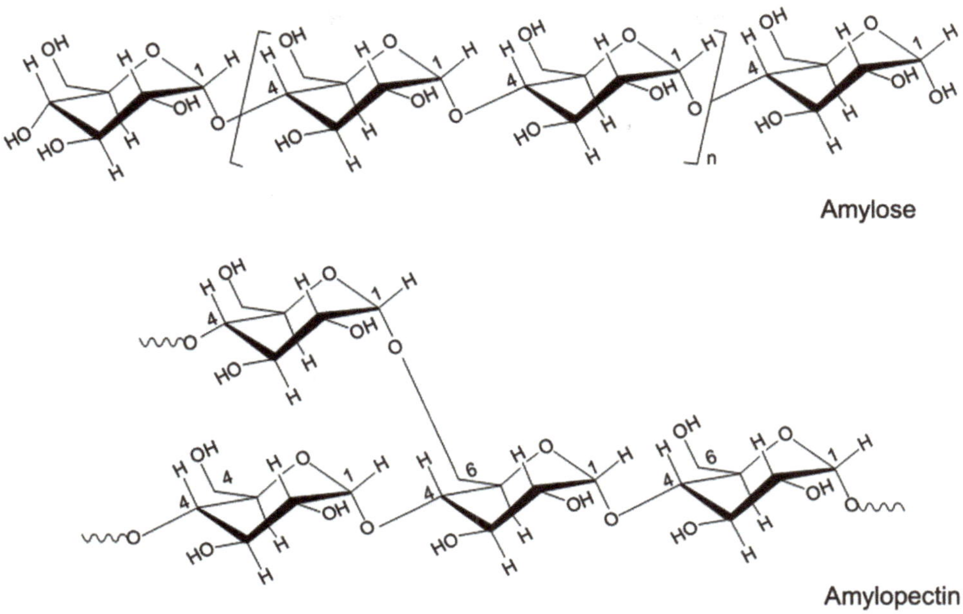

FIG. 1.2 Starch chemical structure of both amylose and amylopectin.

they are undesirable as they possess fast degradability (AL-Oqla et al., 2018a; Aridi et al., 2016a,b; Fares et al., 2019). Furthermore, as the petroleum prices are continually vibrating, and lately, they are reaching unprecedented levels. There is a strong economic need to search for other alternatives to the traditional materials available (AL-Oqla and Sapuan, 2018a).

Extensive efforts have been spent in the last few years in developing such alternatives from renewable resources. The United States is leading the world in this field as it is the largest producer of ethanol from biomass. Ethanol, in turn, is used in producing two important nondegradable bioplastics: these are the biopolyethylene and bio-PET. The production pathway of the biopolyethylene is usually achieved by several steps as first, ethanol is dehydrated to ethylene, and then, ethylene is polymerized using one of many available mechanisms with the help of catalysts. Another example of developing alternative products from renewable resources is the butyl rubber (a copolymer of isoprene and isobutylene) that will soon be commercialized. First, isobutanol is fermented from cellulose sugar or from starch. Then, isobutanol is converted into isobutylene. Combining the derived isobutylene and the renewable isoprene produces the butyl rubber, which is called the natural/synthetic rubber. One more example of developing alternative products from renewable resources is the 3-hydroxypropionic acid (3-HP). This monomer can be dehydrated to produce renewable acrylic acid (used in paints and in superabsorbants), or it can be polymerized to produce biopolyester poly(3-HP). Nylons, on the other hand, have desirable properties beside their toughness. However, they show relatively high prices. In the past, Nylon-11 has been derived from castor oil. Nowadays, many researchers are trying to develop other grades of Nylon from renewable resources. As an example, Nylon-4 can now be derived from monosodium glutamate. The biotechnological industry is confident to reveal new organisms with engineered metabolism to innovate new pathways for other polyamide precursors.

7 SUSTAINABILITY

In order to enhance the properties of plastics, chemical additives are commonly used. unfortunately, these additives have harmful impact on the environment. The microparticles in plastics usually reach the ecosystem by wind, water, animals, and organisms. Hence, they combine with the food chain causing hazardous health problems that may lead to injuries or even death of the organisms. In order to limit the bad effect of plastics, new aware trends should be developed in societies according the sustainable environment. An example to this is by replacing the plastic bags commonly used by bags made of natural fibers. The harmful components in the plastic products can cause many complex

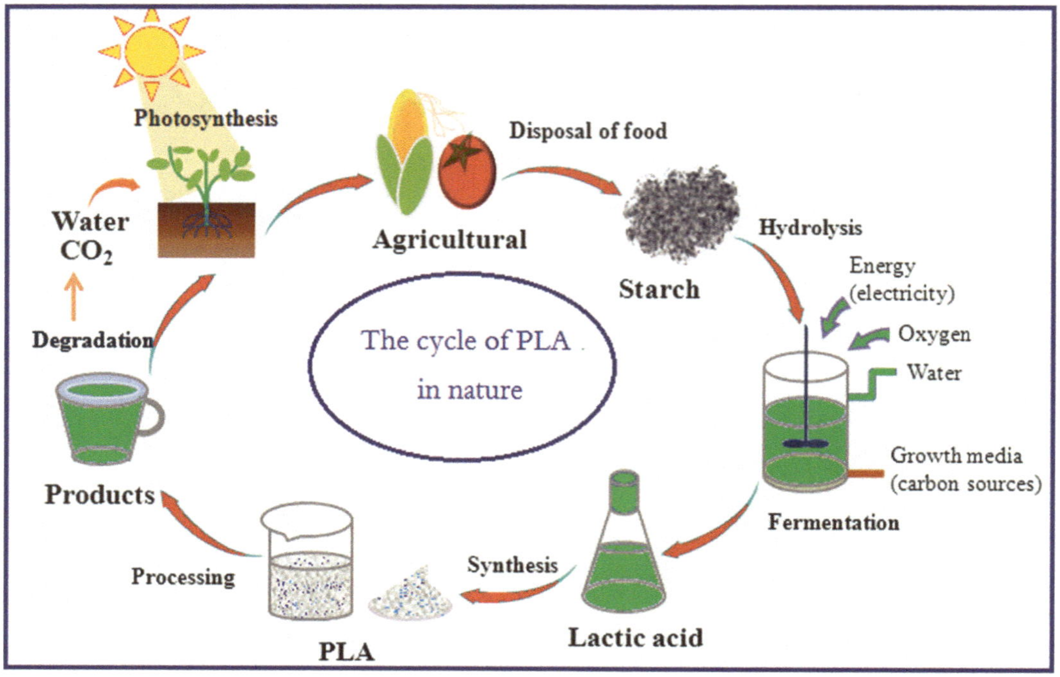

FIG. 1.3 End of life of bioplastics to enhance sustainability.

problems as reducing the quality of air, water, and the whole surrounding. For a product to be claimed that it is sustainable, it should fulfill all the requirements of healthy environment without harming the ecosystem. Fig. 1.3 highlights the closing loop of end life of bioplastics (PLA as an example) to enhance the sustainability. For bioplastics, the plants are grown in farms, then, they are polymerized and converted to the intended products, and then, they are transported to markets until they reach the consumers. At the end of these bioplastics' life, they are composted or recycled (degradation occurs for degradable polymers) without leaving any harmful or toxic components in the environment.

But unfortunately, fossil fuel is required during the manufacturing steps of the bioplastic, as well as during transporting the plants to the manufacturers or the final products to the retailers. However, a partial solution to this issue comes from the fact that many manufacturers are moving toward using clean renewable energy in their factories. Bioplastics are categorized according to many parameters: origin, composition, synthesis process, and application. Fig. 1.4 demonstrates the classifications of biodegradable polymers, and Fig. 1.5 illustrates the bioplastics and the three distinct biopolymer groups with respect to degradability and renewability aspects.

7.1 Types of Bioplastics to Develop Sustainable Industry

7.1.1 Starch-based plastics

Starch-based plastics are the most commonly used among all other bioplastics. It is dominating about 50% of the bioplastic market. Simple starch-based plastics can be produced at home with simple tools. Starch is good in absorbing humidity, and thus, it was a suitable choice for drug capsules as well as medical applications. Fig. 1.6 is an illustrative diagram for the applications of biopolymers in nerve repairing.

Additives like glycerine and sorbitol can work as plasticiser and flexibilizer when added to starch to enhance its thermal characteristics. These thermal characteristics can be tailored to the required needs by controlling the amount of the additives in the starch-based plastic (called thermoplastical starch). Starch is commonly mixed with biodegradable polyesters in order to produce varieties of compounds, such as starch/polylactic acid, starch/Ecoflex, or starch/polycaprolactone. These compounds are compostable, and they are used for many applications. Other compounds are also developed as starch/polyolefin blends. They are not biodegradable, but they reveal lower carbon footprint comparing to petro-based plastics for same applications. Starch is cheap, renewable, and

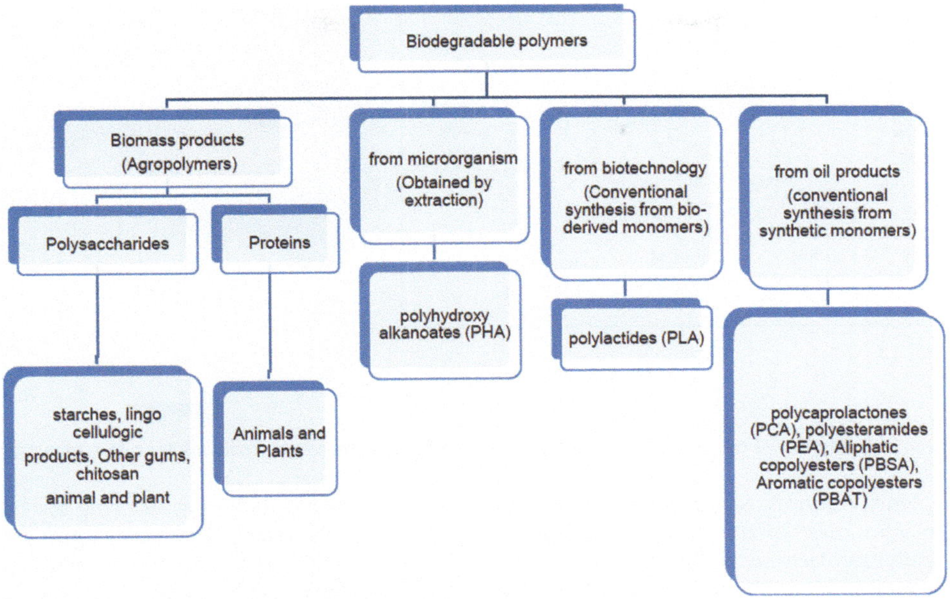

FIG. 1.4 Biodegradable polymers.

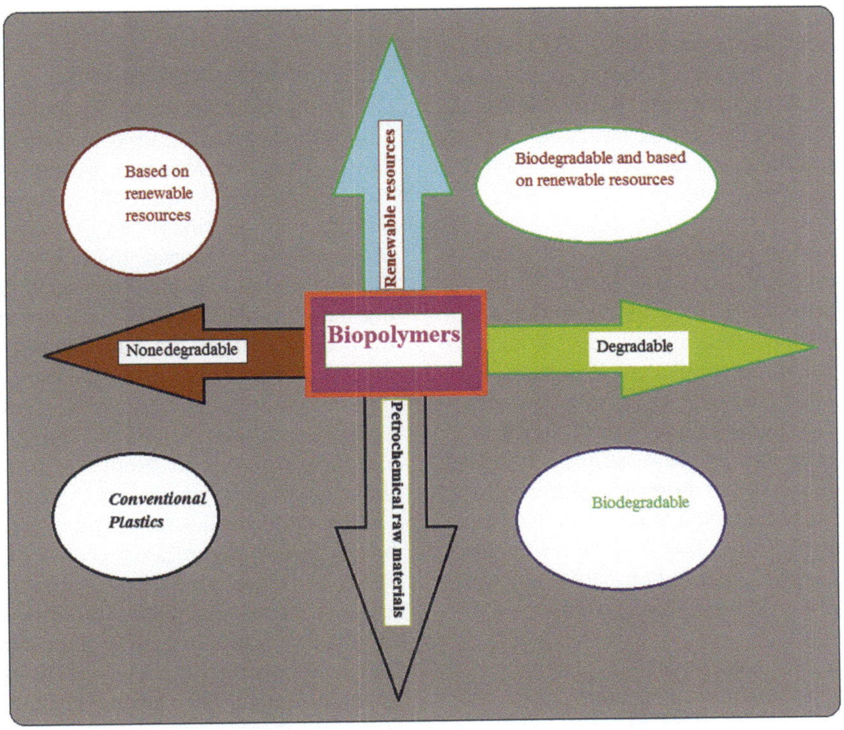

FIG. 1.5 Illustration of the bioplastics and the three distinct biopolymer groups.

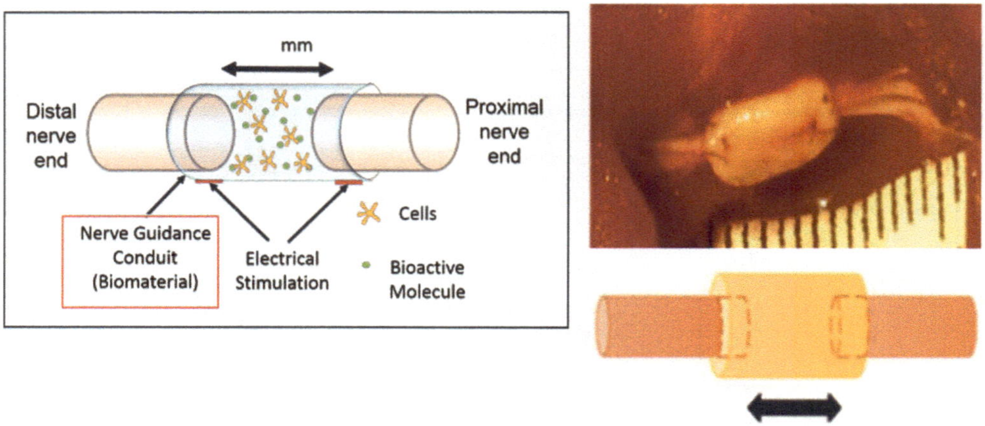

FIG. 1.6 Illustrative diagram for the applications of biopolymers-based materials for nerve repairing.

available in plenty amounts. Starch-based plastics are composed of complex blends with compostable plastics as polycaprolactone, polylactic acid, PBS, polyhydroxyalkanoates, and polybutylene adipate terephthalate. This combination will enhance the performance properties of the plastic, as well as improve its water resistivity. Bioplastic films are produced mainly from blending starch with biodegradable polyesters; hence, these films are biodegradable and compostable. Their applications are mainly in packaging: food packaging as in bakery or fruit and vegetable bags and goods packaging where bubble films are commonly used. Newly developed starch-based films by the Agricultural Research Service scientists have the potential to be used as papers.

7.1.2 Cellulose-based plastics

The cellulose-based bioplastics commonly used are the cellulose esters (including cellulose acetate and nitrocellulose and their derivatives). Cellulose acetate is commonly used in packaging blisters. It is possible to modify the cellulose-based plastics to become thermoplastic materials such as the cellulose acetate. However, the cellulose acetate is expensive relative to other plastics, as well as it requires extensive modifications. Therefore, its use in packaging is not common. On the other side, adding cellulosic fibers to starch can enhance its characteristics as the mechanical properties, the water resistance, and the gas permeability.

7.1.3 Protein-based plastics

Proteins derived from different sources have been already used in plastics to form biodegradable bioplastics. For example, soy proteins have been used in bioplastics since early 1900s. Ford company had used soy-based plastics in their automobiles long time ago. There are some challenges in using soy-based plastics due to their cost and their water sensitivity. However, this can be overcome by blending soy protein with some existing biodegradable polyesters in order to decrease the cost and improve the water sensitivity. Other proteins are being investigated for potential blending. For example, wheat gluten and casein revealed very potential properties to be used as raw materials for varieties of bioplastics.

7.2 Environmental Impact

Cellulose, starch, sugar, wood, and many other renewable resources can be used as sustainable alternatives to the fossil fuel resources in the plastic production. The bioplastics produced from these renewable resources are sustainable in comparison to the conventional plastics. However, the environmental impact of bioplastics should be investigated. This is conducted using many metrics such as the water and energy usage, the biodegradation, the deforestation, etc. The environmental impacts of bioplastics are categorized as the nonrenewable energy usage, the eutrophication, the acidification, and the climate change. The nonrenewable energy usage in bioplastic production is lower than the required energy for the conventional plastics for the same applications. The emission of the greenhouse gases is significantly reduced in bioplastic production. Hence, governments and organizations can increase the sustainability of the environment by adopting the bioplastics instead of the conventional plastics, and by applying suitable regulations to achieve this goal. Although the environmental impact metrics regarding energy and greenhouse gases encourage using bioplastics instead of conventional plastics, other metrics should also be investigated. Bioplastics have higher potentials of eutrophication than conventional plastics.

Eutrophication is the water richness of nutrients. It is considered a serious threat to water resources as it ends the life of the aquatic organisms, harms the freshwater, and causes harmful algal blooms. The reason of the eutrophication is that during production of the biomass, phosphate, and nitrate filtrate into the freshwater reservoirs. Another metric is the acidification. In fact, this environmental impact is negative in bioplastics as they increase the acidification. The high harmful increases of both metrics (acidification and eutrophication) are not only during the bioplastic production but also during planting and growing the raw materials where the chemical fertilizers are used.

7.3 Biodegradation of Bioplastics

The biodegradation in bioplastics is occurred by depolymerizing the polymeric materials using inherent enzymes. The biodegradation starts at the solid/liquid interface where the enzymes are in contact with the solid polymers. Some conventional plastics can also be biodegradable if they contain biodegradable additives. Bioplastics are able to biodegrade under extreme conditions of heat and hydration. The biodegradation speed of these plastics is altered by the environmental conditions, where in the presence of soil and compost, the biodegradation becomes faster and more efficient as rich microbial diversity will exist. The efficiency of the biodegradability in compost environments can be increased more by increasing the temperature and by adding soluble sugar. Beside increasing the biodegradation efficiency, composting is able to incredibly reduce the emission of greenhouse gases. On the other side, biodegradation in soil environments will provide higher diversity of microbes; making it easier for biodegradation to take place. However, soil environments take longer time to biodegrade the bioplastics and require higher temperatures. Other environments can increase the efficiently of the biodegradation as the aquatic environment. However, using the aquatic environments in order to biodegrade the bioplastics is risky and not recommended; it should be avoided as it harms the freshwater and the ecosystems and ends the life of the aquatic organisms. Last parameters affecting the biodegradation speed are the composition and the structure of the bioplastics, and in developing the bioplastics, the researchers put great efforts to optimize the biodegradation speed rate to meet the applications needs.

8 CONCLUSIONS

The social and the environmental demands in one hand, as well the late achievements of the biotechnology and the nanotechnology on the other, are dramatically affecting our view for plastics and how they should be produced. These reasons are pushing researchers to accelerate in enhancing the bioplastic materials to higher levels of performance and in commercializing their processes. Starch-based plastics are the most commonly used among all other bioplastics. The emission of the greenhouse gases is significantly reduced in bioplastic production. Hence, governments and organizations can increase the sustainability of the environment by adopting the bioplastics instead of the conventional plastics, and by applying suitable regulations to achieve this goal. Nowadays, great successes are being achieved in new bioplastic materials commercially. These achievements are directly reflected from the successful applications of the industrial biotechnology.

REFERENCES

AL-Oqla, F.M., 2017. Investigating the mechanical performance deterioration of Mediterranean cellulosic cypress and pine/polyethylene composites. Cellulose 24, 2523–2530.

AL-Oqla, F.M., El-Shekeil, Y., 2019. Investigating and predicting the performance deteriorations and trends of polyurethane bio-composites for more realistic sustainable design possibilities. Journal of Cleaner Production.

AL-Oqla, F.M., Omar, A.A., Fares, O., 2018a. Evaluating sustainable energy harvesting systems for human implantable sensors. International Journal of Electronics 105, 504–517.

AL-Oqla, F.M., Omari, M.A., 2017. Sustainable biocomposites: challenges,Potential and barriers for development. In: Jawaid, M., Sapuan, S.M., Alothman, O.Y. (Eds.), Green Biocomposites: Manufacturing and Properties. Springer International Publishing (Verlag), cham, switzerland, pp. 13–29.

AL-Oqla, F.M., Salit, M.S., 2017. Materials Selection for Natural Fiber Composites. Woodhead Publishing, Elsevier, Cambridge, USA.

AL-Oqla, F.M., Sapuan, M.S., Ishak, M.R., Aziz, N.A., 2014. Combined multi-criteria evaluation stage technique as an agro waste evaluation indicator for polymeric composites: date palm fibers as a case study. BioResources 9, 4608–4621.

AL-Oqla, F.M., Sapuan, M.S., Ishak, M.R., Nuraini, A.A., 2015a. Decision making model for optimal reinforcement condition of natural fiber composites. Fibers and Polymers 16, 153–163.

AL-Oqla, F.M., Sapuan, S., 2018. Investigating the inherent characteristic/performance deterioration interactions of natural fibers in bio-composites for better utilization of resources. Journal of Polymers and the Environment 26, 1290–1296.

AL-Oqla, F.M., Sapuan, S., 2018b. Natural fiber composites. In: Sapuan, S.M., Ishak, M.R., Sahari, J., Sanyang, M.L. (Eds.), Kenaf Fibers and Composites. CRC Press.

AL-Oqla, F.M., Sapuan, S., Anwer, T., Jawaid, M., Hoque, M., 2015b. Natural fiber reinforced conductive polymer composites as functional materials: a review. Synthetic Metals 206, 42–54.

AL-Oqla, F.M., Sapuan, S., Fares, O., 2018b. Electrical–Based Applications of Natural Fiber Vinyl Polymer Composites, Natural Fibre Reinforced Vinyl Ester and Vinyl Polymer Composites. Elsevier, pp. 349–367.

AL-Oqla, F.M., Sapuan, S., Ishak, M., Nuraini, A., 2015c. Predicting the potential of agro waste fibers for sustainable automotive industry using a decision making model. Computers and Electronics in Agriculture 113, 116–127.

AL-Oqla, F.M., Sapuan, S., Jawaid, M., 2016. Integrated mechanical-economic–environmental quality of performance for natural fibers for polymeric-based composite materials. Journal of Natural Fibers 13, 651–659.

AL-Oqla, F.M., Sapuan, S.M., 2014a. Date Palm Fibers and Natural Composites. In: Postgraduate Symposium on Composites Science and Technology 2014 & 4th Postgraduate Seminar on Natural Fibre Composites 2014, 28/01/2014, Putrajaya, Selangor, Malaysia.

AL-Oqla, F.M., Sapuan, S.M., 2014b. Enhancement selecting proper natural fiber composites for industrial applications. In: Postgraduate Symposium on Composites Science and Technology 2014 & 4th Postgraduate Seminar on Natural Fibre Composites 2014, 28/01/2014, Putrajaya, Selangor, Malaysia.

AL-Oqla, F.M., Sapuan, S.M., 2014c. Natural fiber reinforced polymer composites in industrial applications: feasibility of date palm fibers for sustainable automotive industry. Journal of Cleaner Production 66, 347–354.

Alaaeddin, M., Sapuan, S., Zuhri, M., Zainudin, E., Al-Oqla, F.M., 2019a. Photovoltaic applications: status and manufacturing prospects. Renewable and Sustainable Energy Reviews 102, 318–332.

Alaaeddin, M., Sapuan, S., Zuhri, M., Zainudin, E., AL-Oqla, F.M., 2019b. Physical and mechanical properties of polyvinylidene fluoride-Short sugar palm fiber nanocomposites. Journal of Cleaner Production 235, 473–482.

Alaaeddin, M., Sapuan, S., Zuhri, M., Zainudin, E., AL-Oqla, F.M., 2019c. Polymer matrix materials selection for short sugar palm composites using integrated multi criteria evaluation method. Composites Part B: Engineering 107342.

Alaaeddin, M., Sapuan, S., Zuhri, M., Zainudin, E., AL-Oqla, F.M., 2019d. Polyvinyl fluoride (PVF); its properties, applications, and manufacturing prospects. In: IOP Conference Series: Materials Science and Engineering. IOP Publishing, p. 012010.

Aridi, N., Sapuan, S., Zainudin, E., AL-Oqla, F.M., 2016a. Investigating morphological and performance deterioration of injection-molded rice husk–polypropylene composites due to various liquid uptakes. International Journal of Polymer Analysis and Characterization 21, 675–685.

Aridi, N., Sapuan, S., Zainudin, E., AL-Oqla, F.M., 2016b. Mechanical and morphological properties of injection-molded rice husk polypropylene composites. International Journal of Polymer Analysis and Characterization 21, 305–313.

Black, M., Whittaker, C., Hosseini, S., Diaz-Chavez, R., Woods, J., Murphy, R., 2011. Life cycle assessment and sustainability methodologies for assessing industrial crops, processes and end products. Industrial Crops and Products 34, 1332–1339.

Deffeyes, K.S., 2008. Hubbert's Peak: The Impending World Oil Shortage, New Edition. Princeton University Press.

Fares, O., AL-Oqla, F.M., Hayajneh, M.T., 2019. Dielectric relaxation of mediterranean lignocellulosic fibers for sustainable functional biomaterials. Materials Chemistry and Physics.

Halley, P., Dorgan, J.R., 2011. Next-generation biopolymers: advanced functionality and improved sustainability. MRS Bulletin 36, 687–691.

Iannotti, G., Fair, N., Tempesta, M., Neibling, H., Hsieh, F.H., Mueller, R., 2018. Studies on the Environmental Degradation of Starch-Based Plastics, Degradable Materials. CRC Press, pp. 425–446.

Jakubowicz, I., 2003. Evaluation of degradability of biodegradable polyethylene (PE). Polymer Degradation and Stability 80, 39–43.

Khabbaz, F., Albertsson, A.-C., Karlsson, S., 1999. Chemical and morphological changes of environmentally degradable polyethylene films exposed to thermo-oxidation. Polymer Degradation and Stability 63, 127–138.

Sadrmanesh, V., Chen, Y., Rahman, M., AL-Oqla, F.M., 2019. Developing a decision making model to identify the most influential parameters affecting mechanical extraction of bast fibers. Journal of Cleaner Production 117891.

Valerio, O., Pin, J.M., Misra, M., Mohanty, A.K., 2016. Synthesis of glycerol-based biopolyesters as toughness enhancers for polylactic acid bioplastic through reactive extrusion. ACS Omega 1, 1284–1295.

CHAPTER 2

Processing of Thermoplastic Starch

RIDHWAN JUMAIDIN • SITI NURAISHAH MOHD ZAINEL • S.M. SAPUAN

1 INTRODUCTION

Due to environment and manageability issues, this century has seen exceptional accomplishments in green innovation in the material science field through the improvement of biocomposites (Joshi et al., 2004). Society's expanding familiarity on the significance of ecological safeguarding has driven tremendous measures for exploring developments of more natural materials. The advancement of elite material produce using normal assets is present around the world. The improvement of regular fiber composites is becoming a more genuine consideration due to their promising properties. Abundant agricultural waste makes them an easy source for material enhancement. The enhancement of bio-based polymers as alternative matrices for petroleum-derived polymers further provides another green perspective for composites. Adapting natural fiber as the reinforcement appear as a practical solution, particularly in automotive, food packaging, infrastructure, and building item applications (Scholten et al., 2014). Among biopolymers, starch stands out as the most encouraging since it is easily accessible, simple, abundant, sustainable, and biodegradable.

2 BIOPOLYMERS

Biopolymers are polymers that occur in nature. Starches and proteins, for instance, are biopolymers. Numerous biopolymers are now financially delivered at a vast scale, despite the fact that they will not be used often for the creation of plastics. Only a small portion of delivered biopolymers are employed in the creation of plastics. By utilizing them more, it would fundamentally diminish our reliance on fabricated, nonsustainable assets. Other than being accessible, biopolymers have a few monetary and ecological points of interest. Biopolymers demonstrate an advantage for handling waste. For instance, replacing polyethylene by a biopolymer could help remove plastic pieces found in compost. Coalition between synthetic polymers, common polymers, and biodegradable polymers can form novel materials since they can consolidate the process capacity with biodegradation and utilization of sustainable crude materials. In any case, customary polymers typically display abnormal amounts of recyclability and the recycling procedure can be influenced by the presentation of a system containing common polymers (Peres et al., 2016).

2.1 Categorization of Biopolymers

Biopolymers can be obtained from microbial frameworks extracted from higher life forms such as plants or integrated synthetically from essential biological buildings. Biopolymers are developed for various applications ranging from packaging, medical, pharmaceutical, etc. A previous study (Scholten et al., 2014) had investigated the role of biopolymer composites for emulsions and gels in food engineering. Biopolymers are further developed for use in clothing textures, water treatment chemicals, modern plastics, sponges, biosensors, and even information stockpiling components. The categorization of biopolymers is provided in Table 2.1.

Biopolymers have their own properties. They are inexhaustible, maintainable, biodegradable, safe, nonimmunogenic, non-cancer-causing, nonthrombogenic, and carbon; for example, chitin/chitosan. Chitin exists in the skeletal system of creatures and has a white appearance (Asghari et al., 2017). Inexhaustible resources are utilized progressively in the creation of polymers (specifically, monomers). For example, carbon dioxide, terpenes, vegetable oils, and sugars can be applied as feed stocks for the fabrication of a variety of manageable materials and products, including elastomers, plastics, hydrogels, adaptable gadgets, resins, engineering polymers, and composites (Zhu et al., 2016). Biodegradable materials are employed as part of bundling, farming, prescription, and distinguished areas. As of late, the enthusiasm for biodegradable polymers is rising. There are two classes of biodegradable polymers present: engineered and characteristic polymers. These are polymers made from stocks

Advanced Processing, Properties, and Applications of Starch and Other Bio-Based Polymers. https://doi.org/10.1016/B978-0-12-819661-8.00002-0

TABLE 2.1
Categorization of Biopolymers.

Polyesters	**Polysaccharides (lant/ Algal)**
Polyhydroxyalkanoates	Starch (amylose/ amylopectin)
Polylactic acid	Cellulose
Proteins	Agar
Silks	Alginate
Collagen/Gelatin	Carrageenan
Elastin	Pectin
Resilin	Konjac
Adhesives	Various gums (e.g., guar)
Polyamino acids	**Polysaccharides (animal)**
Soy, zein, wheat gluten, casein	Chitin/chitosan
Serum albumin	Hyaluronic acid
Polysaccharides (bacterial)	**Lipids/Surfactants**
Xanthan	Acetoglycerides, waxes, surfactants
Dextran	Emulsan
Gellan	**Polyphenols**
Levan	Lignin
Curdlan	Tannin
Polygalactosamine	Humic acid
Cellulose (bacterial)	**Specialty polymers**
Polysaccharides (fungal)	Shellac
Pullulan	Poly-γ-glutamic acid
Elsinan	Natural rubber
Yeast glucans	Synthetic polymers from natural fats and oils; nylon from castor oil

acquired from either oil assets (noninexhaustible assets) or organic assets (sustainable assets).

3 STARCH

Starch is hydrophilic in nature (Zhang et al., 2014). As a completely biodegradable polysaccharide that is biosynthesized by numerous plants, starch is one of the most abundant renewable resources known to man (Mohammadi Nafchi et al., 2013). Starch is the predominant resource of carbohydrates devoured by humans, providing around 66% of the required day-to-day calories. It assumes a crucial part in our regular daily existence and is a vital plant-derived material generally used as food (Riyajan, 2015), in packaging products, and in the manufacturing of nonnourishment products. It is not only a major dietary carbohydrate but is also employed to manufacture household products such as pharmaceuticals, paper, and textiles (Seung et al., 2015). Starch is the second largest source for biomass (inexhaustible) on this planet, providing an eco-friendly approach to produce enormous assortments of materials when mixed with other biodegradable polymers. Moreover, it is a familiar polymer accessible in large amounts from various sources (Pasquini et al., 2010). The degradation of starch at night distributes carbohydrates to fuel respiration and growth when photosynthesis is nonviable.

Starch, in its local granular shape, is sometimes utilized as nourishment. In its local state, starch comprises of semicrystalline granules that are insoluble in water (Pasquini et al., 2010). Starch, for the most part, is prepared by heat which prompts gelatinization and subsequently crumbles into essential sugar segments, amylose, and amylopectin (Pasquini et al., 2010). Other refinements of starch include the utilization of enzymatic or chemical treatments. High-control, low-frequency ultrasound handling has the additional capability of hydrolyzing starch particles, with the benefit of being a physical technique (Kang et al., 2016). To date, starch is generally employed as nourishment (Riyajan, 2015), in building materials (Teodoro et al., 2015), and in the creation of paper (Petersen et al., 2013). It is an easy, inexhaustible asset (Riyajan, 2015).

Starch fundamentally exists as efficient granules in which amylopectin exhibits nonrandom appropriation of liner chains and a congregate arrangement of branch linkages, giving rise to a high degree of structural organization. Starch is a semicrystalline polymer made from direct polysaccharide molecules (amylose) and stretched particles (amylopectin). It is a distinctive polymer characterized as accessible granule material (Peres et al., 2016). The preserved design of amylopectin is in charge of the semicrystalline, water-insoluble starch granules. Starch granules are the most important energy reserve in plants and can be found in tubers, cereals, roots, sorghum, and stem of a plant, i.e., barley, wheat, oat, corn, sago, green pea, cassava, sugar palm, and potato (Sahari et al., 2014).

Starch granules consist of two major types of polysaccharides: amylose and amylopectin. Each differs in atomic mass, level of expansion, and chemical properties

(Bai et al., 2017). The minor components such as lipid, protein, and phospholipids are found at the starch's surface or connected with polysaccharide chains.

For potato starch, the shape changes from oval to circular as the width of the granules diminishes. Potato is a tuberous harvest that contains a high measure of starch and generally utilized as part of the food industry, material industry, adhesives, and paper industry. In addition, the state of potato starch granules is typically represented as lenticular and their normal size can change between 5 and 100 μm (Bai et al., 2017). Other than that, it can be applied as thickeners, coating, gelling, bonding, and unique casings for drug release. In various mechanical applications, potato starch is used for nourishment solutions, operators for paper and materials, biodegradable plastics, and pharmaceuticals. Fig. 2.1 displays the potato starch granules.

The granule size of starch could reflect the biosynthetic age. Small sized granules are considered as prepubescent granules that cannot develop into full-estimate granules, while extensive granules are completely developed granules (Wang et al., 2016).

3.1 Application of Starch

In general, starch is a common source of carbohydrates for the human diet. It is consumed in the form of staple foods such as rice, pasta, breads, cereals, and vegetables. Starch is also used in food preparation processes such as a thickening agent for gravy, puddings, etc (Jumaidin et al., 2018). In the pharmaceutical industry, starch is employed as a drug excipient that exhibits the slow release characteristics (Le Bail et al., 1999). It is reported that high amylose starch displays good swelling and drug release behavior when formed as a tablet. In papermaking, starch is applied to enhance the strength of paper. It is also used for adhesive materials in the manufacturing of glue and other types of adhesives. Starch is further used in cosmetics, pet food, aquatic food, and clothing.

3.2 Thermoplastic Starch

Starch has unique characteristics where it can be transformed into thermoplastics when subjected to high temperature, shear, and with the presence of a plasticizer. In general, the role of the plasticizer is to disrupt starch granules by breaking its intermolecular and intramolecular hydrogen bonds. The starch-plasticizer interaction is formed by eliminating the starch-starch interaction. This process is accompanied by partial depolymerization of the starch backbone and a decrease in the melting temperature of starch (Stepto, 2003). The introduction of a plasticizer and heat with raw starch is termed as plasticization. The role of the plasticizer is to reduce the brittleness of starch and increase flexibility for future applications. The unique characteristics of thermoplastic starch allow this biomaterial to melt and harden repeatedly, making it suitable for various fabrication processes for conventional plastics. Hence, this feature can be included in the positive attributes of the material, apart from being biodegradable and renewable.

3.3 Processing of Thermoplastic Starch

As previously mentioned, starch requires the presence of heat, shear, and plasticizer to transform into thermoplastic starch. In general, there are several types of plasticizers: glycerol, sorbitol, fructose, urea, water, etc. Two main processes can be applied to produce this green material. Hot press and solution casting are the most common methods employed in reported studies on thermoplastic starch development.

3.3.1 Hot press

The hot press method refers to a process where the starch mixture is pressed at a certain temperature; the

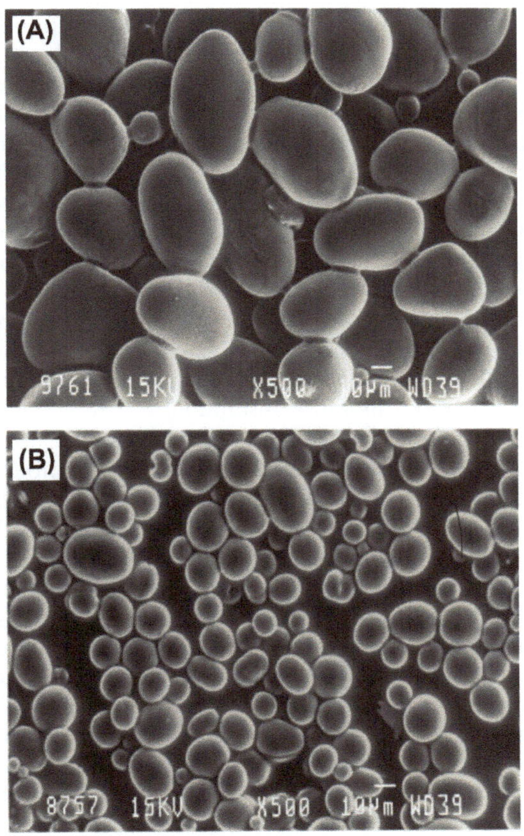

FIG. 2.1 **(A)** The native large potato starch granules by SEM and **(B)** the native small potato starch granules by SEM (Sandhu et al., 2015).

pressure is applied by using hydraulic press equipped with a heating element. To employ the hot press method, several stages must be accomplished. Firstly, native starch should be mixed with a plasticizer at a predetermined ratio, i.e., 20 to 30 wt%. A previous study had reported that mixing can be conducted in a plastic bag until a homogeneous mixture is attained (Lomelí Ramírez et al., 2011). Another study reported that a mechanical stirrer can be used to complete this process (Sahari et al., 2013). It was also mentioned that a high-speed mixer can be employed to ensure good mixing is achieved at this stage (Jumaidin et al., 2016). The homogenous starch/plasticizer mixture can be hot pressed at certain temperature ranges, from 130 to 180°C with pressure ranging between 10 and 20 tons. However, some studies did report using the melt-mixing process with a polymer melt-mixer before the hot pressing process (Lopez et al., 2014; López et al., 2015; Salaberria et al., 2014; Taghizadeh and Favis, 2013). The melt-mixing process is accomplished to obtain early plasticization of the starch, while acquiring a homogeneous starch/plasticizer mixture. This can be achieved more easily during the melt condition. The product from melt-mixing will be crushed into small pellets before placed onto a steel mold for the hot pressing process. Here, the thermoplastic starch will possess similar characteristics to conventional polymer, i.e., polypropylene or polyethylene. The pressing of crushed TPS pellets may be conducted at temperatures between 130 and 180°C with a pressure of 10–30 tons, depending on the study. Removal agent should be applied onto the mold's surface in order to facilitate the removal of samples after hot pressing. Since thermoplastic starch is known to be moisture sensitive, the samples obtained from hot pressing should be cooled down in a desiccator to avoid moisture absorption, which will affect the mechanical properties of the material. The advantage of the hot pressing method is the ability to produce a rigid material and product that is not possible for solution casting. The limitation of this process is the variation in the physical properties of the product due to a nonhomogeneous structure, void, cracks, etc., formed when samples are subjected to high pressure. An accurate amount of the raw mixture is quite difficult to obtain since low amounts will result in unmelted samples at the corner, while excessive amounts will result in increasingly thick samples. To accomplish this process, multiple trials should be conducted in order to obtain the ideal temperature, pressure, time, and amount of raw materials to be placed onto the mold. These parameters are highly dependent on the types of starch, plasticizer, and dimension of the mold.

3.3.2 Solution casting

Solution casting is a process of producing thermoplastic starch by using water solution, plasticizer, heat, and stirring. In general, this method is also used for other types of biopolymers such as agar, alginate, carrageenan, chitosan, chitin, etc. (Haciu et al., 2013; Pereira et al., 2003; Wu et al., 2009). The process of producing thermoplastic starch using solution casting is relatively easier than the hot pressing method. In general, a hot plate or water bath, as well as distilled water and mechanical/magnetic stirrer, is required. Film-forming solution that consists of starch, plasticizer, and distilled water should be prepared for this process. Prior to heating, the amount of starch and plasticizer should be weighed according to the desired ratio to obtain the film-forming solution. Edhirej et al. employed a film-forming solution containing 5 g of cassava starch/100 mL distilled water (Edhirej et al., 2016). Fructose was then used as the plasticizer, at concentrations of 0.30 g/g of dry starch. The film-forming solution containing all necessary materials is then heated to 80°C in a thermal bath and kept at this temperature for 20 min under constant stirring. During stirring, bubbles tend to form within the solution; however, according to Edhirej et al. it can be removed by placing the film-forming solution into a desiccator under a vacuum. The solution should be kept in the vacuum desiccator until no bubbles are visible. The removal of bubbles from the solution is crucial in order to ensure a homogeneous structure of the thermoplastic starch. The existence of bubbles may lead to the formation of a void in the final film that could interfere with the mechanical properties of the resultant film.

The bubble-free solution was then poured onto circular glass plates with 10 cm diameters. Next, the film-forming solution was dried at 50°C in an air circulating oven. The dried films were slowly removed by peeling the film off from the plates. The resultant film is then kept in a zip-locked plastic bag prior to characterization (Edhirej et al., 2016).

Solution casting is a simple method that produces thermoplastic starch that does not require heavy equipment, i.e., heated hydraulic press (hot press method). The advantages of this method is that only minimum amounts of film-forming material are required to produce the film. Hence, multiple productions can be carried out, especially for materials that are difficult to extract or high in cost (such as nanoscale materials). Since there is no pressure applied during the fabrication of the film, the structure of any additional materials (such as natural fiber) is not susceptible to damage as in the hot press method. However, the limitation of this method is that the resultant product is only in the form of a thin film. Hence, fabrication of other shapes is not possible, and the samples can only undergo tensile testing for the characterization of mechanical properties. Thermoplastic starch produced from the hot press method can undergo flexural and impact testing. The film formed from solution casting is only suitable for experimental purposes and is not ready for actual production processes. The hot press method is much more similar to the actual production process for conventional plastics and similar data can be applied for production using extrusion or injection molding.

3.3.3 Injection molding

Injection molding is a type of manufacturing process commonly employed for creating thermoplastic materials in rapid production. This method includes injecting molten materials into a mold, which are then cooled and hardened inside the mold cavity. Apart from plastic, this method is employed for glass, metals (known as die-casting), and elastomers.

Rosa et al. fabricated thermoplastic corn starch by using injection molding (Rosa and Andrade, 2004). Prior to injection molding, the preparation of a plasticized mixture is a key step for this process. To obtain a homogeneous mixture, Rosa et al. prepared a constant weight of starch, glycerol (15 wt%), and water (15 wt %). This mixture was injection molded into ASTM D638-72 Type I specimens, approximately 2 mm thick, using a PicBoy 15/42 Petersen Irmaos Machine (Sao Paulo, Brazil). This was provided with three electrically heated zones, maintained at 130 and 145°C from the feed zone to the die end. The mixtures were manually fed into the machine, and the injection pressure was kept at 113 bar. The injection mold was cooled at 30−40°C by a refrigeration system and kept closed at 1275 bar.

Another study (Avérous, 2004) used extrusion and injection molding to fabricate thermoplastic wheat starch. The authors fabricated two types of plasticized starch matrix: TPS1 and TPS2. TPS1 was prepared with a combination of 70 wt% starch, 18 wt% glycerol, and 12 wt% water; while TPS2 was prepared with 65 wt% starch and 35 wt% glycerol. The processing of TPS began by weighing the starch and glycerol, followed by mixing at high speed (2000 rpm). Next, the starch was heated at 170°C for 45 min in a vented oven to allow glycerol diffusion into the starch granules as well as water volatilization from the mixture. The mixture was then cooled and became a dry blend. The dry blend was added with water at a certain formulation that was not mentioned specifically by the authors. The mixture was subjected to dispersion in a mixer. It then underwent high-speed mixing (2500 rpm) to obtain the final mixture. The powder was extruded with a single screw extruder and granulated. The pellets were extruded for the second time in order to improve dispersion. The final pellets were equilibrated at 50% RH for 8 days prior to the injection molding process. Next, the injection molding machine with a clamping force of 50 tons was used to mold standard dumbbells. The parameters for the injection molding process are as follows: screw barrel temperature (100−130°C), mold temperature (20−25°C), holding pressure (1000 bars), holding time (20 s), and cooling time (10 s). The resultant dumbbell-shaped product was conditioned at a temperature and humidity-controlled room (23°C, 54%RH) during 2, 4, or 6 weeks.

Injection molding is the most common process used for the production of conventional plastics. However, there are some limitations to this process as compared with others. In terms of cost, injection molding requires high-end equipment in order for the process to be carried out. This process will require huge amounts of materials to be used, which can be unlikely depending on the material involved.

4 PROCESSING OF THERMOPLASTIC STARCH COMPOSITES

Even though thermoplastic starch seems to be a promising material for replacing conventional petroleum-based polymers, there are some limitations present for this material, i.e., poor mechanical properties, low thermal stability, high moisture sensitivity, and low dimensional stability. These limitations restrict the material's potential of being used as an alternative to conventional plastics. Hence, various studies have been carried out to improve the properties of thermoplastic starch. This modification includes the incorporation of natural fibers at the macro, micro, and nano levels as well as blending TPS with other polymers that have better physical properties.

Ibrahim et al. investigated the effect of date palm and flax fiber on the behavior of thermoplastic starch composites (Ibrahim et al., 2014), where thermoplastic starch composites were fabricated using the hot press method. Prior to that, native corn starch was mixed with 30 wt % glycerin and 20 wt% distilled water at a temperature between 60 and 80°C. It was reported that the gelatinization process of starch can be enhanced with the presence of water, which could improve the tensile strain of samples without significant effect on tensile strength. Adding glycerin may enhance the processability while reducing embrittlement by inhibiting the retrogradation procedure after processing. The preparation of composites was accomplished by using a positive type mold coated with steric acid as the releasing agent. Natural fiber (i.e., date palm and flax fiber) was chopped and distributed in the mold. Next, thermoplastic starch was emulsified in water with TPS: water ratio of 1:1, 1:2, 1:3, 1:4, and 1:8 for 0, 20, 40, 50, 60, and 80 wt% fiber contents, respectively. The samples were preheated at 140 ± 3°C for 30 min in a hydraulic press to remove excess water from the emulsion. This was followed by hot pressing at 5 MPa and 160 ± 3°C for 30 min and then cooled at a rate of about 2°C/min. The findings indicate that the incorporation of both date palm and flax fiber did improve the mechanical properties and thermal stability, while reducing the water uptake. Fig. 2.2 shows the treated and untreated date palm fiber and the SEM micrograph of starch and the thermoplastic starch.

Jumaidin et al. investigated the effect of agar on the thermal, mechanical, and moisture absorption behaviors of thermoplastic sugar palm starch (Jumaidin et al., 2016). The composites were developed by using a combination of melt-mixing and hot pressing methods. Prior to their fabrication, sugar palm starch was manually extracted from the sugar palm tree by using extraction, washing, sedimentation, and drying processes. The TPS matrix was prepared by using 30 wt% glycerol as the plasticizer. The mixture was then mixed using a high-speed mixer at 3000 rpm for 5 min. The well-mixed sample underwent a melt-mixing process using Brabender plastograph at 140°C and rotor speed of 20 rpm for 10 min. The resultant product from the melt-mixing approach was granulated by using a blade mill with 2 mm mesh to produce small pellets. Lastly, the pellets were pressed in a mold at 140°C for 10 min under the load of 10 tons. The same process was adopted for the composite preparation of agar (10, 20, 30, and 40 wt%) introduced during the high-speed mixing process. The authors reported that the incorporation of agar did improve the thermal stability and mechanical properties, while increasing the moisture uptake and thickness swelling of the composites.

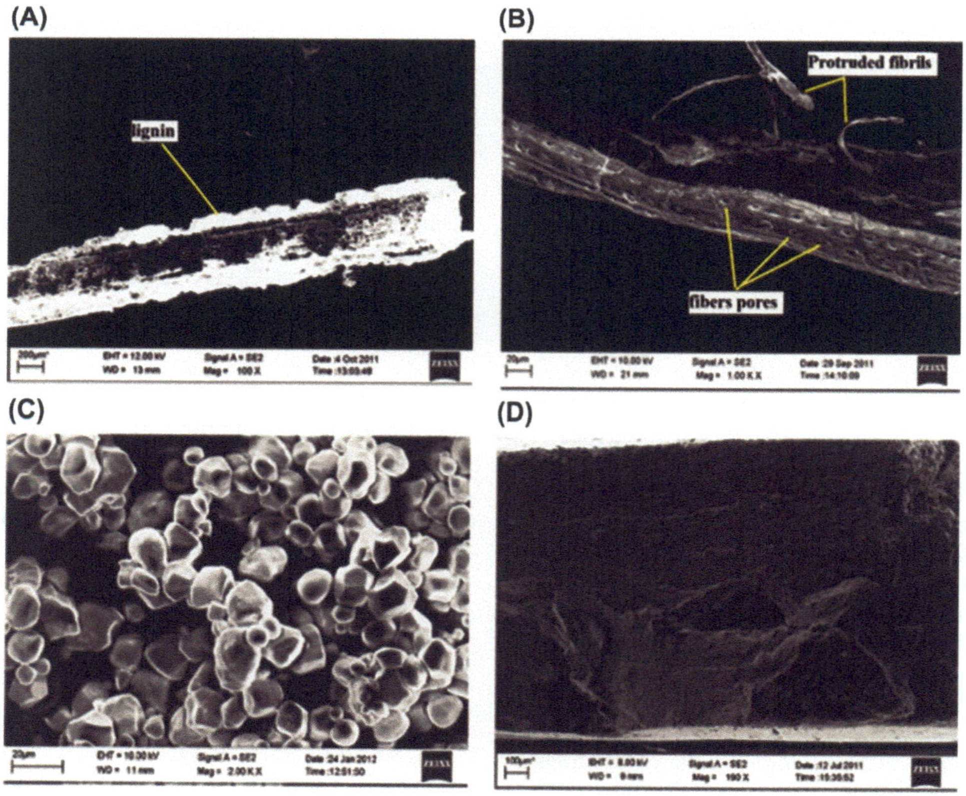

FIG. 2.2 Date palm fibers and starch SEM investigation: **(A)** raw date palm fibers are covered with much lignin; **(B)** NaOH-treated date palm fibers with clean surface and protruded fibrils; **(C)** overnight-stored plasticized corn starch; and **(D)** compression molded thermoplastic starch by emulsion technique (Ibrahim et al., 2014).

Fig. 2.3 shows the SEM micrograph of fracture surface of thermoplastic SPS blended with different ratio of agar (A) 0 wt%, (B) 10 wt%, (C) 20 wt%, (D) 30 wt%, and (E) 40 wt%.

Another recent study had reported the development of thermoplastic starch/natural keratin fiber composites for flame-retardant application (Rabe et al., 2019). In this study, the authors used thermoplastic starch (namely, Mater-Bi EF05B) procured from Novamont S.p.A (Novara, Italy). The keratin fibers were obtained as waste from the beamhouse stage of a Mexican tannery, while the coconut fibers were obtained from the husk of coconut fruits. The natural fibers underwent several procedures prior to the mixing process with TPS. The TPS composites were prepared by the extrusion of TPS and fiber mixture in a twin screw extruder L/D = 32 with a diameter of 27.0 mm and eight heating zones using a counterrotating intermeshing mode. The temperature profile for the compounding process (from feed to die) were as follows: 130/135/135/140/145/150/145/140°C. The rotational speed of the twin screws was set at 120 rpm. The authors adopted the double extrusion method (repeated) in order to ensure better dispersion of fibers and flame-retardant properties. For the second extrusion, the TPS and composites were dried at 105°C in model 30 low pressure dryer with 80 psi (0.5516 MPa) maintained during heating. After each extrusion step, the composites were granulated in a 7.5 HP granulator using a screen with a 5 mm mesh. The findings indicate that the incorporation of fiber did improve the flame-retardant properties of the composites. This was indicated by the lower heat release rate and total heat evolved from the TPS composites.

5 CONCLUSIONS

In general, this chapter has shown that thermoplastic starch is a versatile material that can be processed using

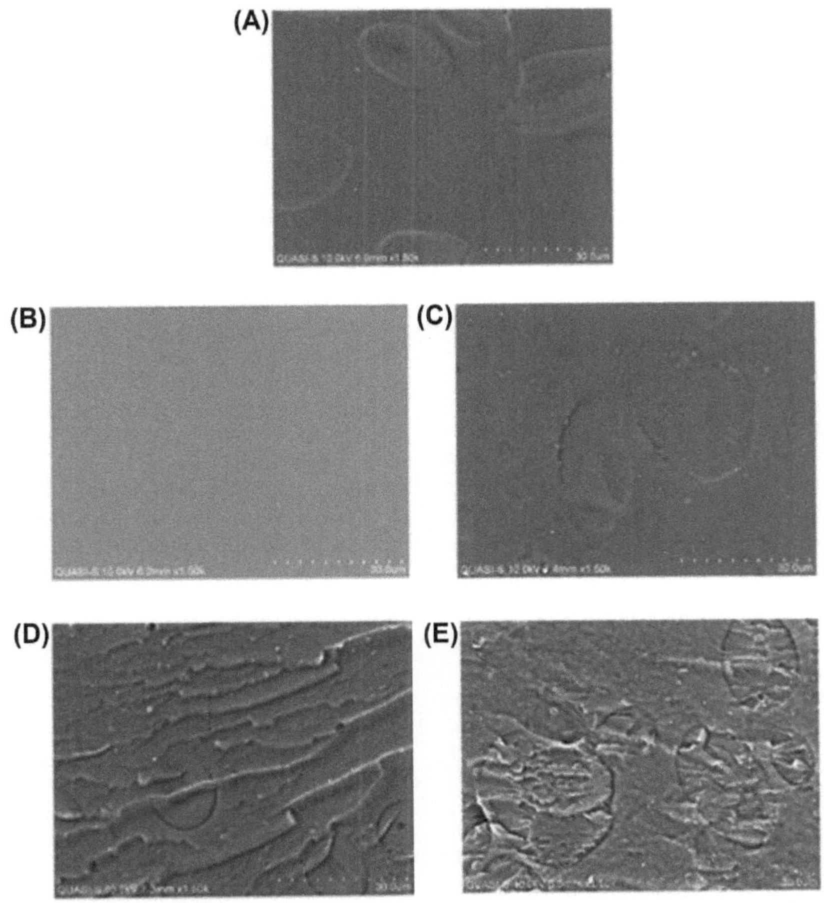

FIG. 2.3 SEM micrograph of fracture surface of thermoplastic SPS blended with different ratio of agar **(A)** 0 wt%, **(B)** 10 wt%, **(C)** 20 wt%, **(D)** 30 wt%, and **(E)** 40 wt%.

various methods. TPS can be fabricated using existing equipment for the manufacturing of commercial plastic products in the plastic industry. Hence, the promising characteristics of TPS provide great opportunities for this material to become a sustainable alternative to synthetic nonbiodegradable polymers. More studies regarding the feasibility of TPS for various applications, especially for high moisture environment, must be explored by researchers to further enhance the readiness of TPS for wider commercial applications.

REFERENCES

Asghari, F., Samiei, M., Adibkia, K., Akbarzadeh, A., Davaran, S., 2017. Biodegradable and biocompatible polymers for tissue engineering application: a review. Artificial Cells, Nanomedicine and Biotechnology 45 (2), 185–192. https://doi.org/10.3109/21691401.2016.1146731.

Avérous, L.,B.N., 2004. Biocomposites based on plasticized starch: thermal and mechanical behaviour. Carbohydrate Polymers 56, 111–122. https://doi.org/10.1016/j.carbpol.2003.11.015.

Bai, W., Hébraud, P., Ashokkumar, M., Hemar, Y., 2017. Investigation on the pitting of potato starch granules during high frequency ultrasound treatment. Ultrasonics Sonochemistry. https://doi.org/10.1016/j.ultsonch.2016.05.022.

Edhirej, A., Sapuan, S.M., Jawaid, M., Zahari, N.I., 2016. Preparation and characterization of cassava starch/peel composite film. Polymer Composites 1–12. https://doi.org/10.1002/pc.

Haciu, D., Saner, S., Türdoğru, O., Ünal, U., 2013. Study of antibacterial effects of a self-standing agar based film incorporated with ZnO. Frontiers in Science 3 (3), 96–101. https://doi.org/10.5923/j.fs.20130303.03.

Ibrahim, H., Farag, M., Megahed, H., Mehanny, S., 2014. Characteristics of starch-based biodegradable composites reinforced with date palm and flax fibers. Carbohydrate Polymers 101, 11–19. https://doi.org/10.1016/j.carbpol.2013.08.051.

Joshi, S.V., Drzal, L.T., Mohanty, A.K., Arora, S., 2004. Are natural fiber composites environmentally superior to glass fiber reinforced composites? Composites Part A: Applied Science and Manufacturing. https://doi.org/10.1016/j.compositesa.2003.09.016.

Jumaidin, R., Sapuan, S.M., Jawaid, M., Ishak, M.R., Sahari, J., 2016. Characteristics of thermoplastic sugar palm starch/agar blend: thermal, tensile, and physical properties. International Journal of Biological Macromolecules 89, 575–581. https://doi.org/10.1016/j.ijbiomac.2016.05.028.

Jumaidin, R., Sapuan, S.M., Jawaid, M., Ishak, M.R., Sahari, J., 2018. Starch: renewable source for thermoplastic. In: Mishra, M. (Ed.), Encyclopedia of Polymer Applications, first ed. Taylor & Francis, pp. 2461–2489. https://doi.org/10.1201/9781351019422-120054068.

Kang, N., Zuo, Y.J., Hilliou, L., Ashokkumar, M., Hemar, Y., 2016. Viscosity and hydrodynamic radius relationship of high-power ultrasound depolymerised starch pastes with different amylose content. Food Hydrocolloids 52, 183–191. https://doi.org/10.1016/j.foodhyd.2015.06.017.

Le Bail, P., Morin, F.G., Marchessault, R.H., 1999. Characterization of a crosslinked high amylose starch excipient. International Journal of Biological Macromolecules 26, 193–200. https://doi.org/10.1016/S0141-8130(99)00082-3.

Lomelí Ramírez, M.G., Satyanarayana, K.G., Iwakiri, S., De Muniz, G.B., Tanobe, V., Flores-Sahagun, T.S., 2011. Study of the properties of biocomposites. Part I. Cassava starch-green coir fibers from Brazil. Carbohydrate Polymers 86, 1712–1722. https://doi.org/10.1016/j.carbpol.2011.07.002.

Lopez, O., Garcia, M.A., Villar, M.A., Gentili, a., Rodriguez, M.S., Albertengo, L., 2014. Thermo-compression of biodegradable thermoplastic corn starch films containing chitin and chitosan. Lebensmittel-Wissenschaft und -Technologie-Food Science and Technology 57 (1), 106–115. https://doi.org/10.1016/j.lwt.2014.01.024.

López, O.V., Ninago, M.D., Lencina, M.M.S., García, M.a., Andreucetti, N.a., Ciolino, A.E., Villar, M.a., 2015. Thermoplastic starch plasticized with alginate–glycerol mixtures: melt-processing evaluation and film properties. Carbohydrate Polymers 126, 83–90. https://doi.org/10.1016/j.carbpol.2015.03.030.

Mohammadi Nafchi, A., Moradpour, M., Saeidi, M., Alias, A.K., 2013. Thermoplastic starches: properties, challenges, and prospects. Starch - Stärke 65, 61–72. https://doi.org/10.1002/star.201200201.

Pasquini, D., Teixeira, E. de M., Curvelo, A.A. da S., Belgacem, M.N., Dufresne, A., 2010. Extraction of cellulose whiskers from cassava bagasse and their applications as reinforcing agent in natural rubber. Industrial Crops and Products. https://doi.org/10.1016/j.indcrop.2010.06.022.

Pereira, L., Sousa, A., Coelho, H., Amado, A.M., Ribeiro-Claro, P.J.A., 2003. Use of FTIR, FT-Raman and 13C-NMR spectroscopy for identification of some seaweed phycocolloids. Biomolecular Engineering 20, 223–228. https://doi.org/10.1016/S1389-0344(03)00058-3.

Peres, A.M., Pires, R.R., Oréfice, R.L., 2016. Evaluation of the effect of reprocessing on the structure and properties of low density polyethylene/thermoplastic starch blends. Carbohydrate Polymers 136, 210–215. https://doi.org/10.1016/j.carbpol.2015.09.047.

Petersen, H., Radosta, S., Vorwerg, W., Kießler, B., 2013. Cationic starch adsorption onto cellulosic pulp in the presence of other cationic synthetic additives. Colloids and Surfaces A: Physicochemical and Engineering Aspects 433, 1–8. https://doi.org/10.1016/j.colsurfa.2013.04.060.

Rabe, S., Sanchez-Olivares, G., Pérez-Chávez, R., Schartel, B., 2019. Natural keratin and coconut fibres from industrial wastes in flame retarded thermoplastic starch biocomposites. Materials 12 (3), 344. https://doi.org/10.3390/ma12030344.

Riyajan, S.A., 2015. Robust and biodegradable polymer of cassava starch and modified natural rubber. Carbohydrate Polymers. https://doi.org/10.1016/j.carbpol.2015.07.038.

Rosa, R.C.R.S., Andrade, C.T., 2004. Effect of chitin addition on injection-molded thermoplastic corn starch. Journal of Applied Polymer Science 92, 2706–2713. https://doi.org/10.1002/app.20292.

Sahari, J., Sapuan, S.M., Zainudin, E.S., Maleque, M.A., 2013. Thermo-mechanical behaviors of thermoplastic starch derived from sugar palm tree (Arenga pinnata). Carbohydrate Polymers 92, 1711–1716. https://doi.org/10.14233/ajchem.2014.15652.

Sahari, J., Salit, M.S., Zainudin, E.S., Maleque, M.A., 2014. Degradation characteristics of SPF/SPS biocomposites fabrication of SPF/SPS biocomposites. Fibres and Textiles in Eastern Europe 22 (5107), 96–98.

Salaberria, A.M., Labidi, J., Fernandes, S.C.M., 2014. Chitin nanocrystals and nanofibers as nano-sized fillers into thermoplastic starch-based biocomposites processed by melt-mixing. Chemical Engineering Journal 256, 356–364. https://doi.org/10.1016/j.cej.2014.07.009.

Sandhu, K.S., Sharma, L., Kaur, M., 2015. Effect of granule size on physicochemical, morphological, thermal and pasting properties of native and 2-octenyl-1-ylsuccinylated potato starch prepared by dry heating under different pH conditions. Lebensmittel-Wissenschaft und -Technologie-Food Science and Technology. https://doi.org/10.1016/j.lwt.2014.11.004.

Scholten, E., Moschakis, T., Biliaderis, C.G., 2014. Biopolymer composites for engineering food structures to control product functionality. Food Structure. https://doi.org/10.1016/j.foostr.2013.11.001.

Seung, D., Soyk, S., Coiro, M., Maier, B.A., Eicke, S., Zeeman, S.C., 2015. PROTEIN TARGETING TO STARCH is required for localising GRANULE-BOUND STARCH SYNTHASE to starch granules and for normal amylose synthesis in arabidopsis. PLoS Biology. https://doi.org/10.1371/journal.pbio.1002080.

Stepto, R.F.T., 2003. The processing of starch as a thermoplastic. Macromolecular Symposia 201, 203–212. https://doi.org/10.1002/masy.200351123.

Taghizadeh, A., Favis, B.D., 2013. Carbon nanotubes in blends of polycaprolactone/thermoplastic starch. Carbohydrate Polymers 98 (1), 189–198. https://doi.org/10.1016/j.carbpol.2013.05.024.

Teodoro, A.P., Mali, S., Romero, N., de Carvalho, G.M., 2015. Cassava starch films containing acetylated starch nanoparticles as reinforcement: physical and mechanical characterization. Carbohydrate Polymers 126, 9–16. https://doi.org/10.1016/j.carbpol.2015.03.021.

Wang, C., Tang, C.H., Fu, X., Huang, Q., Zhang, B., 2016. Granular size of potato starch affects structural properties, octenylsuccinic anhydride modification and flowability. Food Chemistry. https://doi.org/10.1016/j.foodchem.2016.06.006.

Wu, Y., Geng, F., Chang, P.R., Yu, J., Ma, X., 2009. Effect of agar on the microstructure and performance of potato starch film. Carbohydrate Polymers 76 (2), 299–304. https://doi.org/10.1016/j.carbpol.2008.10.031.

Zhang, Y., Rempel, C., Liu, Q., 2014. Thermoplastic starch processing and characteristics-a review. Critical Reviews in Food Science and Nutrition 54 (February), 1353–1370. https://doi.org/10.1080/10408398.2011.636156.

Zhu, Y., Romain, C., Williams, C.K., 2016. Sustainable polymers from renewable resources. Nature 540 (7633), 354–362. https://doi.org/10.1038/nature21001.

CHAPTER 3

Natural Polylactic Acid-Based Fiber Composites: A Review

M.N.M. AZLIN • S.M. SAPUAN • E.S. ZAINUDIN • M.Y.M. ZUHRI • R.A. ILYAS

1 INTRODUCTION

Natural fiber reinforced composite (NFRC) was used in products, for instance, in the aerospace field, marine, transportation, and defense. NFRC is unique because of its strength and stiffness properties (Yahaya et al., 2014). NFRC is a material that is composing fiber as a reinforcement that bound together by a matrix that has a significant influence in terms of mechanical properties (Jumaidin et al., 2019; Liu et al., 2018). Fiber like flax, banana, jute, kenaf, and hemp are among fibers that been proved to be an excellent reinforcement to produce composites (Alavudeen et al., 2015; Misnon et al., 2016; Salman et al., 2015; Tripathi et al., 2018).

The matrices that commonly used are thermoset and thermoplastic. The easiest way to differentiate between thermoset and thermoplastic is based on their behavior when the heat is applied. Each of the matrices will have its melting temperature. Thermoplastic will melt when it is heated beyond a specific temperature, while thermoset will remain in solid state until degradation takes place due to high temperature. A typical example of thermoplastic is polystyrene, polypropylene, polyethylene, poly(vinyl chloride) (Guo, 2017), and polylactic acid (PLA) (Esmaeili et al., 2019), while the example of thermoset is epoxy and unsaturated polyester that are widely used in composite fabrication.

Recently, awareness among people on the usage of natural-based products and eco-friendly products has emerged significantly. Many types of research have been done on NFRC; Moudood et al. (2017) used flax fibers with epoxy to produce NFRC. Baley et al. (2006) also reported that composites reinforced with flax fibers had shown many advantages such as excellent mechanical properties, low density, and biodegradable. Other than flax, fibers, such as hemp (Misnon et al., 2016), kenaf (Aisyah et al., 2019; Asumani et al., 2012; Mazani et al., 2019), jute (Tripathi et al., 2018), banana (Venkateshwaran Narayanan, 2010), sugar palm (Atiqah et al., 2019; Norizan et al., 2020; Nurazzi et al., 2019), pineapple leaf (Indra Reddy et al., 2018), and bamboo (Roslan et al., 2018), have been studied for their mechanical properties and their potential contributions in composite materials. This chapter presents a review of the recent works on NFRC PLA composites.

2 NATURAL FIBERS

2.1 Plant Fibers

The awareness among people in regard to the importance of natural-based products has steered to the increasing usage of natural substances. Nowadays, there are abundantly natural-based fibers that are not yet fully utilized. Plant fibers extracted from stem and leaves have sclerenchyma walls in the form of thickened lignified walls, which make them strong and waterproof. The sclerenchyma cells can elongate and occur in different parts of plants, mainly in the stems and leaves (Carrillo-López and Yahia, 2019; Ilyas et al., 2019a, 2019b; 2020; Lopez and Barclay, 2017). Matured sclerenchyma tissue has dead thickened cell walls containing lignin and high cellulose content (60%–80%) and plays an essential effect by providing structural support for the (Carrillo-López and Yahia, 2019; Lopez and Barclay, 2017) (Table 3.1).

3 POLYLACTIC ACID

Public awareness of the usage of natural-based materials has increased recently. The situation has also influenced the fields such as composite manufacturing to use natural- and biodegradable-based materials. One of the fastest growing and promising composite reinforced materials is PLA. Recently, extensive research efforts have used PLA since biodegradable plastic decreases the adverse impact on the

Advanced Processing, Properties, and Applications of Starch and Other Bio-Based Polymers. https://doi.org/10.1016/B978-0-12-819661-8.00003-2

TABLE 3.1
Chemical Composition and Moisture Content of Plant Fibers (Akil et al., 2011; Alonso Pippo et al., 2011; Bar et al., 2015; Ilyas et al., 2018, 2017; Ilyas et al., 2019a,b,c; Ramamoorthy et al., 2015).

Fibers	Cellulose (%)	Hemicellulose (%)	Lignin (wt%)	Pectin (wt%)	Moisture Content
Abaca	56–63		12–131	1	5–10
Bagasse	32–48	27–32	19–24	15–5	8.8
Bamboo	26–43	15–26	21–31	17–5	8.9
Banana	63–64	10	5	–	10–12
Cereal straw	38–45	15–31	12–20	8	
Cotton	85–90	5.7		0–1	7.85–8.5
Coir	32–43	0.15–0.25	40–45	3–4	8
Flax	71	18.6–20.6	2.2	2.3	8–12
Henequen	77.6	4–8	13.1	–	
Hemp	70–74	17.9–22.4	3.7–5.7	0.9	6.2–12
Jute	61–71.5	13.6–20.4	12–13	0.2	12.5–13.7
Kenaf	45–57	21.5	8–13	3–5	–
Kapok	22–45	22–45	15–22	–	–
Oil palm EFB	65		11	–	–
PALF	70–82		5–12.7		11.8
Ramie	68.6–76.2	13.1–16.7	0.6–0.7	1.9	1.5–17
Sisal	66–78	10–14	10–14	10	10–22
Sugarcane	46	27	23	0	4
Sugar palm	43.9	7.2	33.2	–	8.36

environment due to the accumulation of nonbiodegradable plastic wastes (Hamad et al., 2014).

Instead of PLA, there are many more polymers that are used to produce NFRC, either thermoset or thermoplastic. Thermoplastic is the most frequently used as matric material for NFRC as they are different from thermoset can be remoldable, thus allowing more effective usage of materials for recycling. The properties of PLA also can be tailored during the manufacturing process, such as lowering the glass transition temperature (T_g) and the molecular weight (M_W) (Behera et al., 2018).

4 NATURAL FIBER REINFORCED POLYLACTIC COMPOSITES

4.1 Short Fibers Reinforced PLA Composites

In the world of composite manufacturing, many natural fibers have been used to strengthen PLA as a reinforcement for composite material (NFC). A researcher has used PLA to coat the sisal fiber individually after dissolving PLA in chloroform and it was reinforced into unsaturated polyester (Gupta and Singh, 2018). The incorporation of PLA-coated sisal fibers has increased the flexural and mechanical properties of the composites.

Rajesh and Prasad (2014) have treated jute fibers with NaOH at different concentrations and then mixed with PLA matrix at different weight proportions. The composite sample also was prepared with a various weight proportion of treated and untreated jute fibers PLA matrix. The tensile testing result showed that the treated samples at more fiber loading were improved than untreated samples.

Huda et al. (2008) have fabricated NFRC using kenaf fiber treated with alkaline and silane treatment using PLA as a matrix. The researchers have converted the PLA into sheet form and use it to produce NFRC composites using film stacking method. The result showed that all samples treated with silane (FIBSI) and

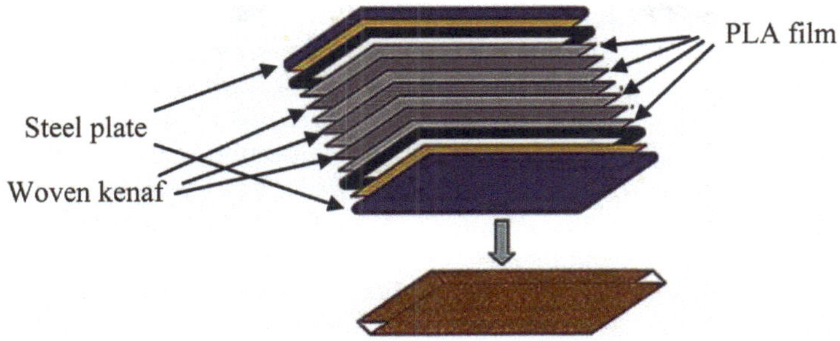

FIG. 3.1 Fabrication procedure of laminated composites (Huda et al., 2008).

alkaline-treated sample (FIBNA) showed higher strength in terms of mechanical properties if compared with NFRC without treatment (FIB) (Fig. 3.1).

Morales et al. (2017) have used the strips of bamboo fiber and bonded with PLA as the matrix. The research used two types of bamboo, which are Moso (*Phyllostachys edulis*), culms, and green bamboo (*Guadua angustifolia* Kunth). Different types of bamboo were used to compare which species can give better strength to the composites for structural application. The bamboo-PLA composite material could be proposed as an alternative to the reference of the study, which is E-glass reinforced epoxy composite.

Agung et al. (2018) also reported that pineapple leaf fiber (PALF) also could be used to mix with PLA to produce green composites. However, in the research, they used maleated anhydride polyethylene as compatibilizing agent. The composite samples were then tested for water absorption, flexural, and tensile tests for the treated PALF and untreated PLAF composites. It can be concluded the hydrophilicity property of natural fiber weakens the flexural and tensile strength of the composites.

Xu et al. (2019) investigated the mechanical properties of hemp/PLA composites. Hemp and PLA fibers were open, cleaned, and mixed using blowing and carding machine to form carded fiber web. The web obtained was later needle punched to get hemp/PLA hybrid fiber mat since the interfacial bonding between both fibers is weak, thus requiring a needle punching process. The hybrid fiber mat and PLA film were oven-dried for 30 min at 105°C to prepare the material for the next process. Finally, the PLA/hemp hybrid fiber mats were hot-pressed with PLA film to form natural reinforced composites.

4.2 Particle Reinforced PLA Composites

Battegazzore et al. (2019) have used side products from agriculture production such as grape stem, alfalfa, and hemp hurd. The morphology, chemical composition, and thermal stability were investigated. The different filler contents (10, 20, 30, 40, and 50 wt%) were melt blended with PLA. They concluded that the best filler among others was hemp hurd in the chip form.

Alias et al. (2019) reported that kenaf core powder also could be used to produce natural fiber reinforced PLA composites. However, in this research, they mixed kenaf core powder and natural rubber using melt-blend and followed by hot pressing. A variation of fiber loading was used, namely 0, 5, 10, 15, and 20 phr. The increment of kenaf loading has increased the water absorption percentages of biocomposite. The poor interfacial adhesion between kenaf core powder and PLA also has weakened the impact strength of the composites.

Nanthakumar et al. (2018) used ground sugarcane leaf fiber (SLF) as a filler to reinforce PLA composites. Five different weight percentages (wt%) of SLF were used, which are 5, 10, 15, and 20. The increasing SLF content at the same time raises Young's modulus and tensile strength of PLA/SLF composites. Nonetheless, the result for elongation at break gave contrary trends to Young's modulus and tensile strength of the composites.

Bamboo fibers also have been reported to be used as filler for PLA matrix reinforced composites (Puspita et al., 2019). The extraction of bamboo microfibrils (BMF) used a combined method of chemical and mechanical treatment. The fiber loading used is 0%, 20%, 30%, 40%, and 50%, hot-pressed at 170°C at the pressure of 10 MPa for 10 min. The elastic modulus and tensile strength showed a significant effect on BMF content.

Wu (2009) investigated the thermal properties of green coconut fiber (GCF) reinforced polylactide (PLA) composites. GCF is a by-product of coconut processing and easily found in Taiwan. The coconut fibers (CF) were blended and followed by drying. The CF then were treated with 0.3% sodium methoxide in a vacuum reactor and quenched with water for 30 min.

4.3 Woven Fabrics Reinforced PLA Composites

Du et al. (2019) studied the properties and potentiality of producing woven jute/PLA composites. Jute woven fabrics were treated with NaOH and combined with PLA by the compression molding method. However, it is reported that alkaline treatment has slightly decreased the strength of the composites and increased the modulus. The small reduction in strength was expected since the decrease in the hemicellulose and lignin. Gunge et al. (2019) investigated the effect of alkaline treatment on mechanical properties of woven banana reinforced polyvinyl alcohol composites.

Kadi et al. (2019) studied the effect of different textile structure to produce biocomposites. Warp knitted, hemp fiber, and woven hemp fabrics were compressed molded for composite production. The tensile modulus and tensile strength results showed the woven hemp/PLA composites gave higher strength as compared with knitted fabrics. They also conclude that the structure of the fabrics does not have a substantial effect on composite strength, but the thickness of the composite has affected the production time.

Durante et al. (2017) studied used different fiber loadings of woven hemp fabric reinforced PLA composites. It was found that the impact strength results showed a higher result with the increment fiber content to the composites.

Rawi et al. (2013) investigated the modulus and tensile strength of composites from bamboo reinforced PLA composites. The impact strength of the woven bamboo/PLA composites in the warp direction is higher than weft direction. The attribute obtained is a result of different breaking forces in different directions and the same findings also can be observed in other research (Porras and Maranon, 2012).

Khan et al. (2016) investigated the effect of woven and nonwoven jute fiber reinforced PLLA composites. The result showed that the impact and flexural and tensile strengths of woven reinforced composites in the warp direction gave the highest results. The results showed a good percentage of improvement for tensile strength (103%), tensile modulus (211%), flexural strength (95.2%), flexural modulus (42.2%), and impact strength (85.9%) in woven reinforced PLLA samples.

4.4 Natural Fiber Reinforced PLA Hybrid Composites

4.4.1 Fiber hybrid composites

There also researches that use several natural fibers such as bamboo (B), coir (C) and kenaf(K) fiber reinforced PLA composites (Yusoff et al., 2016). These fibers were hybridized in three types of unidirectional configuration to reinforce PLA, namely KC/PLA, BC/PLA, and KBC/PLA. The tensile strength of KBC/PLA composites showed the highest result, while KC/PLA revealed the lowest flexural strength (Fig. 3.2).

Siakeng et al. (2018) investigated the properties of coir/pineapple leaf/PLA hybrid composites. The effect of fiber ratio on density, the result for C7P3 samples, which consist of 70% PLA, 21% coir, and 9% of PALF, showed the highest density. Thickness swelling and water absorption result of untreated coir fiber/PALF hybrid composites increased as the coir fiber ratio increased. The C30 (70% PLA/30% coir fiber) showed the highest thickness swelling and water absorption result, whereas P30 (70% PLA/30% PALF) and C1P1 (70% PLA/15% coir fiber/15% PALF) showed least result, respectively.

King et al. (2018) studied the mechanical behavior on bagasse/basalt reinforced PLA hybrid composites.

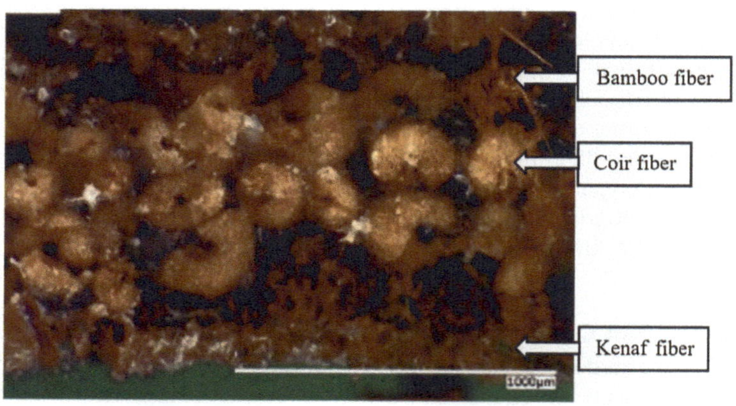

FIG. 3.2 Kenaf/bamboo/coir/coir/bamboo/kenaf (KBCCBK) composites (Yusoff et al., 2016).

The different weights of PLA, BS, and BG were compounded using extruder. The research reveals that 84 wt% of PLA, 12 wt% of basalt fiber, and 4 wt% of bagasse gave the best result as compared with other composites. A good degradation property of the composites has been proven by the water absorption test result with the increment of fiber content, and the water absorption rate also increased.

Asaithambi et al. (2017) investigated the thermal behavior of banana/sisal fibers (BSF) combination reinforced PLA hybrid composites. The fibers were treated with NaOH, followed by benzoyl peroxide (BP). The differential scanning calorimetry test has exhibited that the melting (T_m) and glass transition (T_g) temperatures of the composite samples (virgin PLA, untreated [UT-BSF/PLA], treated [BP-T-BSF/PLA]) have not significantly affected. The adhesion between PLA and BSF for the treated samples showed improvement based on scanning electron micrograph (SEM) micrographs. The effect of treatment has improved the adhesion between BSF/PLA composites due to the elimination of hemicellulose and lignin that affected the tensile properties of BSF. The fiber became stretched and rearranged, thus improving the fiber/matrix adhesion (John et al., 2008).

4.4.2 Woven fabric hybrid composites

Several studies have been done to produce woven hybrid composites. The hybrid can be done using a combination of woven natural and woven synthetic fiber. Manral et al. (2019) studied properties of woven flax/jute/PLA hybrid composites. Hybrid composites of woven flax/jute/PLA composites gave higher modulus and flexural strength, whereas woven flax/PLA gave higher tensile strength. Jute/flax/PLA composites also gave the highest impact strength in comparison with jute/PLA and flax/PLA.

5 PRETREATMENT OF NATURAL FIBERS

5.1 Retting Treatment

5.1.1 Water retting

Water retting is the most conventional method to extract natural fibers for composite fabrication. Fernando et al. (2019) have cut away the top part of the hemp plant and water retted it at 37°C for 5 days. The researcher has compared between dew retted and water retting appearance for hemp fibers and concluded that water-retted hemp produced a softer appearance.

Ruan et al. (2015) studied the characterization of flax water retting of different durations (2, 6, and 10 days). Based on the appearance, water-retted flax with the increment retting duration significantly increased the whiteness of the flax fibers. It is also was found that a retting duration of 6 days is suitable for retting flax due to excellent retting efficiency and fibers of consistent color, fineness, and tensile properties were obtained.

5.1.2 Enzymatic retting

Enzymatic retting is a modification of water retting and sometimes are called bioscouring. Wong et al. (2016) used *Aspergillus fumigatus* R6 pectinase in the kenaf retting process. Diluted pectinase enzyme (*A. fumigatus* R6) was used to treat kenaf fibers and incubated in a box at 30°C for a variation of hours, namely 0, 8, 16, 24, 32, and 40 h. The samples were taken out in 8-hour intervals and washed with tap water. Retting time of 32 h gave the highest tensile strength (458 MPa). The researcher also has compared *A. fumigatus* R6 pectinase enzyme with other sources of pectinase enzyme (*Bacillus subtilis* ADI1 culture filtrate), and the tensile test result displayed no apparent differences between these two enzymes. Therefore, it can be concluded that the retting of kenaf using *A. fumigatus* R6 pectinase was effective.

The influence of enzymatic treatment on flax also has been studied by De Prez et al. (2019). Three enzymes were used, which are Rohapect MPE (a pectin methylesterase), Rohapect PTE (a pectin lyase), and both from AB enzymes and polygalacturonase from *Aspergillus niger*. The treatment was implemented on flax stem, which needs to be dried for 24 h at 105°C before being processed.

5.2 Chemical Treatment

Various methods can be used to improve the fiber properties for composite manufacturing. Some of the researchers have treated the fibers with NaOH (Huda et al., 2008; Li et al., 2007), silane (Agrawal et al., 2000), acetylation (Bogoeva-Gaceva et al., 2007), hydrogen peroxide (Razak et al., 2014), and sodium bicarbonate (Fiore et al., 2016). These treatments will modify the surface of the fibers and at the same time improve the fiber properties.

5.2.1 Alkaline treatment

Rajesh and Prasad (2014) studied the treated jute/PLA composites in terms of tensile properties. The retted fiber was immersed in water for 1 h, dried at 50°C, and cut into 3 mm length before the treatment. The alkaline treatment was done using three different concentrations (5%, 10%, 15%) of NaOH solution for 6 h at 70°C (Li et al., 2007; Liu and Dai, 2007). The different weight proportions were also used (5%, 10%, 15%, 20%, 25%) in PLA matrix.

Islam et al. (2010) studied the effect of alkaline treatment on hemp/PLA composites. The retted hemp fibers were placed into stainless steel canister and the solution of sodium sulphite (Na_2SO_3) followed by the solution of sodium hydroxide (NaOH) were then poured into the canister. The ratio of fibres to 2 wt.% sodium sulphite (Na_2SO_3) and 5 wt.% sodium hydroxide (NaOH) solution was 1:2:10 by weight. The canister was placed into pulp digester at 120°C for 60 min. The fibers were then washed about 45 min until pH 7 was achieved and was oven-dried for 48 h at 70°C. The alkaline treatment has enhanced the interfacial bonding between hemp-PLA, thus increasing the strength of the composites.

5.2.2 Alkali and silane treatment

Yu et al. (2010) treated ramie fiber (RF) with alkaline and silane treatment. The RF was cut into 10 mm in average length. RF was first dipped in a 5% NaOH solution for 3 h at room temperature. The RF was then washed with the mixture of few drops of acetic acid and distilled water. The treated fires were air-dried for 72 h and vacuum-dried for 6 h at 80°C.

Tran et al. (2014) treating husk with combined alkaline and silane surface treatments. Einkorn wheat husks and the long-grain rice were vacuum-dried for 24 h at 100°C. The husks were immersed in a NaOH solution at 2%, 5%, and 10% for 24 h at room temperature. In this study, as for NaOH 5% concentration, both husks were immersed for 6, 12, 18, 24, and 48 h at room temperature. The husks then were washed with 1% concentration of acetic acid and tap water. The husks were washed and dried two times separately. First, the wet husks were air-dried for 3 days and followed by vacuum-dried at 100°C for 6 h. They concluded that composites produced, that is, alkaline-silane-treated husk composites, gave higher stress and bending moduli than untreated husks and silane-treated composites.

5.2.3 Hydrogen peroxide

Marwah et al. (2014) investigated the influence of hydrogen peroxide (H_2O_2) treatment on oil palm empty fruit bunch fiber (OPEFBF)/PLA composites. The oxidizing bleaching agent (H_2O_2) has caused discoloration of the fiber and became brighter. The perhydroxyl ions (HOO^-) have been generated due to the separation of hydrogen peroxide in alkaline solution, causing the fiber to decolourize. The perhydroxyl ions will attack chromophores containing groups of lignin and cellulose. The dissociation of hydrogen peroxide (H_2O_2) in alkaline medium to form perhydroxyl ions is shown in the following equation (Maekawa et al., 2007):

$$H_2O_2 + OH^- \rightarrow H_2O + HOO^-$$

Razak et al. (2014) studied on the bleaching effect of kenaf/PLA composites. The treatment was done at 80°C, pH 11, and for 60 min. Melt blending method was used to compound the bleached kenaf fiber with PLA. It was found that the treatment has enhanced the mechanical properties of the composites. Morphological analysis using SEM also showed that the interfacial adhesion between bleached kenaf fiber and PLA improved.

6 PROCESSING METHODS

The process to form NFRC is quite challenging because of processing different forms of fibers, such as chopped fiber strands, short fiber, long fibers, randomly oriented fiber mat, and woven fabric. Common methods to process NFRC are compression molding (Memon and Nakai, 2013) and injection (Huda et al., 2008; Li et al., 2011).

6.1 Injection Molding

The injection molding process is used to fabricate materials in the form of short to intermediate length fibers and highly dependant on the length of the pellets after the compounding process (Leong et al., 2013). Anuar et al. (2012) have opted for the injection molding method to produce kenaf fiber/PLA composites. Kenaf fiber and PLA were mixed and extruded using twin screw extruder between 180°C and 190°C in all zones. Pelletized kenaf/PLA composite was allowed to crystallize at 120°C for 2 h. The composites were injection molded at a temperature range of 165°C to 190°C.

Kaewpirom and Worrarat (2014) convert PALF fiber and PLA into pellet form using twin screw extruder at 120°C uniform temperature. Type I tensile specimen according to ASTM D638 standard method (2014) used injection molding process. The pellets were dried in a desiccator at 80°C for 4 h before being injection molded. The temperature was set at 120°C for all zones, and the mold surface temperature was set at 50°C and cooled for 25 s. Filling and holding pressures were 130 and 70 MPa correspondingly (Fig. 3.3).

6.2 Compression Molding

The compression molding is a versatile method that can be used for almost all forms of natural fibers. However, the forming process requires a longer time if compared with the injection molding method. Natural fibers in the form of mat and film can be sandwiched between thermoplastic resin and preheated before

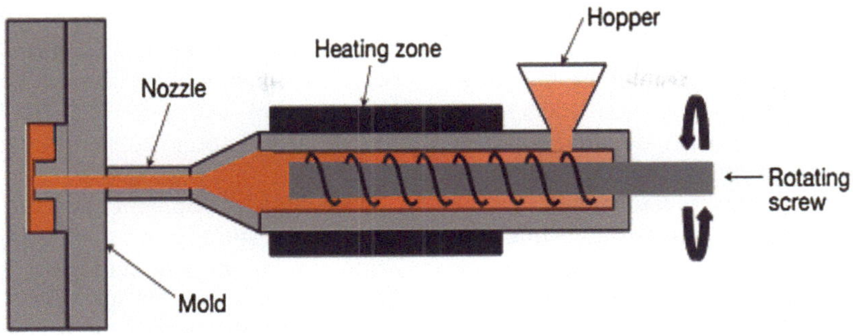

FIG. 3.3 Schematic diagram of injection molding for polymers (Wang et al., 2019).

the compression process. Preheating of the resin is important to ensure the thermoplastic polymer sufficiently melts to avoid the uneven flow of resin of natural fibers. The preheating process is followed by the compression of composites at a certain temperature, pressure, and molding time. The compressed composite samples are then cooled while the pressure is made constant. After cooling, the samples will be demolded, and the excess resin on the sides of the molding will be removed by trimming or grinding (Leong et al., 2013) (Fig. 3.4).

Siengchin and Wongmanee (2014) used a compression molding method to produce 4/4 hopsack and 2/2 twill woven flax fabric reinforced PLA composites. The layers of woven $twill_{2\times2}$ and $hopsack_{4\times4\ flax}$ fabrics were compressed with PLA sheets at 10 m/h of infeed rate. The pressure was set at 20 bars, while the temperature was set at 200°C. Based on the SEM test, it was found that the interfacial gaps between PLA and woven flax became small using the compression method. The result obtained also showed that the hot pressing method is the best method to produce woven flax reinforced PLA composites.

7 MECHANICAL PROPERTIES OF NFR/PLA

A lot of works have been detailed out regarding the mechanical properties of NFR/PLA composites. Mukherjee and Kao (2011) stated that several factors are affecting the mechanical properties of NFR/PLA composites. The mechanical properties of the NFR/PLA composites highly depended on the fiber-matrix adhesion, which related to the processing conditions. Processing condition influences the degree of fiber wetting, crystallinity, and composite porosity. Heinemann and Fritz (2005) assumed that the hydrophilic nature of PLA had generated an excellent adhesion property between PLA and natural fibers. Findings from Mukherjee and Kao (2011) also showed an agreement with what has been reported by Fratzl et al. (2004), concerning the dependency of adhesion between fiber-matrix would affect the mechanical properties of NFR/PLA composites.

Yu et al. (2014) treated short RF with maleic anhydride (MA) to produce ramie/PLA composites. Benzoyl peroxide (BPO) has been used as initiator for MA and PLA. The mechanical property result was improved with the addition of PLA-g-MA, which can give a better adhesion between fiber-matrix. Ramie reinforced PLA composite with 3% of MA showed the highest tensile strength of 64.30 Mpa, flexural strength of 112.4 MPa, and impact strength of 7.1 kJ/m^2. Shih and Huang (2010) used NaOH and silane to produce modified banana fiber (MBF). The MBF later was mixed with PLA and dicumyl peroxide to produce composites. The mechanical test reveals that the flexural and tensile strength of the PLA-g-MA composites hiked with the increment of fiber content that was showing the highest 78.60 and 65.40 MPa, respectively, for 40 phr fiber. Plackett et al. (2003) also

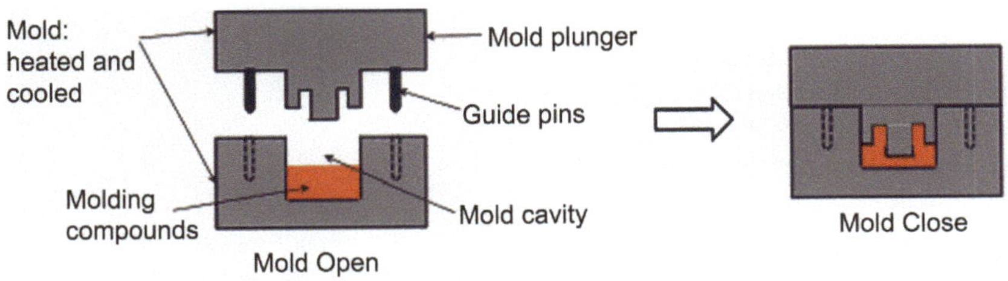

FIG. 3.4 Schematic diagram of compression molding for polymers (Wang et al., 2019).

researched the properties of jute/PLA composites. The finding shows that the tensile strength increased and depended on the heating temperature in the range 210–220°C for the highest tensile strength (100.5 Mpa).

Tayommai and Aht-Ong (2010) also studied about mechanical properties of CF/PLA composites. The increment of Cf content in composites has deteriorated the mechanical properties of the composites while vice versa for impact strength. The weak interfacial adhesion between CF-PLA has been identified to cause the result. Clean edges and large void can be seen between CF-PLA using SEM micrographs.

Works on investigating the mechanical properties of bamboo fiber reinforced PLA composites also have been presented by Maruyama et al. (2019). The researchers used short bamboo fiber (SBF) treated with silane to reinforce PLA composites. The results show that after the composites were immersed in water for 250 h, the tensile strength of all SBF/PLA composites diminished.

As for untreated SBF/PLA composites, the results reduced by 55%, while for silane-treated composites, it decreased by 45% only. The results have proven that the silane treatment has affected the interfacial adhesion for NFRC samples.

8 APPLICATION OF NFRC

The increasing usage of petroleum-based polymers has caused environmental pollution, and at the same time, disturbing the balance of nature. As for the past few decades, the use of conventional polymers can be found in various applications. The situation is due to the advantages of this synthetic polymer, which is versatile, easy to process, lightweight, and durable. However, due to their nonbiodegradable property and also contribution to the accumulation of waste, people are now aware of the importance of reducing the consumption of petroleum-based polymers, reusing, and recycling the material. The biopolymer can be said as a new generation of polymers that are still in the development phase to replace synthetic polymers to produce environmental-friendly products.

8.1 Packaging

Several requirements need to be taken into consideration before the products can be used in the packaging area. The material supposed to have the ability to make the item safe from chemical and physical damage to the products. The requirement also covers the ability of the material to provide simplicity during transportation and handling of the item (Duhovic et al., 2008).

Rawi et al. (2014) have suggested using woven bamboo/PLA composites in the packaging products since they show a better thermal and impact strength in comparison with pure PLA. The impact strength of pure PLA was about 200% lower than bamboo/PLA composites.

Existing development necessity of eco-friendly packaging is stimulated by the rising number of packaging regulations and standards. The application of PLA also has expanded into the NFRC industry and other commodity areas supported by the recent available technology. The advancement of technology has lowered the production cost (Babu et al., 2013; Rose and Palkovits, 2011).

8.2 Structural Application

Morales et al. (2017) investigated the possibility to use bamboo fiber reinforced PLA composites for structural application. The researchers also reported in compression test, and the bamboo reinforced PLA composites showed about 84% of tensile strength, which is comparable with aerospace composites. The results showed that the composites have a good potential to be used for structural application but need to improve the humidity problem by coating to avoid deterioration of mechanical properties of composites (Fig. 3.5).

8.3 Automotive Application

The automotive industry is among the leading end user of raw materials in the manufacturing area. Approximately 50 million vehicles have been produced yearly, including 9 million tons of polymer materials (Cunha et al., 2006). The incorporation of NFRC in the automotive industry

FIG. 3.5 SmartFit PLA Bamboo ("SmartFit PLA - Natural Fibers - Print-Me," 2019).

TABLE 3.2
Applications of Natural Fibers in the Automotive Industry (Jamrichov; and Akov, 2013; Ngo, 2018; Peças et al., 2018).

	Manufacturer	Model	Application
1	Audi	Coupe, Roadster, A2, A3, A4, A6, A8,	Spare tire and boot lining, door panels (back and side), hat rack, and seat backs
2	BMW	3, 5, 7 Series	Headliner, noise insulation and door panels, seat backs, and boot lining
3	Citroën	C5	Door panels (interior)
4	Fiat	Alfa Romeo 146, 156, Punto, Marea, Brava	Door panels
5	Ford	Focus, Cadillac de Ville, Mondeo CD 162	Boot lining, seat parts, and door panels
6	General Motors	Chevrolet Trailblazer	Seat backs and floor mat (car boot)
7	Lotus	Eco Elise	Interior floor mats, seats, body panels, and spoiler
8	Peugeot	406	Rear parcel shelf and seat backs
9	Renault	Twingo, Clio	Rear parcel shelf
10	Rover	2000 and others	Rear storage shelf and insulation,
11	Saab	9s	Seat backs and door panels
12	Toyota	Raum, Brevis, Celsior, Harrier	Spare tire cover, seat backs, and door panels
13	Volkswagen	Bora, Golf, Passat	Seat backs, door panels, boot lid finish panel, and lining
14	Volvo	V70, C70	Boot floor tray, seat padding, and natural foams

by the European car manufacturers can be indicated by the application of car-related parts such as headliners, dashboard and interior parts in their car. The use of natural fibers, such as flax, kenaf, hemp, sisal, and jute fibers, lead to the production of lightweight products (Holbery and Houston, 2006). Jamrichov and Akov (2013) reported that the usage of natural-based fiber in car production is from 80,000 to 160,000 tones per year in Western Europe. The German-based car manufacturer is a leader in using natural fibers in the car parts. Daimler-Chrysler is also reported to initiate programs in the development of natural-based products that will benefit the agriculture industry.

Cunha et al. (2006) reported that Toyota Raum 2003 model is among the cars that used NFRC as a spare tire cover. The cover is made by kenaf fiber reinforced with PLA to produce composites. During EcoInnovAsia 2008, Mazda 5 has been using kenaf/PLA composites as seat covers and claimed that almost 30% of the interior of Mazda 5 components are made from bio-based material (Suddell, 2008) (Table 3.2) .

9 CONCLUSION

NFR/PLA composite has become a popular choice among industry players to use as a basic component in the manufacture of their products. The manufacturing industry has grown rapidly as these composites offer features such as ease of processing, lighter end products, appropriate strength, and renewable properties. Pretreated natural fiber with various surface treatment has improved the strength and interfacial adhesion of the fiber when combined with PLA composites (Islam et al., 2010; Marwah et al., 2014; Rajesh and Prasad, 2014; Tran et al., 2014; Yu et al., 2010). The drawbacks of using natural fibers are hydrophilic nature of natural fiber, which can affect the strength of the composites, and it is suggested to hybrid the composites with other synthetic fibers in the products to reduce this problem. However, with the advancement of technology, the natural-based composites that have gained attention from people hopefully can gradually replace the use of petroleum-based fibers in composite manufacturing for industrial usage.

The situation is a good prospect to the agriculture-based country to grab the chances and also an opportunity for us to reduce the environmental impact issues around the world.

REFERENCES

Agrawal, R., Saxena, N.S., Sharma, K.B., Thomas, S., Sreekala, M.S., 2000. Activation energy and crystallization kinetics of untreated and treated oil palm fibre reinforced phenol formaldehyde composites. Materials Science and Engineering A 277, 77–82. https://doi.org/10.1016/s0921-5093(99)00556-0.

Agung, E.H., Hamdan, M.H.M., Siregar, J.P., Bachtiar, D., Tezara, C., Jamiluddin, J., 2018. Water absorption behaviour and mechanical performance of pineapple leaf fibre reinforced polylactic acid composites. International Journal of Automotive and Mechanical Engineering 15, 5760–5774. https://doi.org/10.15282/ijame.15.4.2018.4.0441.

Aisyah, H.A., Paridah, M.T., Sapuan, S.M., Khalina, A., Berkalp, O.B., Lee, S.H., Lee, C.H., Nurazzi, N.M., Ramli, N., Wahab, M.S., Ilyas, R.A., 2019. Thermal properties of woven kenaf/carbon fibre-reinforced epoxy hybrid composite panels. International Journal of Polymer Science 2019, 1–8. https://doi.org/10.1155/2019/5258621.

Akil, H.M., Omar, M.F., Mazuki, A.A.M., Safiee, S., Ishak, Z.A.M., Abu Bakar, A., 2011. Kenaf fiber reinforced composites: a review. Materials and Design 32, 4107–4121. https://doi.org/10.1016/j.matdes.2011.04.008.

Alavudeen, A., Rajini, N., Karthikeyan, S., Thiruchitrambalam, M., Venkateshwaren, N., 2015. Mechanical properties of banana/kenaf fiber-reinforced hybrid polyester composites: effect of woven fabric and random orientation. Materials and Design 66, 246–257. https://doi.org/10.1016/j.matdes.2014.10.067.

Alias, N.F., Ismail, H., Ishak, K.M.K., 2019. The effect of kenaf loading on water absorption and impact properties of polylactic acid/natural rubber/kenaf core powder biocomposite. Materials Today: Proceedings 17, 584–589. https://doi.org/10.1016/j.matpr.2019.06.338.

Alonso Pippo, W., Luengo, C.A., Alonsoamador Morales Alberteris, L., Garzone, P., Cornacchia, G., 2011. Energy recovery from sugarcane-trash in the light of 2nd generation biofuel. Part 2: socio-economic aspects and techno-economic analysis. Waste and Biomass Valorization 2, 257–266. https://doi.org/10.1007/s12649-011-9069-3.

Anuar, H., Zuraida, A., Kovacs, J.G., Tabi, T., 2012. Improvement of mechanical properties of injection-molded polylactic acid-kenaf fiber biocomposite. Journal of Thermoplastic Composite Materials 25, 153–164. https://doi.org/10.1177/0892705711408984.

Asaithambi, B., Ganesan, G.S., Ananda Kumar, S., 2017. Banana/sisal fibers reinforced poly(lactic acid) hybrid biocomposites; influence of chemical modification of BSF towards thermal properties. Polymer Composites 38, 1053–1062. https://doi.org/10.1002/pc.23668.

ASTM, 2014. D638 – 14 Standard Test Method for Tensile Properties of Plastics, vol. 17. https://doi.org/10.1520/D0638-14.1.

Asumani, O.M.L., Reid, R.G., Paskaramoorthy, R., 2012. The effects of alkali-silane treatment on the tensile and flexural properties of short fibre non-woven kenaf reinforced polypropylene composites. Composites Part A: Applied Science and Manufacturing 43, 1431–1440. https://doi.org/10.1016/j.compositesa.2012.04.007.

Atiqah, A., Jawaid, M., Sapuan, S.M., Ishak, M.R., Ansari, M.N.M., Ilyas, R.A., 2019. Physical and thermal properties of treated sugar palm/glass fibre reinforced thermoplastic polyurethane hybrid composites. Journal of Materials Research and Technology. https://doi.org/10.1016/j.jmrt.2019.06.032.

Babu, R.P., O'Connor, K., Seeram, R., 2013. Current progress on bio-based polymers and their future trends. Progress in Biomaterials 2, 8. https://doi.org/10.1186/2194-0517-2-8.

Baley, C., Perrot, Y., Busnel, F., Guezenoc, H., Davies, P., 2006. Transverse tensile behaviour of unidirectional plies reinforced with flax fibres. Materials Letters 60, 2984–2987. https://doi.org/10.1016/j.matlet.2006.02.028.

Bar, M., Alagirusamy, R., Das, A., 2015. Flame retardant polymer composites. Fibers and Polymers 16, 705–717. https://doi.org/10.1007/s12221-015-0705-6.

Battegazzore, D., Noori, A., Frache, A., 2019. Natural wastes as particle filler for poly(lactic acid)-based composites. Journal of Composite Materials 53, 783–797. https://doi.org/10.1177/0021998318791316.

Behera, K., Sivanjineyulu, V., Chang, Y.H., Chiu, F.C., 2018. Thermal properties, phase morphology and stability of biodegradable PLA/PBSL/HAp composites. Polymer Degradation and Stability 154, 248–260. https://doi.org/10.1016/j.polymdegradstab.2018.06.010.

Bogoeva-Gaceva, G., Avella, M., Malinconico, M., Buzarovska, A., Grozdanov, A., Gentile, G., Errico, M.E., 2007. Natural fiber eco-composites. Polymer Composites 28, 98–107. https://doi.org/10.1002/pc.

Carrillo-López, A., Yahia, E.M., 2019. Morphology and anatomy. Postharvest Physiology and Biochemistry of Fruits and Vegetables 113–130. https://doi.org/10.1016/b978-0-12-813278-4.00006-3.

Cunha, A.M., Campos, A.R., Cristovão, C., Vila, C., Santos, V., Parajó, J.C., 2006. Sustainable materials in automotive applications. Plastics, Rubber and Composites 35, 233–241. https://doi.org/10.1179/174328906X146487.

De Prez, J., Van Vuure, A.W., Ivens, J., Aerts, G., Van de Voorde, I., 2019. Effect of enzymatic treatment of flax on ease of fiber extraction and chemical composition. BioResources 14, 1–16.

Du, S., Peng, X., Gu, H., 2019. Experimental investigation on fabrication and thermal-stamping of woven jute/polylactic acid biocomposites. Journal of Composite Materials 53, 851–861. https://doi.org/10.1177/0021998318792081.

Duhovic, M., Peterson, S., Jayaraman, K., 2008. Natural-fibre-biodegradable polymer composites for packaging. Properties and Performance of Natural-Fibre Composites 301–329. https://doi.org/10.1533/9781845694593.2.301.

Durante, M., Formisano, A., Boccarusso, L., Langella, A., Carrino, L., 2017. Creep behaviour of polylactic acid reinforced by woven hemp fabric. Composites Part B: Engineering 124, 16–22. https://doi.org/10.1016/j.compositesb.2017.05.038.

Esmaeili, M., Pircheraghi, G., Bagheri, R., Altstädt, V., 2019. Poly(lactic acid)/coplasticized thermoplastic starch blend: effect of plasticizer migration on rheological and mechanical properties. Polymers for Advanced Technologies 30, 839–851. https://doi.org/10.1002/pat.4517.

Fernando, D., Daniel, G., Thygesen, A., Meyer, A.S., 2019. Elucidating field retting mechanisms of hemp fibres for biocomposites: effects of microbial actions and interactions on the cellular micro-morphology and ultrastructure of hemp stems and bast fibres. BioResources 14, 4047–4084. https://doi.org/10.15376/biores.14.2.4047-4084.

Fiore, V., Scalici, T., Nicoletti, F., Vitale, G., Prestipino, M., Valenza, A., 2016. A new eco-friendly chemical treatment of natural fibres: effect of sodium bicarbonate on properties of sisal fibre and its epoxy composites. Composites Part B: Engineering 85, 150–160. https://doi.org/10.1016/j.compositesb.2015.09.028.

Francis Luther King, M., Arul Jeyakumar, A., Srinivasan, V., 2018. Investigation of mechanical behaviour on bagasse/basalt reinforced poly lactic acid hybrid composites: tensile, flexural, impact and water absorption. Journal of Advanced Microscopy Research 13, 85–91. https://doi.org/10.1166/jamr.2018.1362.

Fratzl, P., Burgert, I., Gupta, H.S., 2004. On the role of interface polymers for the mechanics of natural polymeric composites. In: Physical Chemistry Chemical Physics, pp. 5575–5579. https://doi.org/10.1039/b411986j.

Gunge, A., Koppad, P.G., Nagamadhu, M., Kivade, S.B., Murthy, K.V.S., 2019. Study on mechanical properties of alkali treated plain woven banana fabric reinforced biodegradable composites. Composites Communications 13, 47–51. https://doi.org/10.1016/j.coco.2019.02.006.

Guo, Q., 2017. Thermosets: Structure, Properties, and Applications, second ed. https://doi.org/10.1016/C2015-0-06205-0

Gupta, M.K., Singh, R., 2018. Flexural and dynamic mechanical analysis (DMA) of polylactic acid (PLA) coated sisal fibre reinforced polyester composite. Materials Today: Proceedings 5, 6109–6114. https://doi.org/10.1016/j.matpr.2017.12.216.

Hamad, K., Kaseem, M., Ko, Y.G., Deri, F., 2014. Biodegradable polymer blends and composites: an overview. Polymer Science - Series A 56, 812–829. https://doi.org/10.1134/S0965545X14060054.

Heinemann, M., Fritz, G.H., 2005. Polylactide structure, properties and applications 19. In: Stuttgater, Plastic Congress.

Holbery, J., Houston, D., 2006. Natural-fiber-reinforced polymer composites in automotive applications. Journal of Occupational Medicine 58, 80–86. https://doi.org/10.1007/s11837-006-0234-2.

Huda, M.S., Drzal, L.T., Mohanty, A.K., Misra, M., 2008. Effect of fiber surface-treatments on the properties of laminated biocomposites from poly(lactic acid) (PLA) and kenaf fibers. Composites Science and Technology 68, 424–432. https://doi.org/10.1016/j.compscitech.2007.06.022.

Ilyas, R.A., Sapuan, S.M., Atiqah, A., Ibrahim, R., Abral, H., Ishak, M.R., Zainudin, E.S., Nurazzi, N.M., Atikah, M.S.N., Ansari, M.N.M., Asyraf, M.R.M., Supian, A.B.M., Ya, H., 2019a. Sugar palm (*Arenga pinnata* [*Wurmb.*] *Merr*) starch films containing sugar palm nanofibrillated cellulose as reinforcement: water barrier properties. Polymer Composites 1–9. https://doi.org/10.1002/pc.25379.

Ilyas, R.A., Sapuan, S.M., Ibrahim, R., Abral, H., Ishak, M.R., Zainudin, E.S., Asrofi, M., Atikah, M.S.N., Huzaifah, M.R.M., Radzi, A.M., Azammi, A.M.N., Shaharuzaman, M.A., Nurazzi, N.M., Syafri, E., Sari, N.H., Norrahim, M.N.F., Jumaidin, R., 2019b. Sugar palm (*Arenga pinnata* (*Wurmb.*) *Merr*) cellulosic fibre hierarchy: a comprehensive approach from macro to nano scale. Journal of Materials Research and Technology 8, 2753–2766. https://doi.org/10.1016/j.jmrt.2019.04.011.

Ilyas, R.A., Sapuan, S.M., Ibrahim, R., Abral, H., Ishak, M.R., Zainudin, E.S., Atikah, M.S.N., Mohd Nurazzi, N., Atiqah, A., Ansari, M.N.M., Syafri, E., Asrofi, M., Sari, N.H., Jumaidin, R., 2019c. Effect of sugar palm nanofibrillated cellulose concentrations on morphological, mechanical and physical properties of biodegradable films based on agro-waste sugar palm (*Arenga pinnata* (*Wurmb.*) *Merr*) starch. Journal of Materials Research and Technology 8, 4819–4830. https://doi.org/10.1016/j.jmrt.2019.08.028.

Ilyas, R.A., Sapuan, S.M., Ibrahim, R., Abral, H., Ishak, M.R., Zainudin, E.S., Atiqah, A., Atikah, N., Syafri, E., Asrofi, M., Jumaidin, R., 2020. Thermal, biodegradability and water barrier properties of bio-nanocomposites based on plasticised sugar palm starch and nanofibrillated celluloses from sugar palm fibres. Journal of Biobased Materials and Bioenergy 14, 1–13. https://doi.org/10.1166/jbmb.2020.1951.

Ilyas, R.A., Sapuan, S.M., Ishak, M.R., 2018. Isolation and characterization of nanocrystalline cellulose from sugar palm fibres (*Arenga pinnata*). Carbohydrate Polymers 181, 1038–1051. https://doi.org/10.1016/j.carbpol.2017.11.045.

Ilyas, R.A., Sapuan, S.M., Ishak, M.R., Zainudin, E.S., 2017. Effect of delignification on the physical, thermal, chemical, and structural properties of sugar palm fibre. BioResources 12, 8734–8754. https://doi.org/10.15376/biores.12.4.8734-8754.

Indra Reddy, M., Anil Kumar, M., Rama Bhadri Raju, C., 2018. Tensile and flexural properties of jute, pineapple leaf and glass fiber reinforced polymer matrix hybrid composites. Materials Today: Proceedings 5, 458–462. https://doi.org/10.1016/j.matpr.2017.11.105.

Islam, M.S., Pickering, K.L., Foreman, N.J., 2010. Influence of alkali treatment on the interfacial and physico-mechanical properties of industrial hemp fibre reinforced polylactic acid composites. Composites Part A: Applied Science and Manufacturing 41, 596–603. https://doi.org/10.1016/j.compositesa.2010.01.006.

Jamrichov, Z., Akov, E., 2013. Mechanical testing of natural fiber composites for automotive industry. University Review 7, 20–25.

John, M.J., Francis, B., Varughese, K.T., Thomas, S., 2008. Effect of chemical modification on properties of hybrid fiber biocomposites. Composites Part A: Applied Science and Manufacturing 39, 352–363. https://doi.org/10.1016/j.compositesa.2007.10.002.

Jumaidin, R., Afif, M., Khiruddin, A., Asyul, Z., Saidi, S., Sapuan, M., Ahmad, R., 2019. Effect of cogon grass fibre on the thermal, mechanical and biodegradation properties of thermoplastic cassava starch biocomposite. International Journal of Biological Macromolecules. https://doi.org/10.1016/j.ijbiomac.2019.11.011.

kadi, N., Baghaei, B., skrifvars, M., 2019. Effect of Textile structure in the process parameters of thermoplastic biocomposite. MATEC Web of Conferences 261, 01005. https://doi.org/10.1051/matecconf/201926101005.

Kaewpirom, S., Worrarat, C., 2014. Preparation and properties of pineapple leaf fiber reinforced poly(lactic acid) green composites. Fibers and Polymers 15, 1469–1477. https://doi.org/10.1007/s12221-014-1469-0.

Khan, G.M.A., Terano, M., Gafur, M.A., Alam, M.S., 2016. Studies on the mechanical properties of woven jute fabric reinforced poly(L-lactic acid) composites. Journal of King Saud University - Engineering Sciences 28, 69–74. https://doi.org/10.1016/j.jksues.2013.12.002.

Leong, Y.W., Thitithanasarn, S., Yamada, K., Hamada, H., 2013. Compression and Injection Molding Techniques for Natural Fiber Composites, Natural Fibre Composites: Materials, Processes and Applications. Woodhead Publishing Limited. https://doi.org/10.1533/9780857099228.2.216.

Li, X., Tabil, L.G., Panigrahi, S., 2007. Chemical treatments of natural fiber for use in natural fiber-reinforced composites: a review. Journal of Polymers and the Environment 15, 25–33. https://doi.org/10.1007/s10924-006-0042-3.

Li, Z., Zhou, X., Pei, C., 2011. Effect of sisal fiber surface treatment on properties of sisal fiber reinforced polylactide composites. International Journal of Polymer Science 2011. https://doi.org/10.1155/2011/803428.

Liu, X.Y., Dai, G.C., 2007. Surface modification and micromechanical properties of jute fiber mat reinforced polypropylene composites. Express Polymer Letters 1, 299–307. https://doi.org/10.3144/expresspolymlett.2007.43.

Liu, Y., Ma, Y., Yu, J., Zhuang, J., Wu, S., Tong, J., 2018. Development and characterization of alkali treated abaca fiber reinforced friction composites. Composite Interfaces 6440, 1–16. https://doi.org/10.1080/09276440.2018.1472456.

Lopez, F.B., Barclay, G.F., 2017. Plant Anatomy and Physiology, Pharmacognosy: Fundamentals, Applications and Strategy. Elsevier Inc. https://doi.org/10.1016/B978-0-12-802104-0.00004-4.

Maekawa, M., Hashimoto, A., Tahara, M., 2007. Effects of pH in hydrogen peroxide bleaching of cotton fabrics pretreated with ferrous sulfate. Textile Research Journal 77, 222–226. https://doi.org/10.1177/0040517507078795.

Manral, A., Ahmad, F., Chaudhary, V., 2019. Static and dynamic mechanical properties of PLA bio-composite with hybrid reinforcement of flax and jute. Materials Today: Proceedings 2–5. https://doi.org/10.1016/j.matpr.2019.07.240.

Maruyama, S., Takagi, H., Nakagaito, A.N., 2019. Influence of silane treatment on water absorption and mechanical properties of Pla/short bamboo fiber-reinforced green composites. Materials and Contact Characterisation IX 1, 101–107. https://doi.org/10.2495/mc190101.

Marwah, R., Ibrahim, N.A., Zainuddin, N., Saad, W.Z., Razak, N.I.A., Chieng, B.W., 2014. The effect of fiber bleaching treatment on the properties of poly(lactic acid)/oil palm empty fruit bunch fiber composites. International Journal of Molecular Sciences 15, 14728–14742. https://doi.org/10.3390/ijms150814728.

Mazani, N., Sapuan, S.M., Sanyang, M.L., Atiqah, A., Ilyas, R.A., 2019. Design and fabrication of a shoe shelf from kenaf fiber reinforced unsaturated polyester composites. In: Lignocellulose for Future Bioeconomy. Elsevier, pp. 315–332. https://doi.org/10.1016/B978-0-12-816354-2.00017-7.

Memon, A., Nakai, A., 2013. Fabrication and mechanical properties of jute spun yarn/PLA unidirection composite by compression molding. Energy Procedia 34, 830–838. https://doi.org/10.1016/j.egypro.2013.06.819.

Misnon, M.I., Islam, M.M., AEpaarachchi, J., 2016. Fabric parameter effect on the mechanical properties of woven hemp fabric reinforced composites as an alternative to wood product. Advance Research in Textile Engineering 1, 1004.

Morales, A.P., Güemes, A., Fernandez-Lopez, A., Valero, V.C., de La Rosa Llano, S., 2017. Bamboo-polylactic acid (PLA) composite material for structural applications. Materials 10, 1–22. https://doi.org/10.3390/ma10111286.

Moudood, A., Hall, W., Öchsner, A., Li, H., Francucci, G., Moudood, A., Hall, W., Öchsner, A., Li, H., Rahman, A., Moudood, A., Hall, W., Öchsner, A., Li, H., 2017. Effect of moisture in flax fibres on the quality of their composites effect of moisture in flax fibres on the quality of their composites. Journal of Natural Fibers 00, 1–16. https://doi.org/10.1080/15440478.2017.1414651.

Mukherjee, T., Kao, N., 2011. PLA based biopolymer reinforced with natural fibre: a review. Journal of Polymers and the Environment 19, 714–725. https://doi.org/10.1007/s10924-011-0320-6.

Nanthakumar, K., Yeng, C.M., Chun, K.S., 2018. Tensile and water absorption properties of solvent cast biofilms of sugarcane leaves fibre-filled poly(lactic) acid. Journal of Thermoplastic Composite Materials 1–16. https://doi.org/10.1177/0892705718805526.

Ngo, T.-D., 2018. Natural fibers for sustainable biocomposites. In: Natural and Artificial Fiber-Reinforced Composites as Renewable Sources. InTech, p. 13. https://doi.org/10.5772/intechopen.71012.

Norizan, M.N., Abdan, K., Ilyas, R.A., Biofibers, S.P., 2020. Effect of fiber orientation and fiber loading on the mechanical and thermal properties of sugar palm yarn fiber reinforced unsaturated polyester resin composites. Polimery 65, 34–43. https://doi.org/10.14314/polimery.2020.2.5.

Nurazzi, N.M., Khalina, A., Sapuan, S.M., Ilyas, R.A., 2019. Mechanical properties of sugar palm yarn/woven glass fiber reinforced unsaturated polyester composites : effect of fiber loadings and alkaline treatment. Polimery 64, 12–22. https://doi.org/10.14314/polimery.2019.10.3.

Peças, P., Carvalho, H., Salman, H., Leite, M., 2018. Natural fibre composites and their applications: a review. Journal of Composites Science. https://doi.org/10.3390/jcs2040066.

Plackett, D., Andersen, T.L., Pedersen, W.B., Nielsen, L., 2003. Biodegradable composites based on L-polylactide and jute fibres. Composites Science and Technology 63, 1287–1296. https://doi.org/10.1016/S0266-3538(03)00100-3.

Porras, A., Maranon, A., 2012. Development and characterization of a laminate composite material from polylactic acid (PLA) and woven bamboo fabric. Composites Part B: Engineering 43, 2782–2788. https://doi.org/10.1016/j.compositesb.2012.04.039.

Puspita, D., Musyarofah, L., Hidayah, E., Sujito, 2019. Fabrication and tensile properties of bamboo micro-fibrils (BMF)/poly-lactic acid (PLA) green composite. Journal of Physics: Conference Series 1217. https://doi.org/10.1088/1742-6596/1217/1/012005.

Rajesh, G., Prasad, A.V.R., 2014. Tensile properties of successive alkali treated short jute fiber reinforced PLA composites. Procedia Materials Science 5, 2188–2196. https://doi.org/10.1016/j.mspro.2014.07.425.

Ramamoorthy, S.K., Skrifvars, M., Persson, A., 2015. A review of natural fibers used in biocomposites: plant, animal and regenerated cellulose fibers. Polymer Reviews 55, 107–162. https://doi.org/10.1080/15583724.2014.971124.

Rawi, N.F.M., Jayaraman, K., Bhattacharyya, D., 2014. Bamboo fabric reinforced polypropylene and poly(lactic acid) for packaging applications: impact, thermal, and physical properties. Polymer Composites 35, 1888–1899. https://doi.org/10.1002/pc.22845.

Rawi, N.F.M., Jayaraman, K., Bhattacharyya, D., 2013. A performance study on composites made from bamboo fabric and poly(lactic acid). Journal of Reinforced Plastics and Composites 32, 1513–1525. https://doi.org/10.1177/0731684413498296.

Razak, N.I.A., Ibrahim, N.A., Zainuddin, N., Rayung, M., Saad, W.Z., 2014. The influence of chemical surface modification of kenaf fiber using hydrogen peroxide on the mechanical properties of biodegradable kenaf fiber/poly(Lactic Acid) composites. Molecules 19, 2957–2968. https://doi.org/10.3390/molecules19032957.

Rose, M., Palkovits, R., 2011. Cellulose-based sustainable polymers: state of the art and future trends. Macromolecular Rapid Communications 32, 1299–1311. https://doi.org/10.1002/marc.201100230.

Roslan, S.A.H., Rasid, Z.A., Hassan, M.Z., 2018. Bamboo reinforced polymer composite - a comprehensive review. IOP Conference Series: Materials Science and Engineering 344. https://doi.org/10.1088/1757-899X/344/1/012008.

Ruan, P., Raghavan, V., Gariepy, Y., Du, J., 2015. Characterization of flax water retting of different durations in laboratory condition and evaluation of its fiber properties. BioResources 10, 3553–3563. https://doi.org/10.15376/biores.10.2.3553-3563.

Salman, S.D., Sharba, M.J., Leman, Z., Sultan, M.T.H., Ishak, M.R., Cardona, F., 2015. Physical, mechanical, and morphological properties of woven kenaf/polymer composites produced using a vacuum infusion technique. International Journal of Polymer Science 2015, 1–10. https://doi.org/10.1155/2015/894565.

Shih, Y.F., Huang, C.C., 2010. Polylactic acid (PLA)/banana fiber (BF) biodegradable green composites. 7th Asian-Australasian Conference on Composite Materials 2010, ACCM 2010 2, 1325–1328. https://doi.org/10.1007/s10965-011-9646-y.

Siakeng, R., Jawaid, M., Ariffin, H., Sapuan, S.M., 2018. Thermal properties of coir and pineapple leaf fibre reinforced polylactic acid hybrid composites. IOP Conference Series: Materials Science and Engineering 368. https://doi.org/10.1088/1757-899X/368/1/012019.

Siengchin, S., Wongmanee, S., 2014. Mechanical and impact properties of PLA/2 × 2 twill and 4 × 4 hopsack weave flax textile composites produced by the interval hot pressing technique. Mechanics of Composite Materials 50, 387–394. https://doi.org/10.1007/s11029-014-9424-0.

SmartFit PLA, 2019. Natural Fibers - Print-Me [WWW Document].

Suddell, B.C., 2008. Industrial fibres: recent and current developments. Proceedings of the Symposium on Natural Fibres 44, 71–82.

Tayommai, T., Aht-Ong, D., 2010. Natural fiber/PLA composites: mechanical properties and biodegradability by gravimetric measurement respirometric (GMR) system. Advanced Materials Research 93–94, 223–226. https://doi.org/10.4028/www.scientific.net/AMR.93-94.223.

Tran, T.P.T., Bénézet, J.C., Bergeret, A., 2014. Rice and Einkorn wheat husks reinforced poly(lactic acid) (PLA) biocomposites: effects of alkaline and silane surface treatments of husks. Industrial Crops and Products 58, 111–124. https://doi.org/10.1016/j.indcrop.2014.04.012.

Tripathi, P., Gupta, V.K., Dixit, A., Mishra, R.K., Sharma, S., 2018. Development and characterization of low cost jute, bagasse and glass fiber reinforced advanced hybrid epoxy composites. AIMS Materials Science 5, 320–337. https://doi.org/10.3934/MATERSCI.2018.2.320.

Venkateshwaran Narayanan, A.E., 2010. Banana fiber reinforced polymer composites - a review. Journal of Reinforced Plastics and Composites 29, 2387–2396. https://doi.org/10.1177/0731684409360578.

Wang, M., Guo, L., Sun, H., 2019. Manufacture of biomaterials. In: Encyclopedia of Biomedical Engineering. Elsevier, pp. 116–134. https://doi.org/10.1016/B978-0-12-801238-3.11027-X.

Wong, L.Y., Saad, W.Z., Mohamad, R., Md Tahir, P., 2016. Efficacy of Aspergillus fumigatus R6 pectinase in enzymatic retting of kenaf. BioResources 11, 10030–10041. https://doi.org/10.15376/biores.11.4.10030-10041.

Wu, C.S., 2009. Renewable resource-based composites of recycled natural fibers and maleated polylactide bioplastic: characterization and biodegradability. Polymer Degradation and Stability 94, 1076–1084. https://doi.org/10.1016/j.polymdegradstab.2009.04.002.

Xu, Z., Yang, L., Ni, Q., Ruan, F., Wang, H., 2019. Fabrication of high-performance green hemp/polylactic acid fibre composites. Journal of Engineered Fibers and Fabrics 14. https://doi.org/10.1177/1558925019834497.

Yahaya, R., Sapuan, S.M., Jawaid, M., Leman, Z., Zainudin, E.S., 2014. Mechanical performance of woven kenaf-Kevlar hybrid composites. Journal of Reinforced Plastics and Composites 33. https://doi.org/10.1177/0731684414559864.

Yu, T., Jiang, N., Li, Y., 2014. Study on short ramie fiber/poly(-lactic acid) composites compatibilized by maleic anhydride. Composites Part A: Applied Science and Manufacturing 64, 139–146. https://doi.org/10.1016/j.compositesa.2014.05.008.

Yu, T., Ren, J., Li, S., Yuan, H., Li, Y., 2010. Effect of fiber surface-treatments on the properties of poly(lactic acid)/ramie composites. Composites Part A: Applied Science and Manufacturing 41, 499–505. https://doi.org/10.1016/j.compositesa.2009.12.006.

Yusoff, R.B., Takagi, H., Nakagaito, A.N., 2016. Tensile and flexural properties of polylactic acid-based hybrid green composites reinforced by kenaf, bamboo and coir fibers. Industrial Crops and Products 94, 562–573. https://doi.org/10.1016/j.indcrop.2016.09.017.

CHAPTER 4

Processing and Characterization of Cornstalk/Sugar Palm Fiber Reinforced Cornstarch Biopolymer Hybrid Composites

M.I.J. IBRAHIM • S.M. SAPUAN • E.S. ZAINUDIN • M.Y.M. ZUHRI • AHMED EDHIREJ

1 INTRODUCTION

The growing mountains of petroleum plastic-based wastes resulted in serious environmental issues. The landfill spaces are limited, and additional combustion capabilities demand further capital financing and lead to more ecological concerns (Ibrahim et al., 2019a). In an effort to alleviate these concerns, scientists and researchers moved in the direction of the production and improvement of eco-friendly materials characterized by biodegradability and recyclability that could sustain the green natural environment.

The most available polymer starch is documented as one of the highly promising raw materials for the fabrication of recyclable and biodegradable items. Starch-based materials are recognized to come up with low tensile efficiency and high water tendency (Avérous and Halley, 2009). As a result, the incorporation of plasticizers to the starch matrix to some extent improved the performance. The primary function of plasticizers is reducing intermolecular attraction of hydrogen bonds and thus increasing the mobility of starch molecules within the structure (Zhong and Li, 2014). Several studies have been stated on the starch-plasticized materials; the results revealed a partial enhancement in the final product (Sanyang et al., 2015, 2016). Recently, new biomaterials derived from natural plants and agronomic residues called natural cellulosic fibers (NCF) have been utilized as a reinforcing filler with starch-based materials (Sahari et al., 2013a).However, the addition of cellulosic fibers to reinforce the starch-based materials has provided significant enhancement characteristic (Kaushik et al., 2010).

Corn (maize) is as ranked the most abundant agriculture cereals planted in the globe. Huge agronomic regions are devoted to corn planting in 2014; the world production of corn reached1.4 Mt (Sandhu et al., 2004). The worth of corn plant remains after harvesting might be handled by converting maize residues into NCFs. Corn stover (residues) typically contains stalk, husk (leaves), and cobs with 50%, 35%, and 15%, respectively (Sokhansanj et al., 2002). Cornstalk indicated to the vertical stem that holds the plant parts; it is a lignocellulosic plant fiber characterized by flexibility, low density, and extendibility, with a high percentage of hemicellulose and acceptable content of cellulose as well as low concentration of lignin and char (Ibrahim et al., 2019b). Biodegradable composite films were fabricated utilizing corn thermoplastic starch matrix and cornstalk fiber (CSF) with 25% (w/w dry starch) fructose plasticizer via solution casting and dehydration technique. The findings indicated that the reinforcing by 6% stalk fiber provided the most effective enhancement content, leading to considerable physical and tensile characteristics of the film produced. Although the final product has shown acceptable properties, it still revealed some drawbacks in terms of water barrier properties. In order to address this defect, hybridization of the cornstalk reinforced corn TPS-based composite with a fiber owning lower hydrophilic nature, such as sugar palm fiber (SPF), is anticipated to overcome such drawbacks. Hybrid composite is that product developed from integrating two or more different fillers into a single matrix to achieve better characteristics than the utilization of individual fiber (Edhirej et al., 2017a).

Advanced Processing, Properties, and Applications of Starch and Other Bio-Based Polymers. https://doi.org/10.1016/B978-0-12-819661-8.00004-4

SPF is a cellulosic plant fiber derived from sugar palm tree characterized by durability, adequate tensile strength, and ability to resist seawater (Ilyas et al., 2017). Moreover, SPF preparation does not require any further treatment or chemical processing (Edhirej et al., 2017a). In the field of biocomposite materials, several studies have been reported about reinforcing polymer matrix by SPF. The results suggested that SPF has the potential to be employed in many composite materials applications, specifically for applications that require water insolubility (WS) and low water holding capacity. Also, SPF has been used as a hybridizing filler with a variety of starches matrices and different natural fibers. For example, Jumaidin et al. (2017) studied the effect of SPF loading on the physical, mechanical, and thermal properties of seaweed/SPF reinforced sugar palm starch/agar biohybrid composite. Also, Edhirej et al. (2017b) explored the effect of various concentrations of SPF on the barrier and tensile characteristics of cassava/SPF reinforced cassava starch biohybrid composites. The authors reported significant results, particularly about water barrier characteristics and mechanical performance. The properties of the produced hybrid composite are affected by many parameters such as the interaction between matrix and fillers, orientation, and percentage of the individual fiber, as well as the magnitude of failure strain for each fiber; the best characteristics of the hybrid composite is attained when the failure strain and orientation of used fibers are compatible (Sreekala et al., 2002).

This chapter focuses on studying the effect of various SPF contents on the thermal, physical, tensile, morphological, and water barrier properties of the cornstalk/SPF reinforced cornstarch (CS) biohybrid composites. It should be noticed that the NCFs employed in the current investigation have not been chemically treated or thermally adapted, resulting in the development of more environmentally efficient and low-cost biomaterials.

2 MATERIALS AND METHODS

2.1 Materials

Thermoplastic CS was derived from grains of corn ear gathered from a local plantation in Perak state, Malaysia, according to the procedures of Ibrahim et al. (2019b). The chemical structure of CS was 24.6% amylose, 74.36% amylopectin, 7.12% crude fats, 10.45% moisture, and 0.61% char. CSF, the stem of the corn plant, was thoroughly washed, dehydrated, and removed the outer skin and then was crushed and shifted via 300 μm autosieve. SPF was obtained from a local farm in Negri Sembilan, Malaysia; it was thoroughly washed, dried, and directly ground to be in powder form. Table 4.1 exhibits the physiochemical structure of SPF and CSF. The evergreen Sdn Bhd, Malaysia, provided a fructose plasticizer.

2.2 Samples Preparation

The preparation of the CS-based biohybrid films was conducted through the solution casting procedure. CS (5 g) was uniformly dispersed in a lab beaker contains distilled water (100 mL). The aqueous dispersion was placed over a magnetic stirrer and subjected to $90 \pm 3°C$ for 20 min. The fructose plasticizer was then combined with a concentration of 25% (w/w dry starch) together with 6% (w/w dry starch) of CSF to the starch solution. The percentages of both fructose and CSF were chosen based on the findings of our earlier research (Ibrahim et al., 2019a). After that, the SPF hybridized agent was loaded at different concentrations (0%, 2%, 4%, 6%, and 8%) w/w starch based. The heating process was maintained with constant stirring for a further 20 min, to ensure good mixing and starch gelatinization. The mixture was then evacuated regularly into a 140 mm diameter Petri plate and directly was placed in an air circulation furnace for 12 h at 50°C. The films produced were softly pulled out from the plates and maintained at room conditions for 7 days before characterizations. The achieved composite films were encoded based on their components and percentage of SPF, as revealed in Table 4.2.

2.3 Thickness and Density (ρ)

The density of the prepared samples was determined from the proposed size (20×15 mm), weight (m), and volume (v) of films. The thickness was calculated by a microelectronic Vernier scale type (Mitutoyo-Co,

TABLE 4.1
Physical and Chemical Properties of Cornstalk and Sugar Palm Fibers.

No.	Content	Cornstalk	Sugar Palm
1	Cellulose (%)	10.8	43.88
2	Hemicellulose (%)	60.3	10.1
3	Lignin (%)	1.98	33.24
4	Ash (%)	1.97	1.01
5	Density (g/cm^3)	1.41	1.28
6	Moisture (%)	11.1	6.45
7	Crystallinity (%)	34.7	35.3

TABLE 4.2
The Percentages of Materials Used in Film Preparation.

Film	Fructose (%) of dry starch	CS g/100 mL Distilled water	CSF (%) of dry starch	SPF (%) of dry starch
CS-film	25	5	0	0
CS-CSF	25	5	6	0
CS-CSF/SPF2%	25	5	6	2
CS-CSF/SPF4%	25	5	6	4
CS-CSF/SPF6%	25	5	6	6
CS-CSF/SPF8%	25	5	6	8

Japan) with ±0.001 mm precision. The density (ρ) was attained from the average of five determination via the equation:

$$\rho = \frac{m}{v} \quad (4.1)$$

2.4 Moisture Content

The initial weight (w_1) of the sample (20 × 15 mm) was recorded before being dried in a laboratory oven at 105°C for 24 h. After that, the sample was reweighed in dry matter (w_2). The weight variation was utilized to estimate the percentage of moisture that was removed from the film through the equation:

$$\text{MC}\ (\%) = ((w_1 - w_2)\ /\ w_1) \times 100 \quad (4.2)$$

2.5 Water Solubility

The WS assay of samples was performed based on the procedure of Shojaee-Aliabadi et al. (2013). A film strip (20 × 15 mm) was dried for 24 h at 105°C and immediately weighted (W_i). The film was then submerged in a laboratory flask filled with extracted water for 6 hours with constant stirring at ambient conditions. The remaining fragments of the film were dried on an air circulation oven at 105°C till a constant weight (W_f) was reached. Finally, the solubility (%) was calculated using the following equation:

$$\text{WS}(\%) = ((W_i - W_f)\ /\ W_i) \times 100 \quad (4.3)$$

2.6 Water Vapor Permeability

The water vapor permeability (WVP) screening quantifies the amount of vapor transmission across the unit thickness of the tested sample. The assay was conducted in accordance with ASTM E96 standard. A cup of 70 mm diameter was loaded with 5 grams of silica gel. A known thickness circular film sample was firmly installed in the top of a cup with the same diameter of the sample and was initially weighted. After that, the testing cup was then kept in a humidity chamber at room temperature and 75% relative humidity. The weight gain in the test cup was periodically recorded until the fixed weight status was achieved. Lastly, the WVP (10^{-1}. mm.g. s^{-1}.m^{-2}. Pa^{-1}) of the specimen is determined using the following formula:

$$\text{WVP} = w \times d/A \times t \times P \quad (4.4)$$

where w (g) is the mass gain, d (mm) is film thickness, A (m^2) is the area of the film, t (s) is the time of permeability, and P (Pa) is the atmospheric pressure.

2.7 Scanning Electron Microscope

The surface fracture of the samples was inspected by the scanning electron microscope (SEM) device type (Hitachi S-3400N, Nara, Japan). The specimen was painted with a gold coating prior to applying a voltage of 20 kV within a high vacuum atmosphere. The applied voltage generates a set of electrons that send signals to convey information about the morphological structure of the sample and produce high-resolution images.

2.8 Fourier-transform Infrared Spectroscopy

The infrared spectrum of samples was tracked by a spectrometer type (Bruker vector 22, Lancashire, UK). The range of frequency was between 4000 and 400 cm^{-1}, and the spectral resolution was set to be 4 cm^{-1}. The test was carried out using 16 scans per sample.

2.9 X-ray Diffraction

A diffractometer 2500 X-ray (Rigaku, Tokyo, Japan) was employed to perform the diffraction analysis of the hybrid films. The applied power throughout the test was kept at 35 mA and 40 kV for current and voltage, respectively. The speed of scattering was at a rate of 0.02 degrees per angle within angular (2θ) range from 5 degrees to 60 degrees. The equation computed the relative crystallinity (Cr):

$$Cr = (Ac / (Ac + Aa)) \times 100 \quad (4.5)$$

where Aa is the amorphous area and Ac is the crystalline area.

2.10 Thermogravimetric Analysis

The thermostability of the samples was achieved using a thermogravimeter analyzer (Q500 V20.13 Build 39, Bellingham, USA). A film sample (10 mm^2) was exposed to a temperature varying from room temperature to 450°C at a constant rate of 10°C/min.

2.11 Tensile Testing

The mechanical performance of the films was estimated using tensile machine type 5KN INSTRON, working based on ASTM D882 standard. A film sample of 10 × 70 mm size was fixed between the machine clamps. The effective length was set to 30 mm, with 2 mm/min crosshead speed. The tensile values considered are the average of five replicates of each sample.

3 RESULTS AND DISCUSSION

3.1 Thickness and Density

The values of thickness and density for the starch-based hybrid composite films were inserted in Table 4.3. As anticipated, the increase in filler concentration increases the thickness and reduces the density compared with the nonhybridized CS film. The obtained findings might be ascribed to the lower density of the reinforcing fillers, 1.28 g/cm^3 for SPF and 1.41 g/cm^3 for CSF (Sahari et al., 2012). Another possible explanation for this phenomenon concerns the function of fillers in reconstructing the composite structure. The higher content of SPF promotes void growth and creates a heterogeneous surface with lower density than the starch matrix, resulting in thick, coarser films (Versino and García, 2014). Comparable findings were also noticed in prior investigations in the same field (Edhirej et al., 2017a; Shakuntala et al., 2014). Nonetheless, the low density of biomasses makes them desirable for biomaterial applications compared with human-made polymer composites, such as glass fiber (2.500 g/m^3) (Mendes et al., 2015).

3.2 Moisture Content

Moisture (water) content is an important parameter for the selection of cellulosic fibers as strengthened fillers for the production of new biomaterials. The lower moisture content (MC) is required because the higher MC affected negatively on the performance and function of the biocomposite produced, especially in terms of porosity creation and dimensional stability and hence mechanical properties (Jawaid and Khalil, 2011). The introduction of SPF as a reinforcing agent has led to an insignificant decrement in the MC compared with the neat CS film. Table 4.3 revealed the effect of hybridization on the starch-based composite. Ilyas et al. (2018a) attributed the reduction in MC to the increased cellulose content within the biohybrid composites. The incorporation of SPF has increased the cellulose content as two fibers existed within the composite; this, in turn, decreased the MC rate, because cellulose acted as a barrier to water penetration due to the existence of a hydroxyl group (OH) in its composition. A similar description was explained by Soykeabkaew et al. (2004); the authors produced a hybrid composite from the tapioca starch matrix reinforced by flax and jute fibers. Furthermore, the work of (Edhirej et al., 2017a) on cassava bagasse/sugar palm

TABLE 4.3
Physical Characteristics of the Samples.

Sample	Thickness (μm)	Density (g/cm^3)	Moisture Content (%)	Water Solubility (%)
CS film	0.195 ± 25.4	1.55 ± 0.08	11.13 ± 0.11	21.38 ± 0.43
CS/CSF	0.205 ± 25.4	1.49 ± 0.07	10.66 ± 0.19	19.61 ± 0.17
CS-CSF/SPF2%	0.245 ± 25.4	1.46 ± 0.04	9.09 ± 0.20	23.38 ± 0.09
CS-CSF/SPF4%	0.280 ± 25.4	1.44 ± 0.06	8.83 ± 0.23	20.61 ± 0.15
CS-CSF/SPF6%	0.295 ± 25.4	1.42 ± 0.09	9.38 ± 0.19	20.28 ± 0.17
CS-CSF/SPF8%	0.315 ± 25.4	1.39 ± 0.07	9.14 ± 0.11	21.11 ± 0.12

fibre reinforced cassava TPS hybrid composites also reveals similar results.

3.3 Water Solubility

This parameter examines the material integrity as affected by water immersion with continuous stirring. Low solubility materials are suitable for applications that require high moisture resistance and less water leaking (Gontard et al., 1993). From the data in Table 4.3, it has been observed that the initial loading of SPF increased the solubility of the film, whereas additional loading resulted in a slight drop in solubility with comparison to the CS film control. The solubility reduced as the content of SPF further increased due to the ability of SPF to enhance composites' integrity through creating a network that strongly maintains the composites' solidity and hindering moisture diffusion, in turn decreasing WS (Jumaidin et al., 2017).

3.4 Water Barrier Properties

The vapor transmission assay estimates the permeability rate of water vapor through the thickness of the known area of the material. The lower values of vapor transmission are much appropriate for the applications that elimination or decrease of moisture passage are required. Fig. 4.1 shows the values of WVP for the control film, CS/CSF composite, and CS-CSF/SPF biohybrid composite. From the achieved results, the neat CS film showed the highest rate of WVP, by 2.28×10^{-10} mm g. $s^{-1}.m^{-2}.\ Pa^{-1}$. This is attributed to the high water propensity of starches as a whole (Wilhelm et al., 2003; Ilyas et al., 2018b). The incorporation of SPF with CS films has improved the water barrier characteristics, as indicated by the decreased values of WVP compared with the control CS film and CS/CSF composite. As the concentration of SPF increased to 2%, the rate of WVP decreased by 66.42%, making it the most efficient loading; this reduction in WVP was ascribed to the firm crystalline structure established after fiber addition formation and the effective distribution of SPF in the starch matrix, which hinders the transmission path of the vapor particles (Slavutsky and Bertuzzi, 2014). Furthermore, the observed reduction could be related to the enhancement of intermolecular bonding between the reinforcing filler and biopolymer matrix, as revealed on FTIR analysis, which eliminates the matrix chain mobilization (Follain et al., 2013).

The additional increase in SPF concentration from 2% to 8% resulted in a slight increasing in the vapor transmission values of the hybrid films from 0.858×10^{-10} mm g. $s^{-1}.m^{-2}.\ Pa^{-1}$ to 1.14×10^{-10} mm g. $s^{-1}.m^{-2}.\ Pa^{-1}$. This is because the higher content of SPF promotes porosity formation and thus increases the vapor transmission rate (Versino and García, 2014).

3.5 Morphological Properties

Fig. 4.2 demonstrates the morphological structure of CS-CSF/SPF biohybrid composites together with nonhybridized control film and CS/CSF composite. The CS-plasticized film exhibited a compact surface with no voids and rather smooth with the existence of

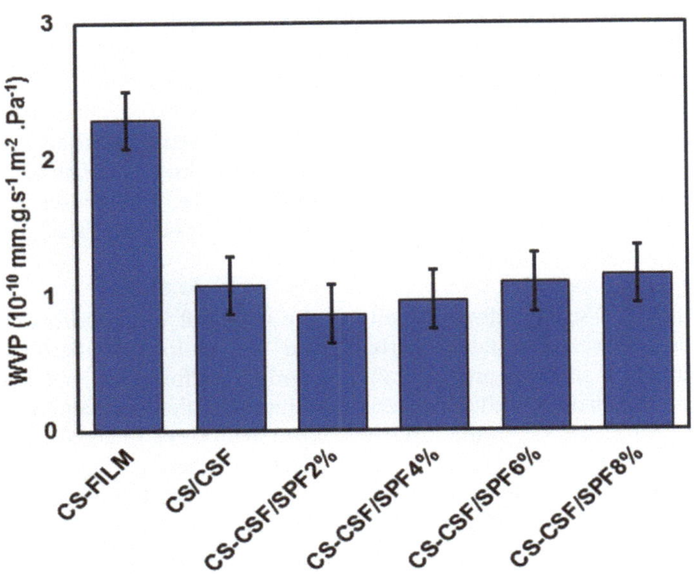

FIG. 4.1 WVP values of CS-CSF/SPF hybrid composites at different SPF loadings.

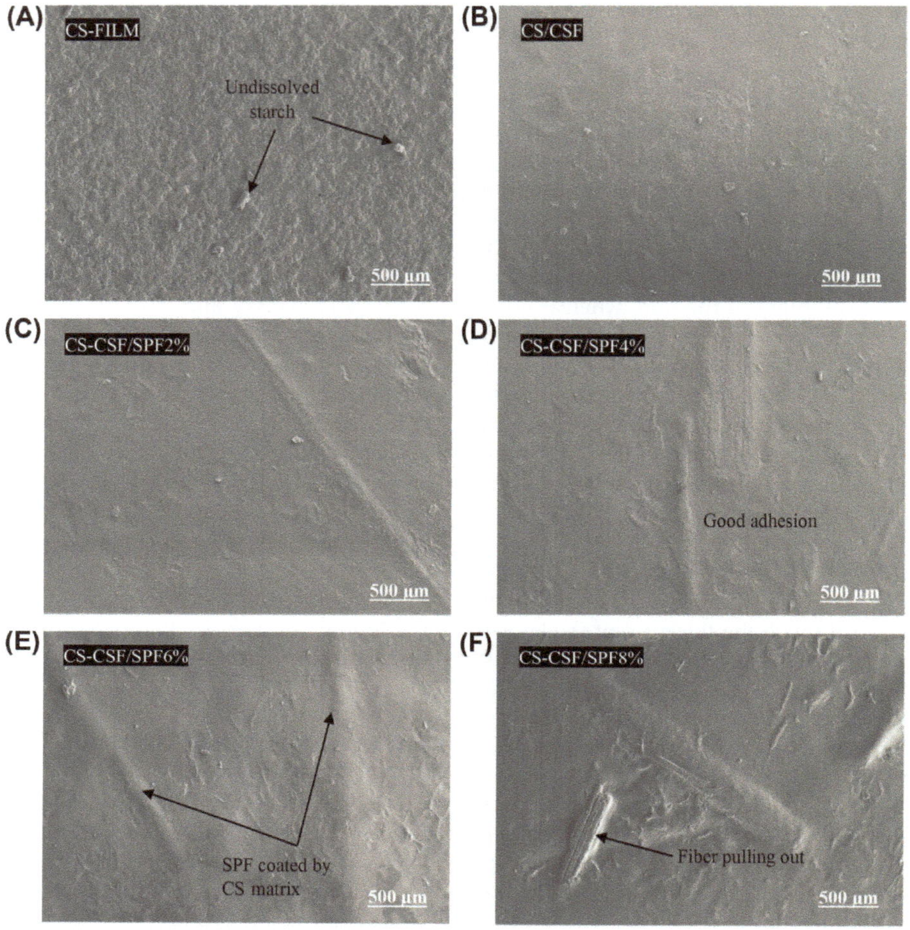

FIG. 4.2 SEM images of CS-CSF/SPS hybrid composite.

nondissolved starch. The addition of CSF to produce a composite (CS/CSF) resulted in a more consistent structure, and the fracture surface turned to be smoother, reflecting the morphological structure of CSF, whereas the incorporation of SPF to produce the hybrid composite showed a structure with coarse surfaces, and the polymer matrix completely covers the filler molecules, with no evidence of clusters or particle aggregations that indicates strong adhesion and good interfacial interaction between the starch matrix and the reinforcing agent; this, in turn, provides an efficient stress transfer and thus better mechanical performance (Avérous et al., 2001). However, the surface micrograph of CS-CSF/SPF hybrid composites (2%, 4%, and 6%) showed consistent dispersion and a more homogenous structure than CS-CSF/SPF8% counterpart. This indicates the lower content of SPF generated more interaction with the starch matrix than the higher content. The highest loading of SPF (CS-CSF/SPF8%) generated a less consistent structure with heterogeneous surfaces, as evidenced by the fiber pulling out from the matrix due to excessive fiber use, which adversely influences the integrity of the hybrid composite (Ludueña et al., 2012).

3.6 FTIR Analysis

The potential interactions between the starch matrix and the reinforcing fillers were evaluated using the Fourier-transform infrared spectroscopy (FTIR) technique. Since the starch and cellulosic fibers were derived from natural raw materials, and their structural composition includes specific components, it was observed that the infrared spectral curve of all film samples showed similar behaviors, as revealed in Fig. 4.3. The bending mode of water particles within the starch created the peaks at 1500–1600 cm^{-1} (Sahari et al.,

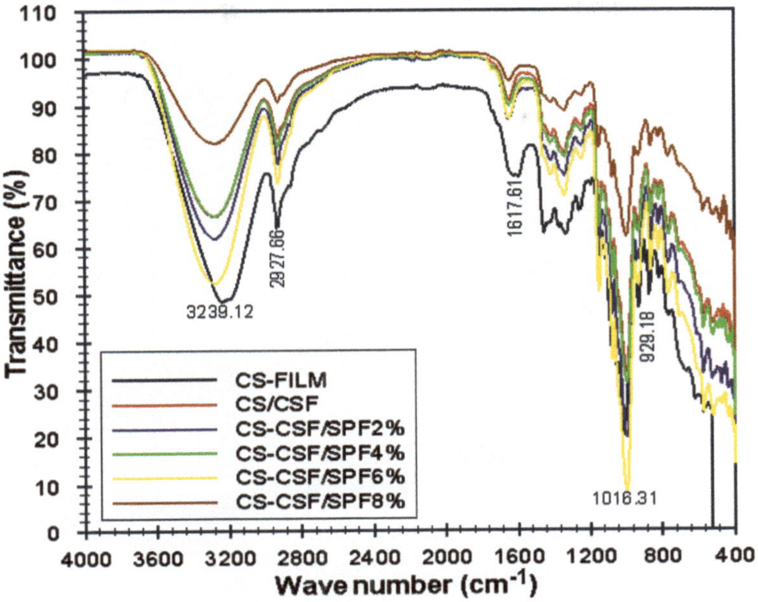

FIG. 4.3 FTIR spectral curves of CS-CSF/SPF at various SPF loadings.

2013b). The observed transmittance bands at the peaks between 950 and 1040 cm^{-1} were assigned to the vibration of the O–C group (Fang et al., 2002), while the vibrational stretching of C–H groups generated the peaks at 2860–3100 cm^{-1} (Prachayawarakorn et al., 2012). The existence of hydroxyl (O–H) groups within the structure of fiber and starch resulted in sharp peaks at 3200–3600 cm^{-1} (Wu et al., 2009). However, the absence of new peaks in the FTIR spectra following the fiber loading suggested no chemical reactions. The potential interactions between the polymer matrix and the reinforcing fiber were determined by tracking the peak shifts after the gradual loading of the fiber. For example, the peaks resulting from the O–H hydroxyl group stretching at 3100–3400 cm^{-1} for the control film were moved to lower wavenumber after fillers addition. According to Wu et al. (2009), peak shifting is a sign for enhancing of intermolecular hydrogen bonding, which means more compatibility and stronger interaction. A comparable explanation was stated by Jumaidin et al. (2017), where the researchers analyzed the structure of sugar palm starch-based composites hybridized with seaweed and sugar palm fiber.

3.7 X-ray Diffraction Analysis

The corresponding data obtained from the diffraction analysis of neat CS film, CS/CSF composites, and CS-CSF/SPF at various loading of SPF are presented in Fig. 4.4. The plotted curve shows that the hybridized samples provided identical patterns to the control film. The only difference was in shifting of the main peaks of scattering angles to higher intensity following the hybridization. With regard to the nonreinforced control film, the retrogradation and gelatinization of starch particles during the preparation result in the crystalline structure of CS film. The main 2θ peaks scattered at angles 17.57 degrees, 20.15 degrees, and 22.48 degrees, which are the typical A-type pattern of the all plant TPS (Zhang et al., 2009). The introduction of fibers into the polymer matrix resulted in an apparent increase in the leading peaks intensity, which in turn improves the relative crystallinity of the hybridized biocomposites. The observed relationship between the addition of reinforcing filler and crystallinity enhancement can be illustrated by the function of cellulose in forming a cross-linked network and promoting the covalent bonding between matrix and fiber (Bledzki and Gassan, 1999). Table 4.4 included the crystallinity index of CS-based films as reinforced by various fillers concentrations. However, NCFs are oriented biomaterials, due to which their crystallinity index is higher than that of starches. Thus, a better crystallinity of biocomposites is anticipated following the hybridization (Ma et al., 2005).

3.8 Thermal Stability Properties

The thermal behavior of CSF/SPF reinforced CS biohybrid composites is displayed in Fig. 4.5. From the

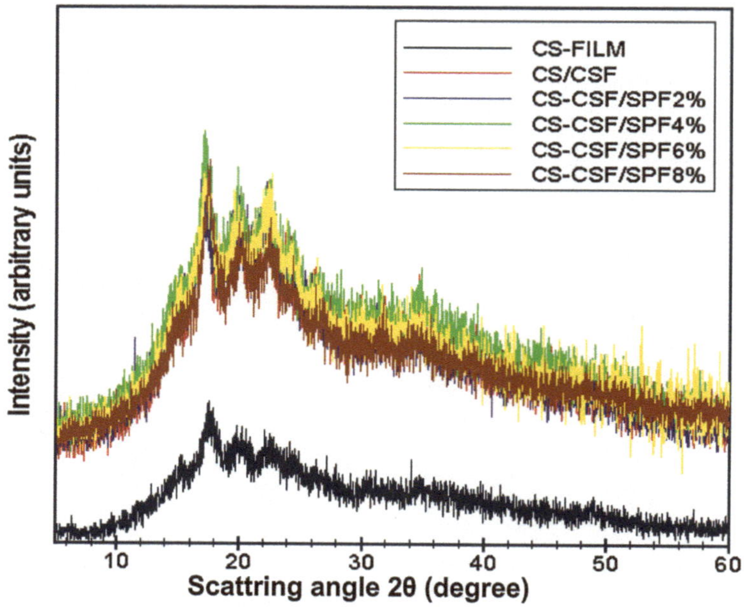

FIG. 4.4 Diffraction curves of the cornstarch-based films with various SPF loadings.

findings, the loss of weight and initial degradation of CS-CSF/SPF biohybrid composites took place in four thermal phases revealed in Table 4.5. The onset degradation events were determined by the prominent peaks in DTG curves, which correspond to a certain loss of mass in the thermogravimetric analysis chart.

The preliminary decomposition and mass loss took place due to the dehydration of water fragments at temperatures below 100°C (Dang and Yoksan, 2015). Films with high MC were exposed to a further heating rate and therefore lost further weight. Additional heating has triggered the subsequent loss of weight at temperatures within 150–200°C; the weight loss in this phase was escribed to the volatilization of plasticizer (fructose).

TABLE 4.4
Relative Crystallinity of CS-CSF/SPF Hybrid Composite Films.

Sample	Relative Crystallinity (%)
CS film	15
CS/CSF	16.4
CS-CSF/SP 2%	17
CS-CSF/SP 4%	18.5
CS-CSF/SP 6%	15.5
CS-CSF/SP 8%	20.3

It is well known that the plasticizer volatilization begins at 150°C (Vega et al., 1996). Further increase in temperature has led to the degradation of starch components, which usually starts with the water-soluble amylopectin degradation (Ibrahim et al., 2019b). The maximum rate of degradation in the final phase was assigned to the degradation of the main constituents of the reinforcing fibers, i.e., cellulose, hemicellulose, and lignin. Based on the reported study of Lomelí-Ramírez et al. (2014), the thermal decomposition of lignocellulosic plant fiber begins with the decaying of hemicellulose at 200°C to 260°C, then cellulose between 230°C to 340°C, and lastly lignin within the range between 270°C to 500°C, which differs according to plant type and the concentrations of the main constituents. Following the complete decomposition of lignin, the remaining constituent is nonorganic particles such as silica, in the form of ash (mass residues). However, the biohybrid films with different SPF concentrations exhibited minimal variations on the degradation temperatures, closed to the decomposition temperature of the single fiber (CSF) biocomposite film, as seen in Table 4.5. Furthermore, following the final degradation, the percentages of mass residues of the biohybrid films showed insignificant differences compared with the starch film and CS/CSF composite film, which indicates that the hybridization does not affect the thermostability of CS-based biohybrid composites. Similar observations were noticed by the researchers Edhirej et al.

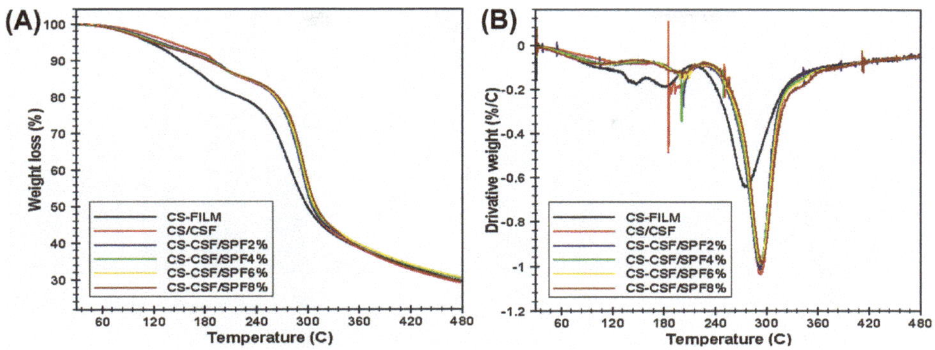

FIG. 4.5 Thermal properties of CS-CSF/SPF hybrid composite. (A) TGA and (B) DTG.

(2017a) when they hybridized the TPS and bagasse of cassava by SPF.

3.9 Tensile Characteristics

The tensile assay was mainly conducted to obtain tensile stress (TS), Young's (elasticity) modulus (E′), and the elongation at break (EB) of CS-CSF/SPF hybrid films. Fig. 4.6 illustrates the results of tensile evaluation before and after hybridization. According to the results achieved, the introduction of SPF with different concentrations resulted in a considerable enhancement in TS and E′, while EB parameter recorded an apparent reduction, indicating that the hybridization has improved the rigidity and stiffness and decreased the flexibility of the films (Dias et al., 2011).

In general, the incorporation of SPF showed considerable enhancement in the tensile characteristics of the biohybrid composites. The film CS-CSF/SPF4% provided the most considerable tensile stress (12.94 MPa) and Young's modulus (1019.21 MPa), more significant than 11.7 MPa achieved by the single fiber (CS/CSF) composite and 6.8 MPa achieved by neat CS film. The hybridization enhanced the tensile strength by 90.29% compared with starch film and by 10.6% compared with CS/CSF composite. This enhancement can be explained through the capability of cellulose to control the mobilization of polymer particles and promotes the interaction between polymer and cellulose fillers, which, in turn, offers an efficient stress transfer (Müller et al., 2009). The enhancement was also related to the crystallinity index as previously mentioned by Salaberria et al. (2014); the authors declared that the increase in the crystallinity index is an indicator for the improvement of the mechanical efficiency of the structure.

Despite this improvement, the loading of SPF beyond 4% has reduced the tensile stress and the modulus of elasticity of the hybridized composites. The observed reduction in tensile performance could be attributed to the excessive use of SPF, which led to pulling out of fiber from the polymer matrix, as seen

TABLE 4.5
Degradation Temperatures of CS-CH/SPF Hybrid Composites.

	ONSET DEGRADATION TEMPERATURE (°C)					
Film Sample	Phase 1 (Moisture)	Phase 2 (Plasticizer)	Phase 3 (Starch)	Phase 4 (Fibers)	Mass Residue (%)	Weight Loss (%)
CS film	82.92	170.13	253.70	–	29.63	50.95
CS/CSF	85.11	184.97	273.92	293.32	29.14	55.96
CS-CSF/SPF2%	73.08	190.64	271.69	292.62	30.44	54.86
CS-CSF/SPF4%	72.48	190.61	272.13	294.45	30.32	55.08
CS-CSF/SPF6%	68.07	184.36	273.06	293.06	30.56	54.90
CS-CSF/SPF8%	65.60	175.40	274.79	294.07	29.53	55.75

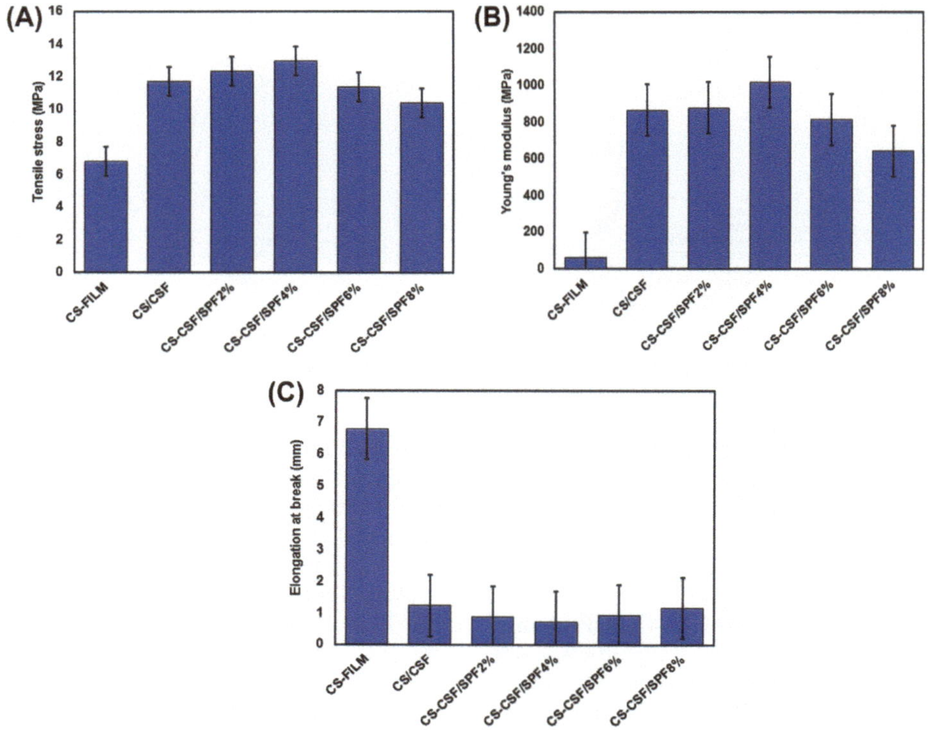

FIG. 4.6 Tensile properties of CS-CSF/SPF hybrid composite. (A) Tensile strength, (B) Young's modulus, and (C) Elongation at break.

in SEM images. Fiber pulling out is an indicator of the disintegration of the structure and poor stress transfer and, therefore, weak tensile characteristics (Islam et al., 2013).

Relating To the elongation of break (EB) for CS-CSF/SPF biohybrid composites, the introduction of SPF and its different loadings showed an adverse effect on elasticity modulus and tensile stress. Increasing of SPF content reduces the elongation of the hybrid composite. The reduction in the EB of the film with additional fiber content is due to the fact that the cellulosic fiber reconstructs the network of the hybrid composites by stimulating the intermolecular attraction of the polymer matrix network. Such reconstruction increases the rigidity and stiffness and hence decreases flexibility (da Rosa Zavareze et al., 2012).

4 CONCLUSIONS

Biohybrid composite films were produced from CS and fibrous residues of CSF and sugar palm (SPF) via the method of solution casting and dehydration. Based on the obtained results, a remarkable enhancement in the general performance of the CS-based biocomposite film was achieved after the hybridization. The results showed that the hybridization had improved the tensile performance of the biocomposite films by 90.29%, with the film having 4% SPF (12.94 MPa), which demonstrated the most effective reinforcing concentration, whereas the water barrier properties of the hybrid composites reduced by 66.42%, which means better resistance to moisture transmission. The crystallinity index increased from 15% to 20.3% after hybridization; this improvement is associated with increased intermolecular bonding, as evidenced by the band shift detected by FTIR, as well as to the stable interfacial adhesion between polymer matrix and fiber and cohesive surfaces as seen in SEM images. With regard to thermal properties, hybridization does not appear to be enhanced film thermal stability due to the absence of any apparent change in degradation temperatures. Overall, hybridization of cornstarch and CSF with SPF enhanced the characteristics of the produced composites, thereby expanding the potential applications of these biomaterials.

REFERENCES

Avérous, L., Halley, P.J., 2009. Biocomposites based on plasticized starch. Biofuels, Bioproducts, and Biorefining 3, 329–343.

Avérous, L., Fringant, C., Moro, L., 2001. Plasticized starch–cellulose interactions in polysaccharide composites. Polymer 42, 6565–6572.

Bledzki, A., Gassan, J., 1999. Composites reinforced with cellulose based fibres. Progress in Polymer Science 24, 221–274.

da Rosa Zavareze, E., Pinto, V.Z., Klein, B., El Halal, S.L.M., Elias, M.C., Prentice-Hernández, C., et al., 2012. Development of oxidised and heat–moisture treated potato starch film. Food Chemistry 132, 344–350.

Dang, K.M., Yoksan, R., 2015. Development of thermoplastic starch blown film by incorporating plasticized chitosan. Carbohydrate Polymers 115, 575–581.

Dias, A.B., Müller, C.M., Larotonda, F.D., Laurindo, J.B., 2011. Mechanical and barrier properties of composite films based on rice flour and cellulose fibers. LWT-Food Science and Technology 44, 535–542.

Edhirej, A., Sapuan, S., Jawaid, M., Zahari, N.I., 2017. Cassava/sugar palm fiber reinforced cassava starch hybrid composites: physical, thermal and structural properties. International Journal of Biological Macromolecules 101, 75–83.

Edhirej, A., Sapuan, S., Jawaid, M., Zahari, N.I., 2017. Tensile, barrier, dynamic mechanical, and biodegradation properties of cassava/sugar palm fiber reinforced cassava starch hybrid composites. BioResources 12, 7145–7160.

Fang, J., Fowler, P., Tomkinson, J., Hill, C., 2002. The preparation and characterization of a series of chemically modified potato starches. Carbohydrate Polymers 47, 245–252.

Follain, N., Belbekhouche, S., Bras, J., Siqueira, G., Marais, S., Dufresne, A., 2013. Water transport properties of bionanocomposites reinforced by Luffa cylindrica cellulose nanocrystals. Journal of Membrane Science 427, 218–229.

Gontard, N., Guilbert, S., CUQ, J.L., 1993. Water and glycerol as plasticizers affect mechanical and water vapor barrier properties of an edible wheat gluten film. Journal of Food Science 58, 206–211.

Ibrahim, M., Sapuan, S., Zainudin, E., Zuhri, M., 2019. Physical, thermal, morphological, and tensile properties of cornstarch-based films as affected by different plasticizers. International Journal of Food Properties 22, 925–941.

Ibrahim, M.I., Sapuan, S.M., Zainudin, E.S., Zuhri, M.Y.M., 2019. Extraction, chemical composition, and characterization of potential lignocellulosic biomasses and polymers from corn plant parts. BioResources 14, 6485–6500.

Ilyas, R., Sapuan, S., Ishak, M., Zainudin, E., 2017. Effect of delignification on the physical, thermal, chemical, and structural properties of sugar palm fibre. BioResources 12, 8734–8754.

Ilyas, R., Sapuan, S., Ishak, M., 2018. Isolation and characterization of nanocrystalline cellulose from sugar palm fibers (*Arenga pinnata*). Carbohydrate Polymers 181, 1038–1051.

Ilyas, R., Sapuan, S., Ishak, M., Zainudin, E., 2018. Development and characterization of sugar palm nanocrystalline cellulose reinforced sugar palm starch bionanocomposites. Carbohydrate Polymers 202, 186–202.

Islam, M.T., Alam, M.M., Zoccola, M., 2013. Review on modification of nanocellulose for application in composites. The International Journal of Innovative Research in Science, Engineering and Technology 2, 5444–5451.

Jawaid, M., Khalil, H.A., 2011. Cellulosic/synthetic fiber reinforced polymer hybrid composites: a review. Carbohydrate Polymers 86, 1–18.

Jumaidin, R., Sapuan, S.M., Jawaid, M., Ishak, M.R., Sahari, J., 2017. Thermal, mechanical, and physical properties of seaweed/sugar palm fiber reinforced thermoplastic sugar palm starch/agar hybrid composites. International Journal of Biological Macromolecules 97, 606–615.

Kaushik, A., Singh, M., Verma, G., 2010. Green nanocomposites based on thermoplastic starch and steam exploded cellulose nanofibrils from wheat straw. Carbohydrate Polymers 82, 337–345.

Lomelí-Ramírez, M.G., Kestur, S.G., Manríquez-González, R., Iwakiri, S., de Muniz, G.B., Flores-Sahagun, T.S., 2014. Bio-composites of cassava starch-green coconut fiber: Part II—structure and properties. Carbohydrate Polymers 102, 576–583.

Ludueña, L., Vázquez, A., Alvarez, V., 2012. Effect of lignocellulosic filler type and content on the behavior of polycaprolactone based eco-composites for packaging applications. Carbohydrate Polymers 87, 411–421.

Ma, X., Yu, J., Kennedy, J.F., 2005. Studies on the properties of natural fibers-reinforced thermoplastic starch composites. Carbohydrate Polymers 62, 19–24.

Mendes, C., Adnet, F., Leite, M., Furtado, C.R.G., Sousa, A., 2015. Chemical, physical, mechanical, thermal and morphological characterization of corn husk residue. Cellulose Chemistry & Technology 49, 727–735.

Müller, C.M., Laurindo, J.B., Yamashita, F., 2009. Effect of cellulose fibers on the crystallinity and mechanical properties of starch-based films at different relative humidity values. Carbohydrate Polymers 77, 293–299.

Prachayawarakorn, J., Limsiriwong, N., Kongjindamunee, R., Surakit, S., 2012. Effect of agar and cotton fiber on properties of thermoplastic waxy rice starch composites. Journal of Polymers and the Environment 20, 88–95.

Sahari, J., Sapuan, S., Ismarrubie, Z., Rahman, M.Z., 2012. Physical and chemical properties of different morphological parts of sugar palm fibers. Fibres and Textiles in Eastern Europe 91, 21–24.

Sahari, J., Sapuan, S., Zainudin, E., Maleque, M., 2013. Mechanical and thermal properties of environmentally friendly composites derived from sugar palm tree. Materials & Design 49, 285–289.

Sahari, J., Sapuan, S., Zainudin, E., Maleque, M., 2013. Thermo-mechanical behaviors of thermoplastic starch derived from sugar palm tree (*Arenga pinnata*). Carbohydrate Polymers 92, 1711–1716.

Salaberria, A.M., Labidi, J., Fernandes, S.C., 2014. Chitin nanocrystals and nanofibers as nano-sized fillers into thermoplastic starch-based biocomposites processed by melt-mixing. Chemical Engineering Journal 256, 356–364.

Sandhu, K.S., Singh, N., Kaur, M., 2004. Characteristics of the different corn types and their grain fractions:

physicochemical, thermal, morphological, and rheological properties of starches. Journal of Food Engineering 64, 119–127.

Sanyang, M., Sapuan, S., Jawaid, M., Ishak, M., Sahari, J., 2015. Effect of plasticizer type and concentration on tensile, thermal and barrier properties of biodegradable films based on sugar palm (*Arenga pinnata*) starch. Polymers 7, 1106–1124.

Sanyang, M.L., Sapuan, S., Jawaid, M., Ishak, M.R., Sahari, J., 2016. Effect of sugar palm-derived cellulose reinforcement on the mechanical and water barrier properties of sugar palm starch biocomposite films. BioResources 11, 4134–4145.

Shakuntala, O., Raghavendra, G., Samir Kumar, A., 2014. Effect of filler loading on mechanical and tribological properties of wood apple shell reinforced epoxy composite. Advances in Materials Science and Engineering 2014.

Shojaee-Aliabadi, S., Hosseini, H., Mohammadifar, M.A., Mohammadi, A., Ghasemlou, M., Ojagh, S.M., et al., 2013. Characterization of antioxidant-antimicrobial κ-carrageenan films containing *Satureja hortensis* essential oil. International Journal of Biological Macromolecules 52, 116–124.

Slavutsky, A.M., Bertuzzi, M.A., 2014. Water barrier properties of starch films reinforced with cellulose nanocrystals obtained from sugarcane bagasse. Carbohydrate Polymers 110, 53–61.

Sokhansanj, S., Turhollow, A., Cushman, J., Cundiff, J., 2002. Engineering aspects of collecting corn stover for bioenergy. Biomass and Bioenergy 23, 347–355.

Soykeabkaew, N., Supaphol, P., Rujiravanit, R., 2004. Preparation and characterization of jute-and flax-reinforced starch-based composite foams. Carbohydrate Polymers 58, 53–63.

Sreekala, M., George, J., Kumaran, M., Thomas, S., 2002. The mechanical performance of hybrid phenol-formaldehyde-based composites reinforced with glass and oil palm fibres. Composites Science and Technology 62, 339–353.

Vega, D., Villar, M.A., Failla, M.D., Vallés, E.M., 1996. Thermogravimetric analysis of starch-based biodegradable blends. Polymer Bulletin 37, 229–235.

Versino, F., García, M.A., 2014. Cassava (*Manihot esculenta*) starch films reinforced with natural fibrous filler. Industrial Crops and Products 58, 305–314.

Wilhelm, H.-M., Sierakowski, M.-R., Souza, G., Wypych, F., 2003. Starch films reinforced with mineral clay. Carbohydrate Polymers 52, 101–110.

Wu, Y., Geng, F., Chang, P.R., Yu, J., Ma, X., 2009. Effect of agar on the microstructure and performance of potato starch film. Carbohydrate Polymers 76, 299–304.

Zhang, L., Xie, W., Zhao, X., Liu, Y., Gao, W., 2009. Study on the morphology, crystalline structure and thermal properties of yellow ginger starch acetates with different degrees of substitution. Thermochimica Acta 495, 57–62.

Zhong, Y., Li, Y., 2014. Effects of glycerol and storage relative humidity on the properties of kudzu starch-based edible films. Starch-Stärke 66, 524–532.

CHAPTER 5

Development and Processing of PLA, PHA, and Other Biopolymers

C.H. LEE • S.M. SAPUAN • R.A. ILYAS • S.H. LEE • A. KHALINA

1 INTRODUCTION

Recently, biodegradable polymers have attracted much attention due to the environmental concerns and sustainability issues associated with petroleum-based polymers (Lee et al., 2014; Ayu et al., 2018). Polylactic acid (PLA) and polyhydroxyalkanoate (PHA) polymer are the most widely used biodegradable thermoplastic polymer. The use of PLA and PHA polymers in various advanced applications has been certified long time ago due to their promising properties. However, a lot more products had refuse to use biopolymers, just because of its processing issue. Biopolymers can perform as or even better than conventional polymer (such as polypropylene). The differences in term of viscosity, melt flow rate, and melt strength has caused the existing processing parameters unsuitable for biopolymers. Different grades of biopolymer have been introduced by manufacturers to apply on varies processing methods. However, thermal degradation phenomena are always observed with deteriorated performances of biopolymer. This is because processing of thermoplastic biopolymer (PLA and PHA polymer), melting at elevated temperature, is the first step of processing, and it is done by screw (single or twin screw) (Lee et al., 2017). The rotating screw conveys the molten polymer toward dies or mold often exerting shear stress for the purpose of homogenous mixing. Yet, this has induced further degradation and lower molecular weight extrudate was obtained (Lee et al., 2018). Therefore, compatibilizers have been applied to enhance the polymer's melt strength or widen its processing limits. Injection molding processing method is always applied on the products that required high outer surface finishing and able are to be produced in a short period of time. Biopolymer extrudates are directly injected into the mold for cooling. The extrudates could also be directed into a blow machine to produce blown films for various applications, especially food packaging films. On the other hand, thermoforming is a process of heating a thermoplastic sheet to its glass transition temperature to stretch and then cool into final products. PLA/PHBV blends have improved the overall satisfaction, yet incorporation with compatibilizers has recorded to dramatically widen its processing time and temperature range. Besides, 3D printing offers many advantages over traditional manufacturing techniques. Utilizing of 3D printing technology has been implemented in many sectors, especially medical applications. Hence, some developments for PLA and PHA polymer's applications (majority of medical developments) are also discussed in this chapter.

2 PROCESSING PROPERTIES OF BIOPOLYMERS

Processing properties of biopolymers are essential before the optimum processing method can be selected for specific biopolymer product. The molecular weight is the crucial property that affects processing properties, including viscosity, melt flow index, and melt strength. The viscosity of a polymer is inversely proportional to the shear stress applied (Fig. 5.1) and can be enhanced by higher molecular weights, the cooperation of additives, fillers, working temperature, and/or pressure.

To analyze plastic's melt index, ASTM D1238 (standard test method for melt flow rates for thermoplastic by extrusion plastometer) is one of the options. Approximately 7 g of plastic materials is loaded into the barrel and heated to the temperature specified for the plastic. A weight specified for the plastic is applied to a plunger so that the molten plastic is forced out through the die. A timed extrudate is collected and weighed. Melt flow rate values are calculated in g/10min according to Eq. (5.1):

$$\text{Flow rate} = \left(\frac{600}{t} \times \text{weight of extrudate}\right) \tag{5.1}$$

where t is time of collection in second.

Advanced Processing, Properties, and Applications of Starch and Other Bio-Based Polymers. https://doi.org/10.1016/B978-0-12-819661-8.00005-6

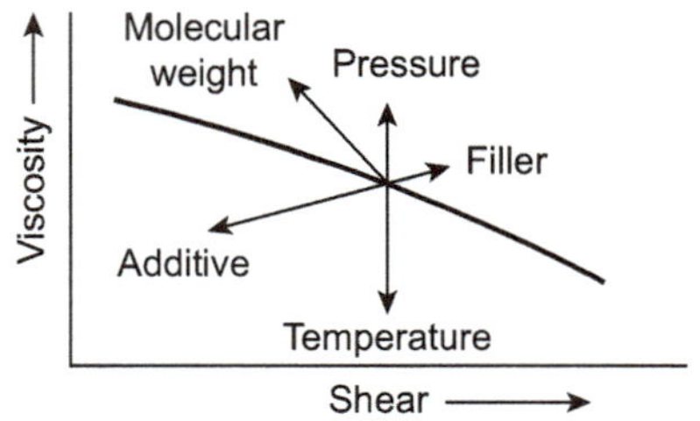

FIG. 5.1 Factors influencing the viscosity of polymer (Rosato, 1998).

Poor melt stability and low melt strength of PLA polymer have made low processability on melt spinning, blow molding, and blown film extrusion processing (Lim et al., 2008). Besides, PLA and PHAs are not preferred to be processed by extrusion because of their sensitivity to moisture and high temperatures.

2.1 PLA

PLA is a crystalline polymer that has low melt viscosity. It has melting temperature and processing temperature ranging from 155 to 185°C and 185 to 190°C, respectively. Working at a higher temperature will cause chain scission reactions to happen; PLA product with lower molecular weight is obtained. Besides, the formation of lactide in molten PLA by a back-biting mechanism can reduce melt viscosity and produce fumes and fouling of equipment under prolonged processing. The polymers would show a significant reduction of viscosity if no prior drying were done before processing. This reduction was due to the hydrolysis reaction and thermal degradation (Speranza et al., 2014).

A typical property of pseudoplastic non-Newton fluids shows evidence for decreased viscosity of PLA polymer with increasing shear stress. This behavior was due to the arrangement of chain segments parallel to the applied shear stress. An inverse relationship between molecular weight and melt flow index for all type of polymers (Djellali et al., 2015). The higher melt index of PLA being recorded is compares with that of the LDPE, which has higher melt viscosity and molecular weight. A similar finding was found by those who studied PLA scrap reprocessing by extrusion compounding. The authors have confirmed that recycled PLA has lower viscosity due to short molecular chains (Peinado et al., 2015).

High activation energy (66.89 kJ/mol) was recorded by PLA polymer at a low shear rate, which is due to strong intermolecular forces in PLA due to its polar nature providing excellent rigidity of macromolecular chains and resistance to flow at low processing shear rate (Djellali et al., 2015). On the contrary, lower activation energy with higher shear rate may be attributed to the higher possibility of entanglement and intermolecular interaction destruction between chain segments, resulting in lower viscosity. A higher concentration of enhancer in PLA reduces the ability of PLA to crystalline, making lower entanglement molecular weights.

Global Biopolymers Co. Ltd. has introduced several grades of PLA polymers for different types of processing to produce the optimum performance of products. The properties of some PLA grades available in the market have listed in Table 5.1.

2.2 PHA

PHAs are a family of bio-based polymer produced by fermentation of renewable feedstock using microbiology. Majority of PHA family have been listed in Table 5.2 with their categories and properties. The most common petrochemical-based biopolymer in the PHA family is poly(3-hydroxybutyrate) (PHB). It is a highly crystalline linear polymer, but its high melting point caused the processing quite problematic and unstable molten. Thermal degradation of PHA is mainly caused by the mechanism of β-elimination and scission, where smaller polymer segments containing carboxylic acid and crotonic end groups are randomly produced (Morikawa and Marchessault, 1981).

ptThe melting temperature of PHA is higher than its decomposition temperature, and it will undergo thermal decomposition during processing and decreased the molecular weight of PHA product as consequences. Therefore, the melting point (180°C) of PHA is the most optimum processing temperature. Like PLA polymer, increasing the temperature from 170 to 210°C observed a change in lower values of viscosity for PHB polymer (El-Hadi et al., 2002). Santos et al. (2018) have recorded 12.28 ± 1.97 g/10min of melt flow index for PHB polymer at 174°C and a load of 2.16 kg under ASTM D1238 (Santos et al., 2018). More than 10 folds of increment on melt property when working temperature has increased to 190°C. Similar findings by Greene (2013) showed a dramatically reduction of viscosity for Tianjin PHA polymers (Greene, 2013). However, Vandi et al. (2018) have revealed a succeed study to process PHBV polymer via extruder with a temperature of 190°C (Vandi et al., 2019). Fig. 5.2 shows the contour mapping of tensile strength versus the processing temperature and screw speed.

The use of PHAs bioplastics is tailorable to the application, depending on the specific combinations of

TABLE 5.1
The Properties of Some PLA Grades Available in the Market.

Company	PLA Grade	Process Suitability	Melt Index, G/10min	Melting Point, °C	Glass Transition Temperature, (T_g), °C
Global Biopolymers Co. Ltd	L100H–GBP20	Thermoforming	2–10[a]	170–180	60–63
	L99L–GBP30	Injection molding	20–40[a]	170–180	60–63
	L96H–GBP70	Stretch blow molding	4–8[a]	145–160	60–63
NatureWorks	Ingeo 2003D	Extrusion	6[b]	145–160	55–60
	Ingeo 2500HP	Extrusion	8[b]	165–180	55–60
	Ingeo 3001D	Injection molding	22[b]	155–170	55–60
	Ingeo 3052D	Injection molding	14[b]	145–160	55–60
	Ingeo 3100HP	Injection molding	24[b]	165–180	55–60
	Ingeo 3251D	Injection molding	80[b]	155–170	55–60
	Ingeo 3260HP	Injection molding	65[b]	165–180	55–60
	Ingeo 3801X	Injection molding	8[b]	160–170	45
	Ingeo 7001D	Stretch blow molding	6[b]	145–160	55–60
	Ingeo 7032D	Stretch blow molding	5–15[b]	155–170	55–60
Total corbion PLA	Luminy L105	Injection molding	30[a]	175	55–60
	Luminy L130	Injection molding	10[a]	175	55–60
	Luminy L175	Thermoforming	3[a]	175	55–60
	Luminy LX175	Thermoforming	3[a]	155	55–60
	Luminy LX575	Extrusion	3[a]	160–170	55–60
	Luminy LX175U	Extrusion, thermoforming	3[a]	145–160	55–60
	Luminy LX130U	Injection molding	7[a]	145–160	55–60

[a] The testing under 190°C and 2.16 kg weight load.
[b] The testing under 210°C and 2.16 kg weight load.

different monomers incorporated into the polymer chain. Cambridge Consultants has developed a series of PLA containers and a bioplastic tray (Rick, 2018). It means the efforts of sorting, washing, and separate bins are no longer needed. All products could decompose in industrial composting facilities or even on side road. Other than that, PHAs have great potential to replace almost all kinds of traditional plastics without causing trouble to end users. Fig. 5.3 has shown an infographic that visually summarizes the physical requirements of some conventional plastic products and potential alternative PHA types.

One of the unique properties of PHA polymer is that it is biodegradable by microorganisms into water-soluble oligomers and monomers, which is nontoxic to human cells and tissues. The details of the biodegradability of PHA materials have been reviewed by previous review paper (Ong et al., 2017). Therefore, PHA polymers are widely used in medical applications. The biosynthesis of PHA for biomedical applications is studied in an earlier review paper (Butt et al., 2018).

3 BIOPOLYMERS PROCESSING AND ITS DEVELOPMENT

3.1 Extrusion

Extrusion is a high-volume manufacturing process in the plastic industry. It is build up by a combination of screw conveyor and compressor. The raw plastic is subjected to an elevated temperature to melt and formed under a continuous cycle. The heated helical screw (single or twin screw) rotates to compress the raw plastics together to homogeneous melt and forces the air out of the barrel, with no damage to the molten. The additives like chain extenders, stabilizers, plasticizers, or/and colorants are used and mixed with the raw plastic in the machine hopper.

The single screw and twin screw have their benefits, depending on the polymer being processed and the products to be fabricated. The differences between

TABLE 5.2
The Categories and Properties of Various PHAs (Joce).

Polymer Code	Polymer Name	Material Class	Properties
P3HB	Poly(3-hydroxybutyrate)	Semicrystalline thermoplastic	• Strong • Brittle • Small thermal processing window • High softening temperature
P4HB	Poly(4-hydroxybutyrate)	Thermoplastic elastomer	• Strong • Flexible • Ductile • High melt viscosity
P(3HB-*co*-4HB)	Poly(3-hydroxybutyrate-*co*-4-hydroxybutyrate)	Semicrystalline thermoplastic/ thermoplastic elastomer	• Strong • Tough • Large thermal processing window • Ductile
PHBV	Poly(3-hydroxyalkanoate-3-hydroxyvalerate)	Semicrystalline thermoplastic	• Strong • Brittle • Large thermal processing window • High softening temperature
PBHH	Poly(3-hydroxybutyrate hexanoate)	Semicrystalline thermoplastic	• Flexible • Ductile • Easy to process low softening and melting temperature

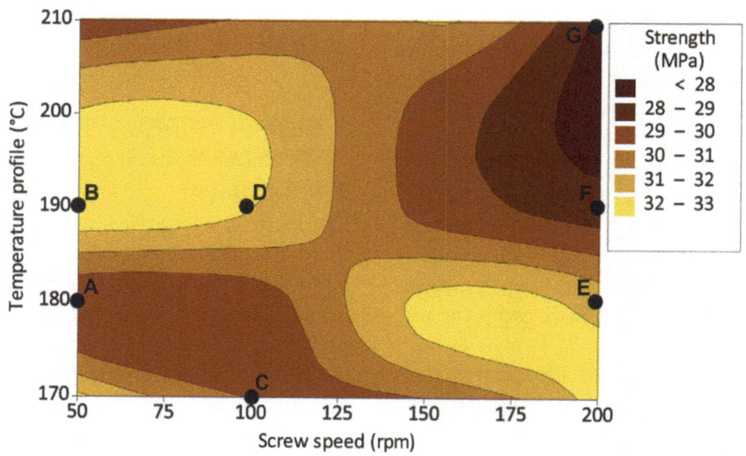

FIG. 5.2 Contour mapping of tensile strength versus the processing temperature and screw speed (Vandi et al., 2019).

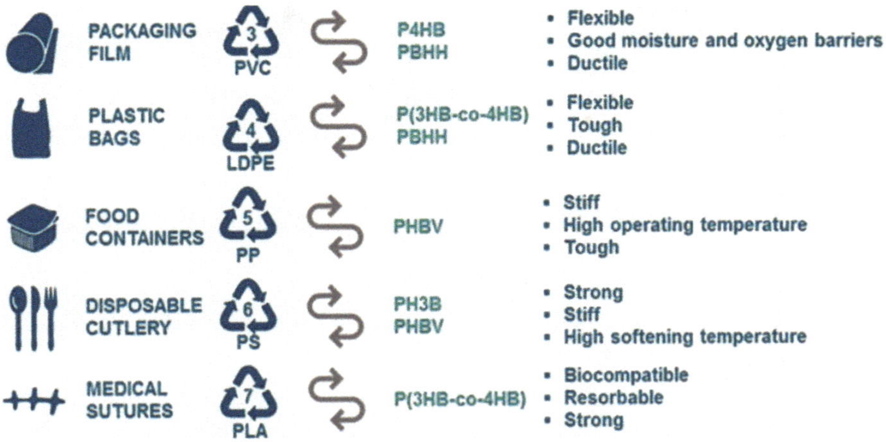

FIG. 5.3 Physical requirements of some conventional plastic products and potential alternative PHA types (Rick, 2018).

single and twin screw are listed in Table 5.3. A typical extrusion screw consists 50%, 25%–30% and 20–25% of its screw length for feed zone, compression zone and metering zone, respectively. The compression ratio (flight depth at feed zone/flight depth at the metering zone) of screw affects the mixing and shear heating properties (Fig. 5.4). The ratio of the screw length to its diameter, L/D ratio, is responsible for homogenous mixing and varies for all polymers. Screws with large L/D ratio provide higher shear heating, better blend, and longer melt residence time in the extruder. Aggressive mixing had additional $7 \times 90°$ elements and featured a reverse flight at the end with 4×30^0 elements at zone 7. The results show that aggressive mixing model has better tensile properties, regardless of the fiber contents and processing temperature.

TABLE 5.3
Comparison of Single- and Twin-screw Extruders (Rosato, 1998).

Criteria	Single-Screw	Twin-Screw
Flow type	Drag	Near positive
Residence time and distribution	Medium and wide	Low and narrow
Effect of back pressure on output	Reduces output	Moderate effect on output
Shear in channel	High (useful for stable polymer)	Low
Maximum screw speed	High (limited by melting and stability of polymer)	Medium
Thrust capacity	High	Low
Mechanical construction	Robust, simple	Complicated
Initial costing	Moderate	High

The speed of the rotating screw and working temperature varies according to the specific type of polymers or mixture of a polymer. During the initial stage of mixing, decrement of barrel temperature was recorded as heat is being absorbed by polymer for melting purpose. The time for the melt temperature to reach the plateau value depends on screw speed. A short plateau time is recorded for higher screw speed due to higher shear heating. On the other hand, the differences in melt temperature were found throughout the cross section with the highest temperature at the center ($r/R = 0.0$) and less pronounced for higher screw rotating speed. High screw rotating speed shows a homogenous cross section of melt molten.

On the other hand, higher working temperature shows increased melt flow capability of the polymer due to weaker macromolecular chain. Besides, Arrhenius expression suggests the shear viscosity of polymer on temperature according to

$$\eta_w = \eta_0 \exp\left(\frac{E}{RT}\right) \tag{5.2}$$

where η_0 is the coefficient related to the viscosity of materials, E is the activation energy of viscous flow, and R is

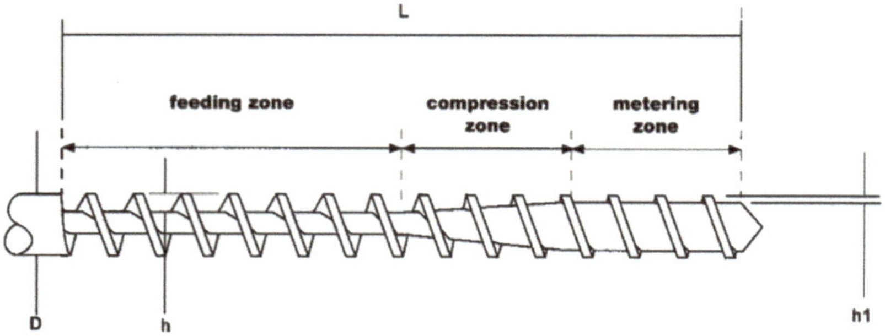

FIG. 5.4 Schematic figure for typical extruder screw (Maier and Calafut, 1998). *D*, diameter of screw; *h*, flight depth at feed zone; *h1*, flight depth at metering zone; *L*, screw length.

the universal gas constant. It was observed that melt viscosity increased with temperature (Liang, 2007).

As for PLA polymer extrusion, hydrolytic degradation is the main drawback and should be dried to 0.025% w/w moisture content before extrusion. Higher working temperature (240°C) must maintain its moisture content below 0.005% w/w to avoid reduction of molecular weight. Drying of PLA resin is a challenging topic as it degrades at elevated temperature and high moisture content is environmental. High moisture contents (80% relative humidity) condition found dramatically molecular weight reduction in less than 30 days for 60°C but 10 days for 80°C, respectively (Henton et al., 2005). On the other hand, high drying temperature permitted fast drying for crystalline PLA resins.

PLA resins are processable by using conventional extruder with general purpose screw (L/D ratio = 24–30) as well as extruder screw specified for PET polymer, which has a low shear rate. Besides, the screw compression ratio is recommended in the range of 2–3. To make sure all PLA crystalline are melted and to achieve an optimal melt viscosity for processing, the heater set point is usually set at 200–210°C. Dicumyl peroxide and triallyl trimesate have been implemented to produce long-chain branching PLA extrudates. The physical properties were studied at a temperature of 60°C, where hydrolytic degradation happens. A significant loss of molar mass was observed in the first 20 days, representing biodegradability is maintained yet long-chain branching approached for various commodity applications.

A PLA/sisal biocomposite has been recycled for eight times by using the extrusion process to study its recyclability (Chaitanya et al., 2019). The performances of specimen dropped significantly after the third recycle processing, responsible by severe damage of fibers and PLA degradation. The processing parameters of 60 RPM screw speed and temperature profile of 150–185°C have induced thermal degradation to the composite. The morphological images have revealed that the increase in the surface roughness was due to the hydrolytic degradation, which lowers its molecular weight and increases its melt flow rate, resulting in improper fiber wetting. Fig. 5.5 shows the factors that influence tensile strength, and it found that fiber ratio has significantly varied the strength of composite rather than processing parameters.

3.2 Injection Molding

Injection molding is one of the most frequently used processing for plastic products. The machine was build up by the injection unit and clamping unit. Injection unit consists of a reciprocal screw that melts polymer with elevated temperature and conveys it to mold that holds in the clamping unit. Then the molten is allowed to cool and solidify in the mold before being ejected. The typical cycle for an injection molding process is shown in Fig. 5.6. Cycle time is an essential parameter for maximum production rate. It is often to transfer partially cooled products to postcooling mold to provide further cooling while it starts the next cycle.

There are currently three types of screw designs available in the marketplace, as listed in Table 5.4. Each design is equipped with specific advantages and disadvantages. An L/D screw ratio of 20:1 is the minimum ratio for injection molding. Longer screw length creates more shear heat for better homogeneity of melt and allows larger products fabrication. However, it also caused higher back pressure that increases melt temperature, lengthens the cycle times, consumes extra energy, causes higher possibility of polymer degradation, and increases wear of barrels.

Injected PLA products are relatively brittle due to rapid physical aging since its glass transition temperature, T_g, is located above ambient temperature. One study concluded that the lower the aging temperature from

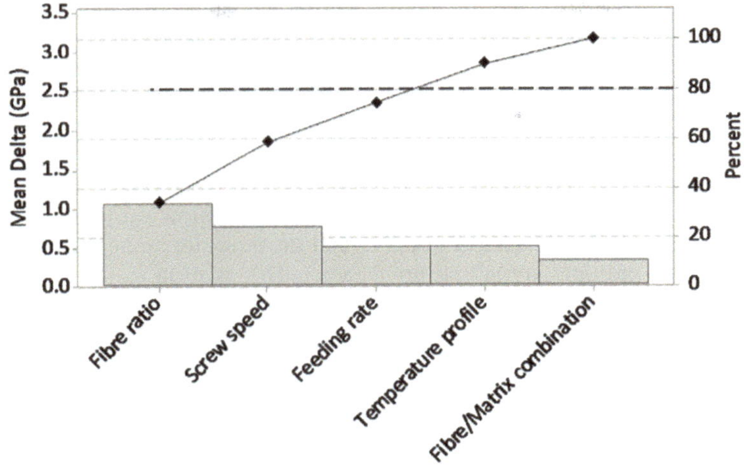

FIG. 5.5 Factors affecting tensile strength (Vandi et al., 2019).

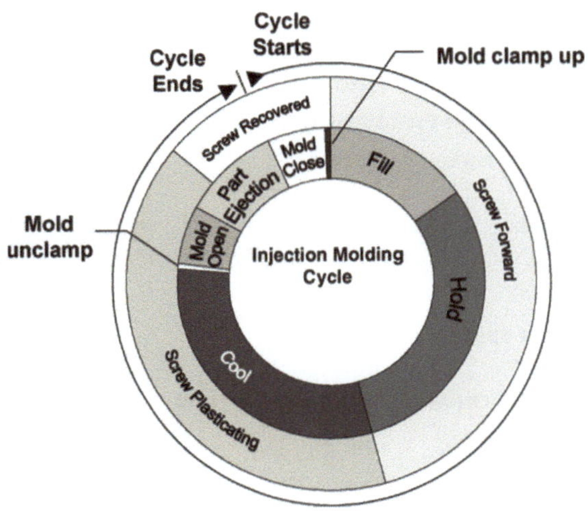

FIG. 5.6 Typical cycle for an injection molding process (Lim et al., 2008).

T_g, the slower the aging rate. On the other hand, no sign of physical aging is taking place for aging temperature higher than T_g of PLA, as recorded from DSC analysis. Therefore, injected products aged at room temperature will more likely become very brittle. PLA polymer that is having high crystallinity could minimize the effect of physical aging since aging below T_g highly depends on its amorphous phase. Besides, mold temperature, back pressure, cooling temperature, and rate are influencing PLA aging properties. Schäfer et al. (2019) has studied the effects of nucleating agents on reducing injection molding cycle times for PLA polymer. The results show the positive impact of bio-based nucleating agents on cycle time reduction. However, fast cooling without posttreatment process recorded lower mechanical properties. Annealed and hot mold cast samples show better properties due to higher crystallinity of PLA polymer.

TABLE 5.4
Advantages and Disadvantages of Screw Types (Grelle, 2005).

Screw Type	Advantages	Disadvantages
Conventional screw	• Low in cost	• Relatively poor mixing quality
Barrier screw	• Better melting rate • Slightly better mixing quality compared with conventional screw	• Higher shear stress • Higher melt temperatures • Prone to solid wedging • Not as forgiving as a conventional screw
Barr ET	• Higher melting efficiency • Better energy utilization • Increased mixing and melt uniformity	• Higher cost due to increased difficulty in manufacturing and design of the screw

There are numerous injection moldable PHA biopolymers available in the market. PHA resins should be dried to 1000 ppm (0.1%) or less. The thermal decomposition temperature of PHA resins was found on 180°C, and hence its processing temperature configuration should range from 176°C at feeding zone to 160°C at the nozzle, with slightly cooling effects throughout the barrel. On the other hand, using a single-flighted, general purpose screw with 2:1 to 2.5:1 compression ratio with 50–150 RPM screw speed is highly recommended. Besides, the mold temperature should fix on 48.9–60°C, to promote crystallization before further cooling is allowed.

Drying of PLA resins at 50°C for 12 h under vacuum condition has been done to prevent hydrolysis degradation. One study using twin-screw microcompounder 100RPM, 180°C and 3 min of mixing period before it directly transferred to the injection molding. It was then processed under 10 bars pressure and injected into a 25°C of mold. A fibrillated surface morphology of PLA was obtained under the injection parameters. On the other hand, to widen the PLA applications, insertion of compatibilizers has been more compatible when PLA blends with PP polymers. However, the properties decreased with an increased portion of PP polymers, since PLA resins hold better performances than PP polymer. Besides, an inorganic filler, hydroxyapatite, has inserted into PLA composites by using injection molding. The parameters of extrusion and injection molding processes are listed in Table 5.5 (Setting 1). On the other hand, injected PLA blend with low crystallinity and amorphous PHA copolymers were demonstrated an improved toughness.

Gunning et al. (2013) has used a 27 mm twin-screw extruder with a 38:1 ratio to mixes short natural fibers into PHB composite pellets (Gunning et al., 2013). The compounds were then processed via injection molding to study its mechanical and biodegradation performances. The injection molding settings were listed in Table 5.5 (Setting 2). The melt flow rate results showed that the dispersion and the size of natural fibers had disturbed the molten's flow properties. However, the presence of fiber agglomerations further restricted the flow of molten and may even form blockages in the die. It will influence the processability of natural fibers reinforced PHA composite in injection molding if the melt flow rate is too low.

3.3 Blow Molding

All plastic bottles are made from thermoplastic materials by using blow molding processing. Three types of blow molding methods are available, injection blow molding, extrusion blow molding, and stretch blow molding.

For Injection blow molding process, the material is injected. The hot injected plastic is then blown into a bottle at blow molding station, allowed to cool (1–4 days), and then ejected at next station. It is often used to produce irregular shapes of bottles, bore necks, or wide necks bottles. A high precision product has produced which does not require after work trimming. However, bottle sizes were limited to a blow-up ratio and ovality ratio lesser than 3:1 and 2:1, respectively. In extrusion blow molding process, plastic is melted and extrudes simultaneously into a hollow tube (parison). This parison is then placed in a cooled metal mold, and the air is blown into the parison, inflating a hollow shape of bottles or containers. The product

TABLE 5.5
Parameters of Extrusion and Injection Molding Processes.

Number of Setting	1	2
Reference	Akindoyo et al. (2017)	Gunning et al. (2013)
EXTRUSION PARAMETERS		
Feeding zone, °C	110	–
Zone 1, 2, 3, 4, 5, 6, 7, °C	165, 165, 170, 170, 175, 175, 170	185, 160, 145, 135, –, –, –
Die, °C	170	–
Screw speed, RPM	200	–
INJECTION MOLDING PARAMETERS		
Feeding zone, °C	50	–
Compression zone, °C	165–185	–
Metering zone, °C	190	–
Nozzle, °C	185	130
Mold, °C	35	25
Screw speed, RPM	150	20
Screw position, RPM	30	–
Injection pressure, bar	550–600	2000
Holding pressure, bar	500–550	900
Injection time, s	0.50	6.50
Cooling time, s	30	20

is then ejected after cooled. The uneven wall thickness of the product is one of the major disadvantages of extrusion blow molding, and a thicker part tends to shrink more. Besides, close dimensional tolerances are nearly hard to achieve, and surface finishing is relatively low. Trimming at the neck and bottom pinch-off areas is needed to remove excess plastic.

On the other hand, stretch blow molding functions as injection blow molding, but in addition, it is stretched throughout the process. The preheated PLA polymers were injected and transferred to a blow molding machine, which will then be stretched in biaxial direction. Biaxial molecular orientation allows better physical properties, clarity, and gas barrier properties. The preform is heated by a few infrared heaters to a suitable temperature to blow. The blow nozzle stretches in the preform by inflating itself. The raised shape will imprint the surface details of the bottles. It has the highest production rate but most expensive to set up.

PLA bottles were reported biodegraded in about 100 days and resides can be recycled as fertilizer or soil enrichments. As bottled drinks demand to increase dramatically worldwide, switching the conventional PET bottles to PLA bottles is a great option to solve environmental issues as well as political and economic aspect. In most cases, there are no or little differences between PET and PLA bottles in terms of capabilities and appearance of the bottle. However, the requirement for PLA blow processing is different from PET blow processing parameters. Therefore, processing flow and parameters adjustments must be done by specialists before changing the bottle materials.

Preheating for the preform polymer before entering the stretch molding is needed for blow processing. The preform heating temperatures of PLA and PET bottle are 75 and 100°C, respectively. Higher PLA preheating (100°C) will cause the polymer to shrink and degrade. On the other hand, different cooling rates of both polymers makes incomplete PLA bottle cool down due to the insufficient time before leaving the molds. New cooling techniques or longer cool down periods must be given to PLA bottles to avoid bottle deformations. Additionally, temperature sensitivity of PLA polymer requires multiple heat tunnels to achieve consistent heating for optimal bottle performance because of different polymer thickness throughout the bottle.

PHAs are valuable for melt blowing at low comonomer content and molecular weight. The low melt strength and low thermal stability produces an inferior grade of blow-molded bottles, while high melt strength is an excellent opaque bottle material for extrusion blow molding process (Greene, 2013). On the other hand, branched PHA compositions can be made by reactive extrusion to improve melt strength. The sufficient extrusion temperature and melting period caused decomposition of a free radical initiator, resulting in a decrease in its molecular weight, while the branching reaction produces an increase in its molecular weight.

The low melt strength of PLA is attributed to its chain scission reactions when subjected to shear and high temperature in an extruder. Melt strength enhancer is commonly applied to allow PLA polymer to undergo processing with higher melt flow index. In this study, PLA was processed at low extrusion temperature resulting in high melt viscosities and has observed a stable blown film production. Another PLA-blend-blown-film has been successfully produced by using blow processing with listed parameters in Table 5.6 (Mallegni et al., 2018). Plasticizer has been applied in the film formulations to enhance tear resistance.

Fig. 5.7 shows the continuous production of film blowing in the study, and the film's thickness was controlled in the range of 180−230 μm. The results showed that the addition of PHA into starch matrix increased the tensile strength, but decreased the modulus, and thermal stability of films.

3.4 Thermoforming

Thermoforming is a process of heating a thermoplastic sheet to its glass transition temperature to stretch and then cool into final products. Elastomers and thermosets are prohibited from the thermoforming process due to nonsoften behavior when heated. Thermoforming is able to create several finishing parts on the same product; therefore, time and cost saving are achievable. However, the extra cost must be considered to fabricate

TABLE 5.6
PLA-blend-blown-film Blow Processing Parameters (Mallegni et al., 2018).

Blown Processing Parameters	Value
Layflat width, mm	410
Bubble diameter, mm	261
Die diameter, mm	55
Die gap, mm	0.8
Film thickness, mm	0.05
Blow-up ratio	4.75
Drawdown ratio	3.37
Forming ratio	0.71

FIG. 5.7 Continuous production of film blowing for starch/PHA composite films (Sun et al., 2017).

an extruded thermoplastic sheet from powder or pellet form. There are three types of thermoforming processing (vacuum thermoforming, pressure thermoforming and mechanical thermoforming) available.

Vacuum thermoforming uses atmospheric pressure to force the heated sheet to deform according to the cavity shape. Low forming pressure, as compared to the injection molding process, is needed in the low-cost tooling processing. Besides, it provides a perfect option for prototype fabrication. Pressure thermoforming uses air pressure to push the soft sheet, while mechanical thermoforming has a core plug and presses the soft sheet into the cavity. Productive Plastics, Inc. has made a comparison between thermoforming and injection molding. The thermoforming processing is preferred, which has short set up time, lowers production, and lowers initial costing.

In general, the working temperature of PLA is much lower than conventional plastics, roughly about 80−110°C. The molds for polypropylene are not useable for PLA thermoforming because polypropylene shrinks more than PLA upon cooling. Besides, higher cooing period is required for PLA products due to lower thermal conductivity and T_g of PLA. An elevated temperature of 90°C must be applied before any trimming work is needed on the PLA sheet. The occurrence of thermal degradation on PLA thermoforming products found when storing it above 40°C.

3.5 3D Printing

3D printing creates parts by building up objects one layer at a time. This method offers many advantages over traditional manufacturing techniques. 3D printing is unlikely to replace many traditional manufacturing methods, yet there are many applications where a 3D printer can deliver a design quickly, with high accuracy from a functional material. Fig. 5.8 is showing the number of 3D printers under $5000 sold globally. The number of printers sold doubled consistently since then.

One of the main advantages of additive manufacture is the speed at which parts can be produced compared with traditional manufacturing methods. Intricate designs can be uploaded from a CAD model and printed in a few hours. The advantage of this is the rapid verification and development of design ideas. Additive manufacturing machines complete a build in one step, with no interaction from the machine operator during the build phase. As soon as the CAD design is finalized, it can be uploaded to the machine and printed in one step in a couple of hours.

The ability to produce a part in one step dramatically reduces the dependence on different manufacturing processes (machining, welding, painting) and gives the designer greater control over the final product. The cost of manufacture can be broken down into three categories: machine operation costs, material cost, and labor costs. Most desktop 3D printers use the same amount of power as a laptop computer. On the other hand, material costs are the most significant contributor to the cost of a part made via additive manufacturing. However, the price of 3D printing materials has dropped exponentially due to the high popularity of 3D printing processing consumed in a massive number of materials. Besides, it requires almost zero labor costs for a 3D printer.

Not only does 3D printing allow more design freedom but it also allows complete customization of designs. Since current additive manufacturing technologies excel in building single parts one at a time, they are

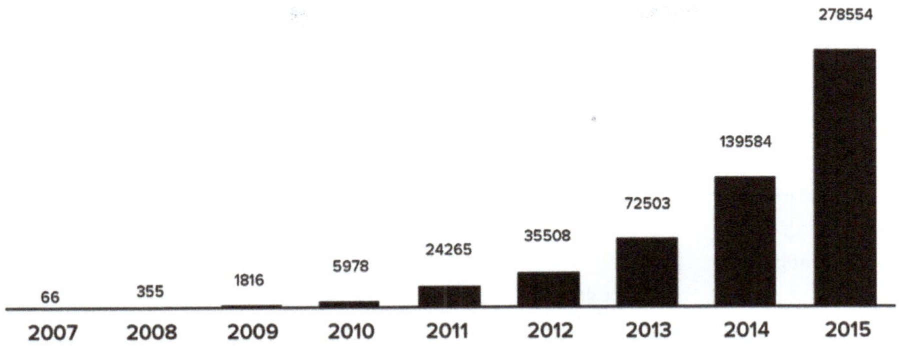

FIG. 5.8 The number of 3D printers under $5000 sold globally (Redwood, 2019).

ideally suited for one off production. Therefore, high customization processing flavored all users.

The purpose of 3D printing product must identify to locate the most optimum type of 3D printing processing. Fig. 5.9 shows the flowchart for the selection of 3D printing processes via material used and performances. 3D printing materials usually come in filament, powder, or resin form depending on the types of 3D printer. Thermoplastic polymers used in fused deposition modeling, FDM, are best suited for functional applications including the manufacturing of end use parts and functional prototypes. A performance cost diagram shown in Fig. 5.10. PLA is considered as the most commodity readily FDM 3D printing material giving biodegradability. However, FDM processing produces low strength product, unless fiber reinforced filaments are being used. High biocompatible of FDM processing allowed PLA or PHA products to be printed in the medical and packaging fields.

Utilizing 3D printing technology in tissue engineering provides a more detailed, complex, and versatile scaffold system. Details of 3D printing technology on ceramic-based scaffolds and bone tissue engineering have have briefly reviewed by Du et al. (2018) (Du et al., 2018). Interesting research on developing 3D printing filaments of composite materials made from maleic anhydride-grafted polyhydroxyalkanoate (PHA-*g*-MA) and coupling agent-treated palm fiber (TPF) showing great performance on water barrier ability and fibroblast cell viability evaluation in water after incubation of the cells for 24 h. These excellent characteristics potentially made the low-cost PHA-*g*-MA/TPF composite as biodegradable 3D printing filaments for tissue scaffold structures. Similar 3D printed scaffolds by using PLA polymer have been done by Farto-Vaamonde et al. (2019) (Farto-Vaamonde et al., 2019). During FDM printing, drugs loaded on the PLA filament may become solubilized in the melted PLA, which causes the drug to be molecularly dispersed in the printed strand matrix. Different drug release profiles have been achieved. The results show the use of commercially available and low-cost 3D printer machine achieving the requirements for bone scaffolds and sustaining more than 8 weeks (Rodrigues et al., 2016).

4 DEVELOPMENTS FOR PLA AND PHA POLYMER'S APPLICATIONS

A harsh global issue on single-use plastic pollution is an emerging new business to reconsider the new phase of plastic products. Bioplastics that are biodegradable have an increasingly feasible technical alternative to conventional polymer, especially in the medical sector.

The medical devices that are used to replace or repair some diseased, damaged, or nonfunctional piece of tissue or bone-like replacement of joints, heart valves, arteries, teeth, tendons, ligaments, ocular lenses, and some other advanced devices to partially or entirely replace or assist in functioning of a vital organ like lung, kidney, liver, and heart, had widely replaced by biopolymers. Besides, advanced and efficient drug delivery systems, PHA capsulate materials made of biopolymers that showing no resistance from human body also widely applied to contain drugs for controlled and sustained release or targeted delivery of drugs.

The scaffold system in biomedical industry is primarily to act as a template for tissue formation and is typically seeded with cells. O'Brien (2011) summarized the biomaterials and scaffolds to develop biological substitutes that restore, maintain, or improve tissue function. PHA scaffold system was developed for rat osteoblasts, the cell that forms new bone matrix. Rough, hydrophilic poly(hydroxybutyrate) (PHB) and

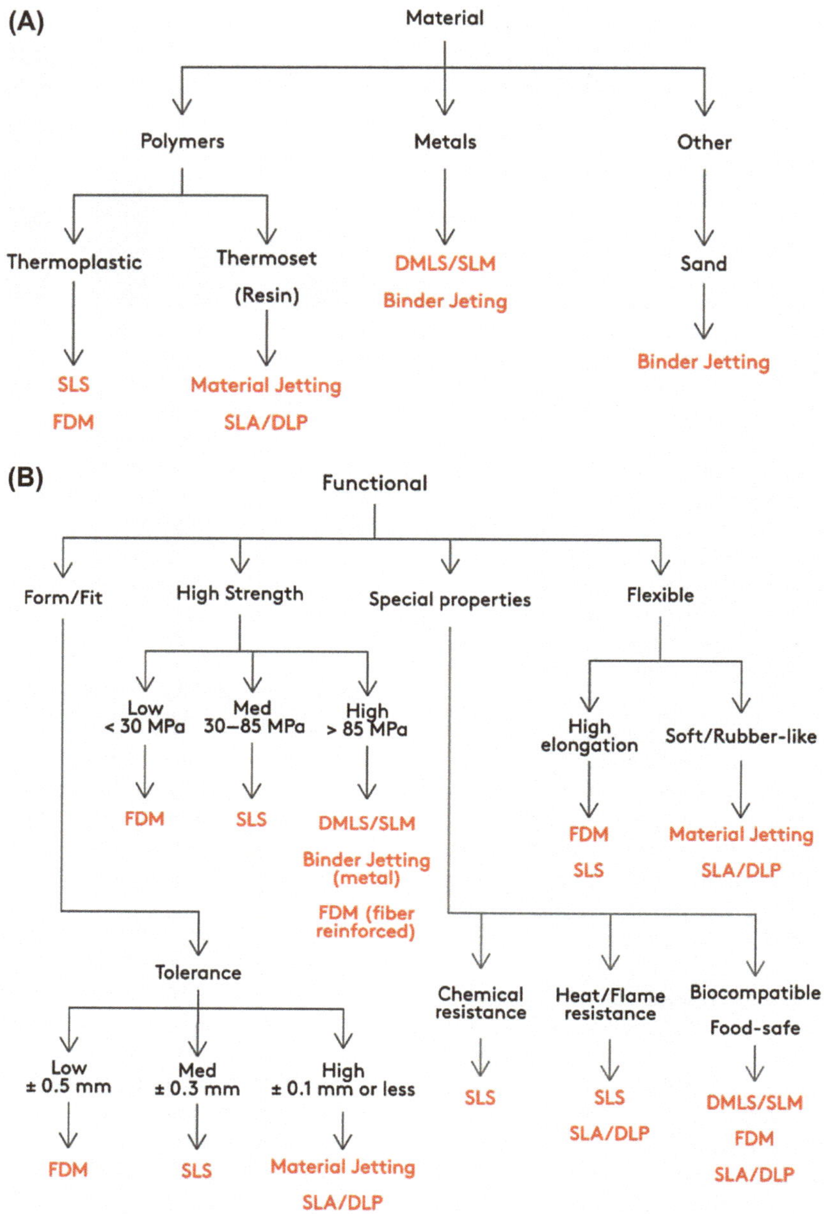

FIG. 5.9 Flowchart for the selection of 3D printing processes via **(A)** material used and **(B)** performances (Alkaios Bournias, 2019).

poly(hydroxybutyrate-*co*-hydroxyvalerate) (PHBV) films incubated comparable viable osteoblast cells growth rate among four PHB copolymer films in the study (Wang et al., 2013). In contrast, smoother and more hydrophobic poly(3-hydroxybutyrate-*co*-3-hydroxyvalerate-*co*-3-hydroxyhexanoate) and poly (hydroxybutyrate-co-hydroxyhexanoate) (PHBHHx) surface encourage the growth of fibroblast, the cell that synthesizes extracellular matrix and collagen and plays a critical role in wound healing (Wang et al., 2013).

A three-dimensional PHA scaffold system is developed to support the joint. Chondrocytes, a cell which secreted the matrix of cartilage, has grown in large amounts on the surface of PHBHHx/PHB scaffolds and slowly forms confluent cell multilayers at

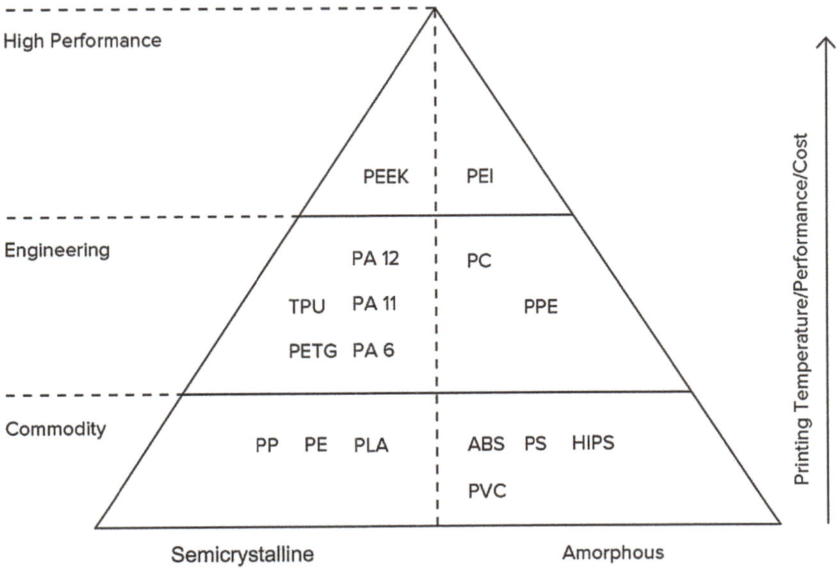

FIG. 5.10 Performance-cost diagram (Alkaios Bournias, 2019).

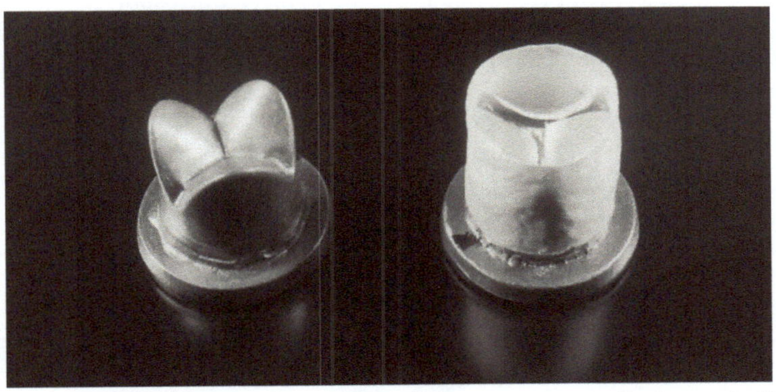

FIG. 5.11 Trileaflet heart valve scaffold made of PHA polymers (Sodian et al., 2000).

14–28 days (Deng et al., 2002). It is an excellent achievement for articular repair sectors for the regeneration of tissue upon isolated polymeric scaffolds. Similar finding from Deng et al. (2003), showing that higher mRNA level of type II collagen of chondrocytes on PHBHHx/PHB scaffold system compared to pure PHB scaffold (Deng et al., 2003). It is because PHBHHx/PHB scaffold provides better surface properties for collagen filaments anchoring. The blend scaffold system has recorded almost fourfold of collagen productions (742.1 mg/g) than in the PHB scaffold system. The PHA-based scaffold systems for bone tissue applications have been intensively review by Lim et al. (2017).

Another innovative development using PHA polymer is to create a scaffold for trileaflet heart valve (Sodian et al., 2000). Then the heart valve scaffold was tested in a pulsatile flow bioreactor, and it was opened and closed like a standard heart valve. Fig. 5.11 shows the trileaflet heart valve scaffold made of PHA polymers.

One study has developed PHB nanohydroxyapatite (HAP) scaffold with improvements on compressive elastic modulus, maximum stress, and osteoblast responses, including cell growth and alkaline phosphatase activity. Besides, the presence of nanobioglass in the PHB scaffolds has developed an osteoblast-like MG63 cell proliferation compared to pure PHB polymers. PHB scaffold promoted attachment, proliferation, and survival of adult Schwann cells and supported marked axonal regeneration within the graft. Xu et al. (2010) compared PHB, P3HB4HB, and PHBHHx regarding their ability to support the central nervous system (NSC) growth and differentiation both on their 2D films and 3D matrices (Xu et al., 2010).

Drug deliveries have become essential tools in the medical field and have been extensively investigated because of the strong demand for the controlled delivery of pharmacologically active materials to cells, tissue, and organs. Hydrophobic micro- or nanoparticle PHA is suitable for carrying hydrophobic drugs as PHA granules with a multifunctional surface displaying both specific binding sites for certain inorganic substances and IgF were produced (Fig. 5.12). PHA has been utilized for the delivery of a range of drug substances from anticancer drugs to hormones and antibiotics. The details of the drug delivery system by PHA nanoparticles are reviewed by Shrivastav et al. (2013).

A specific drug delivery system consisting of PHA nanoparticles, PhaP, and polypeptide or protein ligands fused to PhaP was developed (Natureworks, 2005). The ligand-PhaP-PHA particular drug delivery system was proven effective both in vitro and in vivo for targeting cancer cells (Yao et al., 2008). Similarly, the specificity of the PHB-PhaC-GFP-A33scFv nanoparticle toward the colon cancer cell lines SW1222 (A33+) and HT29 (A33−) was confirmed when PHB nanoparticles were attached with engineered PHA synthase fused with a green fluorescent protein (GFP) and a single-chain variable fragment antibody (A33scFv) specific to colon cancer (Kwon et al., 2014). Bansal et al. (2011) has provided a comprehensive outlook for the advanced drug delivery system approaches by using curcumin as a model agent to introduce highly potent cancer chemoprevention.

Controlled release and accurate targeting are of major importance in drug delivery. The release of antitumor drug (2′,3′-diacyl-5-fluoro-2′-deoxyuridine) from the spheres misting of low molecular weight polymer (molecular weights = 65,000) was faster than that from the spheres of higher molecular weight (molecular weights = 135,000 or 450,000). Besides, the PHB sphere showed low toxicity to and good biocompatibility with mice and rats.

PHA-based polyamidoamine dendrimer (PAMAM) was developed on the transdermal drug delivery system, a route of administration wherein active ingredients are delivered across the skin. PAMAM acted as a polymeric penetration enhancer and enhanced the permeation of drug by pretreatment. Stirring rate, emulsifier concentration, and polymer/solvent ratio reported to the size of PHBHx microsphere. The authors have recorded more than 90% of drug released within the first 24 h and yet the rate of release can be adjusted by changing initial drug/polymer and particle size distributions.

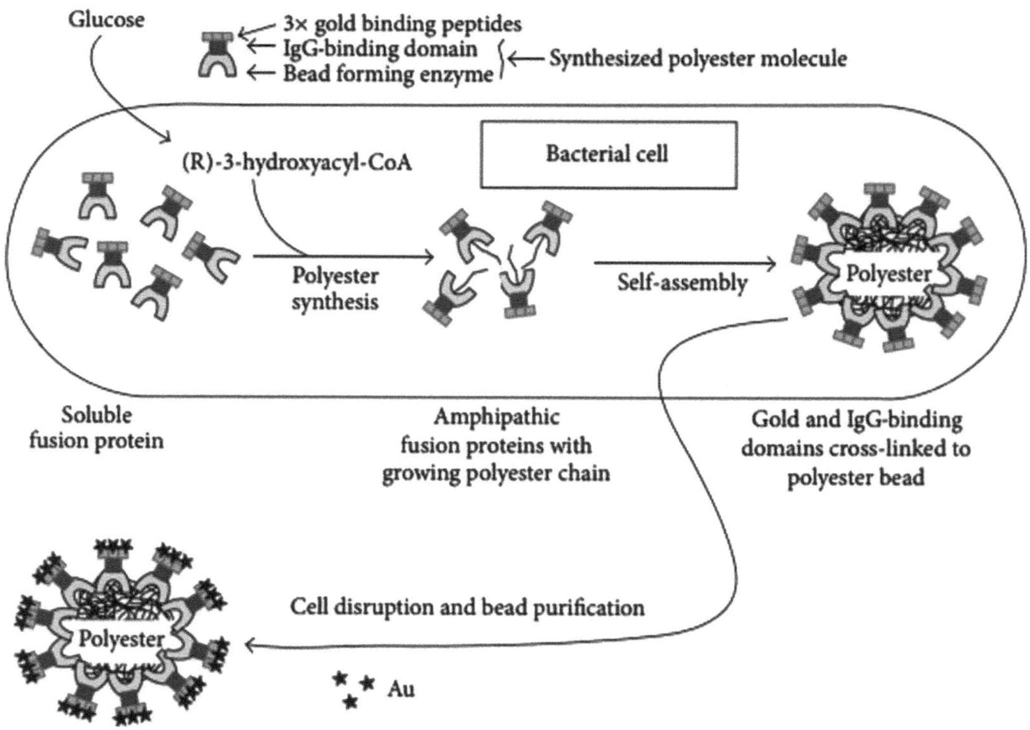

FIG. 5.12 Schematic overview of PHA granules with a multifunctional surface displaying both specific binding sites for certain inorganic substance gold and for IgG (Shrivastav et al., 2013).

Hydrophobic PHA can also be modified to encapsulate hydrophilic drugs. Hydrophilic insulin blending with phospholipid was successfully loaded on PHBHHx nanoparticles (INS-PLC-NPs). Only 20% of encapsulated insulin was released in vitro within 31 days. The hypoglycaemic effect in STZ-induced diabetic rats lasted for more than 3 days after the subcutaneous injection of INS-PLC-NPs, significantly prolonging the therapeutic effect compared with the administration of insulin solution (Peng et al., 2012). PLA nanoparticles were produced to coencapsulate a hydrophilic drug (theophylline) and a lipophilic drug (budesonide) for pulmonary drug delivery. At the end of 24 h, 10%–15% of theophylline was released from both mono- and coencapsulated nanoparticles, while for budesonide, the drug release from mono- and coencapsulated nanoparticles was between 3% and 7% at the end of 24 h.

PHAs are an excellent material to process excellent packaging films by thermoforming, and it can be done by sole PHA polymer or PHA composites. Highly hydrophobicity of PHA shows extreme water vapor and gas barrier behavior, which is suitable for eco-friendly packaging. One study has concluded that increasing the contents of PHBV in starch/PHBV polymer blend was beneficial to the reduction of water vapor transmission rate (Malmir et al., 2018). Keskin et al. (2017) has reviewed on PHA's gas barrier properties, with a specific focus on potential applications of PHBV in packaging. Improvement of barrier properties for organic and inorganic fillers on PHAs polymer has been highlighted in the review paper.

On the other hand, the promising barrier properties make PHAs as substitute materials for liquid foods and CO_2 containing liquids bottle. Besides, high UV barrier is desirable to protect especially unsaturated lipid components in food from the formation of radicals and to reduce the shelf life of food. Khosravi-Darani and Bucci (2015) have discussed the developments (copolymers, blending, and nanoscale) on PHA's properties (mechanical properties, permeability, thermal stability, food contact, and biodegradability) for food packaging.

The famous global company, Nestlé, has announced the development of a biodegradable water bottle on January 2019 (Rick, 2019). Nestlé decided to achieve 100% of its packaging recyclable or reusable by 2025. Although PET bottles are widely recycled, only 29.9% and 9% of PET bottles are recycled in the United States and globally, respectively.

One innovative research created a PHA-modified PLA container by using thermoforming processing (Brum, 2016). The container maintains high biocontent and composability of PLA resin. The wide processing temperature is one of the processing features as PHA polymer has a narrow temperature window. Besides, the clarity is maintained and results in a clear, transparent packaging product.

The use of wood-plastic composites (WPC) is one of the famous options for the furniture industry. It is estimated that a production of 6 million tons of WPC will be produced in the year 2020 (Analysis, 2015). The end-of-life disposal of WPC by using petroleum-based plastic can easily be solved by fully biodegradable PHA matrix. Table 5.7 lists the mechanical properties of PHA/wood composites, compared with commercially available WPC. The results show comparable tensile strength and tensile moduli, and it is expanding the use of PHA/wood composites, allowing them to gain a significant market share.

5 CONCLUSION

PLA and PHA polymer are the most popular biodegradable thermoplastic polymer currently due to their

TABLE 5.7
Summary of Mechanical Properties Achieved for PHA/Wood Composites and Commercial WPC (Vandi et al., 2018).

	Tensile Strength, MPa	Tensile Modulus, GPa	Strain at Failure, %	Impact Strength, kJ/m^2
PHBV ENMAT Y1000	30	2.8	8.0	
PHBV + 50 wt% wood content	27	6.1	1.0	4.0
FKUR-Firolon P7550 (PP + 50wt% wood)	22	3.3	3.0	7.9
FKUR-Firolon P8530 (PLA + 50wt% wood)	34	3.8	3.8	11.7
Jeluplast-PP H50-500-14 (PP + 50wt% wood)	32	4.5	1.9	10.6

promising performances. Processing methods of biodegradable thermoplastic polymer are like conventional thermoplastics but required adjustment on processing parameters or incorporations of compatibilizers. The differences in term of viscosity, melt flow rate, and melt strength have caused the existing processing parameters unsuitable for biopolymers. Ease of thermal degradation phenomena is always an issue needed to deal with biopolymer processing. This is because their processing temperature is near to degradation temperature.

The use of extruder is the most common procedure when processing biopolymers. The rotating screw conveys the molten polymer toward dies or mold often exerting shear stress for the purpose of homogenous mixing. Yet, this has induced further degradation, and lower molecular weight extrudate was obtained. Injection molding or blow machine are the destination of molten polymer. On the other hand, thermoforming is a process of heating a thermoplastic sheet to its glass transition temperature to stretch and then cool into final products. Besides, 3D printing offers many advantages over traditional manufacturing techniques. Utilizing of 3D printing technology has been implemented in many sectors, especially medical applications.

REFERENCES

Akindoyo, J.O., et al., 2017. Effects of surface modification on dispersion, mechanical, thermal and dynamic mechanical properties of injection molded PLA-hydroxyapatite composites. Composites Part A: Applied Science and Manufacturing 103, 96–105.

Alkaios Bournias, V. Selecting the Right 3D Printing Process. Available from: https://www.3dhubs.com/knowledge-base/selecting-right-3d-printing-process/(cited 2019).

Analysis, Z.R., 2015. Global Wood-Plastic Composites Market Set for Rapid Growth, to Reach Around Usd 6.0 Billion by 2020. Available from: http://www.marketresearchstore.com/news/globalwood-plastic-composites-market-set-for-rapid-101 (cited 2019 21th March).

Ayu, R.S., et al., 2018. Effect of modified tapioca starch on mechanical, thermal, and morphological properties of PBS blends for food packaging. Polymers 10 (11), 1187.

Bansal, S.S., et al., 2011. Advanced drug delivery systems of curcumin for cancer chemoprevention. Cancer Prevention Research (Philadelphia) 4 (8), 1158–1171.

Brum, L., 2016. PHA-modified PLA Thermoform Materials for Sustainable Packaging Applications (cited 2019).

Butt, F.I., et al., 2018. Recent progress in the utilization of biosynthesized polyhydroxyalkanoates for biomedical applications – Review. International Journal of Biological Macromolecules 120, 1294–1305.

Chaitanya, S., Singh, I., Song, J.I., 2019. Recyclability analysis of PLA/Sisal fiber biocomposites. Composites Part B: Engineering 173, 106895.

Deng, Y., et al., 2002. Study on the three-dimensional proliferation of rabbit articular cartilage-derived chondrocytes on polyhydroxyalkanoate scaffolds. Biomaterials 23 (20), 4049–4056.

Deng, Y., et al., 2003. Poly(hydroxybutyrate-co-hydroxyhexanoate) promoted production of extracellular matrix of articular cartilage chondrocytes in vitro. Biomaterials 24 (23), 4273–4281.

Djellali, S., et al., 2015. Viscosity and viscoelasticity measurements of low density polyethylene/poly(lactic acid) blends. Polymer Bulletin 72 (5), 1177–1195.

Du, X., Fu, S., Zhu, Y., 2018. 3D printing of ceramic-based scaffolds for bone tissue engineering: an overview. Journal of Materials Chemistry B 6 (27), 4397–4412.

El-Hadi, A., et al., 2002. Effect of melt processing on crystallization behavior and rheology of poly(3-hydroxybutyrate) (PHB) and its blends. Macromolecular Materials and Engineering 287 (5), 363–372.

Farto-Vaamonde, X., et al., 2019. Post-manufacture loading of filaments and 3D printed PLA scaffolds with prednisolone and dexamethasone for tissue regeneration applications. European Journal of Pharmaceutics and Biopharmaceutics 141, 100–110.

Greene, J., 2013. PHA biodegradable blow-molded bottles: compounding and performance. Plastics Engineering 69, 16–21.

Grelle, P.F., 2005. Injection molding. In: Harper, C.A. (Ed.), Handbook of Plastic Processes. Wiley.

Gunning, M.A., et al., 2013. Mechanical and biodegradation performance of short natural fibre polyhydroxybutyrate composites. Polymer Testing 32 (8), 1603–1611.

Henton, D., et al., 2005. Polylactic Acid Technology.

Joce, C. PHAs: Market Opportunities and Technical Possibilities. Cambridge Consultants.

Keskin, G., et al., 2017. Potential of polyhydroxyalkanoate (PHA) polymers family as substitutes of petroleum based polymers for packaging applications and solutions brought by their composites to form barrier materials. In: Pure and Applied Chemistry, p. 1841.

Khosravi-Darani, K., Bucci, D.Z., 2015. Application of poly(hydroxyalkanoate) in food packaging: improvements by nanotechnology. Chemical and Biochemical Engineering Quarterly 29 (2), 275–285.

Kwon, H.-S., et al., 2014. A multi-functional polyhydroxybutyrate nanoparticle for theranostic applications. Journal of Materials Chemistry B 2 (25), 3965–3971.

Lee, C.H., Salit, M.S., Hassan, M.R., 2014. A review of the flammability factors of kenaf and allied fibre reinforced polymer composites. Advances in Materials Science and Engineering 2014, 8.

Lee, C.H., Sapuan, S.M., Hassan, M.R., 2017. Mechanical and thermal properties of kenaf fiber reinforced polypropylene/magnesium hydroxide composites. Journal of Engineered Fibers and Fabrics 12 (2), p. 155892501701200206.

Lee, C.H., Sapuan, S.M., Hassan, M.R., 2018. Thermal analysis of kenaf fiber reinforced floreon biocomposites with magnesium hydroxide flame retardant filler. Polymer Composites 39 (3), 869–875.

Liang, J.Z., 2007. Effects of extrusion rate, temperature, and die diameter on melt flow properties during capillary flow of low-density-polyethylene. Polymer-Plastics Technology and Engineering 46 (3), 245–249.

Lim, L.T., Auras, R., Rubino, M., 2008. Processing technology for poly(lactic acid). Progress in Polymer Science 33, 820–852.

Lim, J., et al., 2017. Emerging bone tissue engineering via Polyhydroxyalkanoate (PHA)-based scaffolds. Materials Science and Engineering: C 79, 917–929.

Maier, C., Calafut, T., 1998. 16 - Extrusion. In: Maier, C., Calafut, T. (Eds.), Polypropylene. William Andrew Publishing, Norwich, NY, pp. 205–221.

Mallegni, N., et al., 2018. Poly(lactic acid) (PLA) based tear resistant and biodegradable flexible films by blown film extrusion. Materials 11 (1).

Malmir, S., et al., 2018. Effects of poly (3-hydroxybutyrate-co-3-hydroxyvalerate) microparticles on morphological, mechanical, thermal, and barrier properties in thermoplastic potato starch films. Carbohydrate Polymers 194, 357–364.

Morikawa, H., Marchessault, R.H., 1981. Pyrolysis of bacterial polyalkanoates. Canadian Journal of Chemistry 59 (15), 2306–2313.

Natureworks, 2005. Processing guide for thermoforming articles. Natureworks LLC, Minnetonka, MN.

O'Brien, F.J., 2011. Biomaterials & scaffolds for tissue engineering. Materials Today 14 (3), 88–95.

Ong, S.Y., Chee, J.Y., Sudesh, K., 2017. Degradation of polyhydroxyalkanoate (PHA): a review. Journal of Siberian Federal University. Biology 10, 21–225.

Peinado, V., et al., 2015. Effect of extrusion on the mechanical and rheological properties of a reinforced poly(lactic acid): reprocessing and recycling of biobased materials. Materials 8 (10), 7106–7117.

Peng, Q., et al., 2012. A rapid-acting, long-acting insulin formulation based on a phospholipid complex loaded PHBHHx nanoparticles. Biomaterials 33 (5), 1583–1588.

Redwood, B. The Advantages of 3D Printing. Available from: https://3dinsider.com/3d-printing-advantages/ (cited 2019).

Rick, L., 2018. PHA bioplastics a 'tunable' solution for convenience food packaging. Packaging Sustainability, Recycling, Resin Pricing, Consumer Products, Materials.

Rick, L., 2019. Nestlé taps Danimer Scientific PHA for biodegradable water bottle development. Packaging Sustainability, Recycling, Materials. Available from: https://www.plasticstoday.com/packaging/nestl-taps-danimer-scientific-pha-biodegradable-water-bottle-development/183546001060374/page/0/1 (cited 2019).

Rodrigues, N., et al., 2016. Manufacture and characterisation of porous PLA scaffolds. Procedia CIRP 49, 33–38.

Rosato, D.V., 1998. Extruding Plastics-A Practical Processing Handbook. Springer, US.

Santos, E.B.C., et al., 2018. Rheological and thermal behavior of PHB/piassava fiber residue-based green composites modified with warm water. Journal of Materials Research and Technology.

Schäfer, H., Pretschuh, C., Brüggemann, O., 2019. Reduction of cycle times in injection molding of PLA through bio-based nucleating agents. European Polymer Journal 115, 6–11.

Shrivastav, A., Kim, H.-Y., Kim, Y.-R., 2013. Advances in the applications of polyhydroxyalkanoate nanoparticles for novel drug delivery system. BioMed Research International 2013, 581684.

Sodian, R., et al., 2000. Fabrication of a trileaflet heart valve scaffold from a polyhydroxyalkanoate biopolyester for use in tissue engineering. Tissue Engineering 6 (2), 183–188.

Speranza, V., De Meo, A., Pantani, R., 2014. Thermal and hydrolytic degradation kinetics of PLA in the molten state. Polymer Degradation and Stability 100, 37–41.

Sun, S., et al., 2017. Effects of low polyhydroxyalkanoate content on the properties of films based on modified starch acquired by extrusion blowing. Food Hydrocolloids 72, 81–89.

Vandi, L.-J., et al., 2018. Wood-PHA composites: mapping opportunities. Polymers 10, 751.

Vandi, L.-J., et al., 2019. Extrusion of wood fibre reinforced poly(hydroxybutyrate-co-hydroxyvalerate) (PHBV) biocomposites: statistical analysis of the effect of processing conditions on mechanical performance. Polymer Degradation and Stability 159, 1–14.

Wang, Y., et al., 2013. Induced apoptosis of osteoblasts proliferating on polyhydroxyalkanoates. Biomaterials 34 (15), 3737–3746.

Xu, X.Y., et al., 2010. The behaviour of neural stem cells on polyhydroxyalkanoate nanofiber scaffolds. Biomaterials 31 (14), 3967–3975.

Yao, Y.C., et al., 2008. A specific drug targeting system based on polyhydroxyalkanoate granule binding protein PhaP fused with targeted cell ligands. Biomaterials 29 (36), 4823–4830.

CHAPTER 6

Nanocellulose/Starch Biopolymer Nanocomposites: Processing, Manufacturing, and Applications

R.A. ILYAS • S.M. SAPUAN • MOHD NOR FAIZ NORRRAHIM • TENGKU ARISYAH TENGKU YASIM-ANUAR • ABUDUKEREMU KADIER • MOHD SAHAID KALIL • M.S.N. ATIKAH • RUSHDAN IBRAHIM • MOCHAMAD ASROFI • HAIRUL ABRAL • A. NAZRIN • R. SYAFIQ • H.A. AISYAH • M.R.M. ASYRAF

1 INTRODUCTION

The rapid increment of human population encourages the increase of plastics production to meet the current market demand. Plastics have excellent properties and benefits for packaging, and they have been implemented for vast applications for over the years. It is also one of the abundant materials in the world today because of the low production cost and good mechanical and thermal abilities (Camann et al., 2010). Plastics are used in several fields and products such as for food packaging, grocery bag, toys, electronic devices, cooking appliances, and others. Polyolefins are the most versatile material in the production of plastics (Norrahim et al., 2013). Among the polyolefins, PP and PE are the most widely used. However, single use of PP and PE contributes to the environmental issue upon its disposal at landfill as they are not biodegradable. Another issue associated with the use of these plastics is its production from nonrenewable petroleum resource. Even though there are numerous efforts being taken to produce PE production by producing PE from bio-based resources through bioethanol pathway (Babu et al., 2013), however, bio-based PE suffers from high production cost due to the high cost of fermentation substrate. The impact of polyolefins on the environment has grown concerns among consumers, which led researchers to find the alternatives to polyolefins.

The use of biodegradable polymers for packaging application is an option. Biodegradable plastics can be defined as plastics that can be degraded biologically and by microorganisms such as bacteria and fungus. According to Verhoogt et al. (1994), this term can be applied if the polymer is readily degraded to carbon dioxide (CO_2) and water (H_2O). There have been many developments on biopolymers, which can be degraded in the environment upon discarding. Table 6.1 shows the list of biodegradable and nonbiodegradable polymers.

As seen in Table 6.1, not all bio-based polymers are biodegradable, and vice-versa. In developing sustainable packaging plastics especially for short-term use, bio-based, biodegradable plastics are favorable. Attentions have been focused on biopolymers that can be produced from natural resources using various techniques and methods (Corre et al., 2012; Ilyas et al., 2018f; Jumaidin et al., 2019a; Nurazzi et al., 2019). Several innovative techniques such as bilayer or the use of multicomponent films with good mechanical properties have been developed to make these biopolymers applicable for packaging application (Camann et al., 2010). For example, an innovative technology was developed for continuous production of laminated PVC with films containing 50% or more starch that could reduce the water sensitivity of such film.

Polyhydroxyalkanoate, cellulosics, polylactic acid, starch-based plastics, proteinaceous plastics, polybutylene succinate, polycaprolactone (PCL), polybutylene adipate-co-terephthalate, polyvinyl alcohol (PVOH), polytrimethylene terephthalate, bio-based epoxy, bio-based polyurethane, polyglycolic acid, and polyethylene oxide are among the examples of biodegradable polymers. Starch-based polymer received numerous attention due to several advantages such as low

Advanced Processing, Properties, and Applications of Starch and Other Bio-Based Polymers. https://doi.org/10.1016/B978-0-12-819661-8.00006-8

TABLE 6.1
Biodegradable and Nonbiodegradable of Bioplastics (Hassan et al., 2013; Ilyas et al., 2016; Ilyas et al., 2018d; Reddy and Misra, 2012; Sapuan and Ilyas, 2017).

	Biodegradable	Nonbiodegradable
Bioplastics	• Polyhydroxyalkanoate (PHA) • Cellulosics • Polylactic acid (PLA) • Starch-based plastics • Proteinaceous plastics • Poly(butylene succinate) (PBS) • Polycaprolactone (PCL) • Polyvinyl alcohol (PVOH) • Poly(butylene adipate-co-terephthalate) (PBAT) • Polytrimethylene terephthalate (PTT) • Bio-based epoxy • Bio-based polyurethane	• Polyamide 11 (PA11) • Bio-derived polyethylene • Polyethylene terephthalate (PET or PETE) • High-density polyethylene (HDPE) • Polyvinyl chloride (PVC) • Low-density polyethylene (LDPE) • Polypropylene (PP) • Polypropylene (PP) • Nylon (PA) • Acrylonitrile butadiene styrene (ABS) • Polycarbonate (PC)

production cost, transparent, renewable in nature, and easy availability (Gadhave et al., 2018). However, practically bioplastics including starch need to be improved before being applied as packaging plastics since some characteristic of these bioplastics do not match with commercial plastics characteristics (Mohammadi et al., 2012). The current trends showed a substantial development of biodegradable polymer especially in design strategies and engineering in order to offer advanced polymers with comparably good performance (Luckachan and Pillai, 2011).

Reinforcement of nanocellulose as a nanoscale filler in composites has been extensively studied for many uses and applications in the field of advanced biomedical material and food packaging material (Norrrahim et al., 2018a; Ahmad et al., 2019). Due to the high stiffness of nanocellulose, it can be implemented to increase the mechanical strength of biodegradable polymers. Several research on nanocellulose composites had shown a great improvement on behavior of mechanical and water barrier (Ariffin et al., 2017). The nanocellulose formed of nanostructured cellulose is a nanoscale size fiber that possesses lightweight as well as high strength, great durability, good biodegradability, and renewability (Zhu et al., 2011). Nanocellulose can be applied either for consumer products or high-tech industrial products due to its high mechanical performance.

2 NANOCELLULOSE

Since the rising interest on environmental pollution and limited resource of petroleum for various manufacturing industries, specific consideration has been given to create value-added bioproducts based on renewable resources (Sreekala et al., 1997). Besides being able to use for various bioproducts manufacturing, the utilization of biomass is still limitedly focused on certain bioproducts especially biocomposites (Aisyah et al., 2019; Norizan et al., 2020; Shinoj et al., 2011). Hence, the manufacturing of nanocellulose, one of the demanded and valuable nanomaterials, can broaden the utilization of biomass (Norrrahim et al., 2018b).

Nanocellulose can be defined as a material having a dimension of 100 nm or less with extremely high specific area, high porosity with excellent pore interconnectivity, lightweight, and high biodegradability (Hernandez et al., 2018; Ilyas et al., 2018f, 2020; Tian et al., 2016). Due to its superior properties, nanocellulose has received tremendous attention to be used for broad applications such as for materials, construction, packaging, automobile, transportation, and biomedical fields (Lee et al., 2014).

3 CLASSIFICATION OF NANOCELLULOSE

In general, nanocellulose can be categorized into cellulose nanofiber (CNF), cellulose nanocrystals (CNCs), and bacterial nanocellulose (BNC) according to its dimensions (Fig. 6.1). Both CNF and CNC are considered as plant-based nanocellulose, and BNC is considered as microbial-based nanocellulose. For CNF and CNC production, the fabrication methods involved the disintegration of plant cellulose using mechanical or chemical methods, while for BNC, it involved the bioformation of cellulose by bacteria (Khalil et al., 2014;

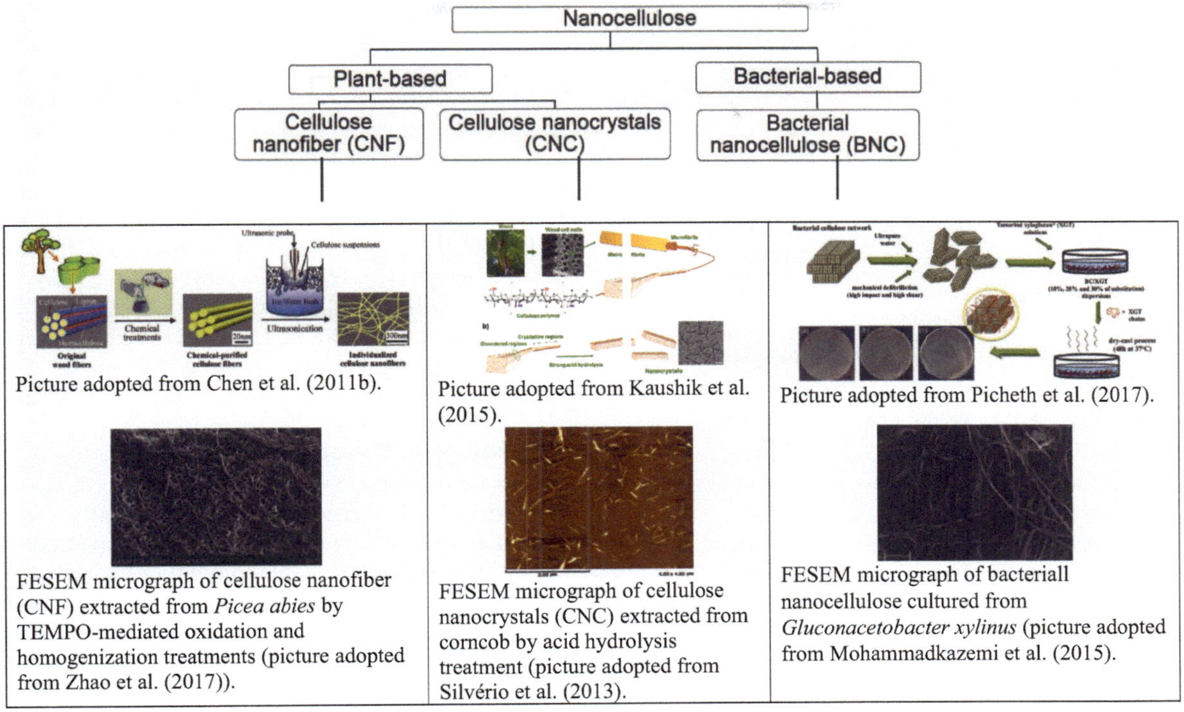

FIG. 6.1 Classification of nanocellulose.

Klemm et al., 2009). All CNF, CNC, and BNC are in nanoscale size, but they are different in shape, size, and composition (Xu et al., 2013).

In past few decades, there are various sources of cellulose fiber that had been used to produced nanocellulose, such as sugar palm (Ilyas et al., 2018a), water hyacinth (Asrofi et al., 2018c), oil palm empty fruit bunch (OPEFB) (Lani et al., 2014), ramie (Syafri et al., 2018a), pineapple leaf (Mahardika et al., 2018), kenaf bast (Sabaruddin and Paridah, 2018), roselle (Kian et al., 2018), cotton (Li et al., 2014), arecanut husk (*Areca catechu*) (Julie Chandra et al., 2016), and sugarcane bagasse (Bhattacharya et al., 2008). The modification of nano-sized cellulose is one way of enhancing its performance. Several methods can be used to obtain nanocellulose, one of which provides a combination between chemical (pulping, bleaching, acid hydrolysis) and mechanical treatments (ultrasonication, high shear homogenization, high-pressure homogenization).

3.1 Cellulose Nanofiber

CNF is made up of flexible and long-linked fibers between 20 and 100 nm wide and several micrometers in length (Tibolla et al., 2014). The nanofiber is made up of alternating crystalline and amorphous domains. These domains are stabilized laterally by hydrogen bonds among the hydroxyl groups and the oxygen atoms of adjacent molecules (Kim et al., 2015). Besides that, the ratio L/d of CNF is very high, which endows it with a very low percolation threshold (Lavoine et al., 2012).

The CNF can be produced from any lignocellulosic materials by several approaches of mechanical methods as indicated in Table 6.2. The selection of fibrillation treatments and lignocellulosic types plays major roles in determining the CNF properties. For example, CNF isolated from cotton linters by grinding process has a width of approximately 10–30 nm, with crystallinity index around 82.3%–85.7%, and thermally degrades at around 320–350 °C (Chen et al., 2011b), while CNF isolated from rice straw by TEMPO-mediated oxidation is having diameter approximately between 2 and 20 nm with crystallinity index around 60%–65% and is degraded at 210–310 °C (Jiang and Hsieh, 2013).

3.2 Cellulose Nanocrystals

CNC is rodlike or whisker-shaped particles with size approximately between 5 and 70 nm wide and less or

TABLE 6.2
The Example of Isolation for Cellulose Nanofiber (CNF) Production.

Methods	Source of Biomass	References
Sonication	Poplar trees	Chen et al. (2011c)
Sonication	Rice straw	Lu and Hsieh (2012)
Electrospinning	Pine needle powder	Xiao et al. (2015)
Electrospinning	Maize	De Oliveira Mori et al. (2014)
High-pressure homogenization	Yellow lobster wastes	Salaberria et al. (2015)
High-pressure homogenization	Wheat straw pulp	Zimmermann et al. (2010)
High-pressure homogenization	Sugar palm fiber	Atikah et al. (2019), Ilyas et al. (2019, 2019, 2019c, 2020), Ilyas et al. (2018g, 2019a, 2019c)
Grinding and tempooxidation	Wood pulp	Chen et al. (2014)
Cryocrushing	Wheat straw and soy hulls	Alemdar and Sain (2008)

more than 100 nm in length (Kaboorani and Riedl, 2015). The major difference structure between both CNF and CNC is due to the isolation method applied to the cellulose. In contrast to the CNF, the CNC can solely be produced by acid hydrolysis from any cellulose materials such as wood fibers, plant fibers, microfibrillated cellulose, microcrystalline cellulose, or CNF.

The acid hydrolysis process is able to chemically reduce the amorphous regions of the long-chain glucose compound, hence reducing the degree of polymerization (Rohaizu and Wanrosli, 2017). The amorphous regions have lesser density compared with the crystalline regions because they consist of amorphous and crystalline regions. Later, the amorphous regions are broke apart and release the individual crystallites after the cellulose is subjected to an acid treatment (Fahma et al., 2010; Lamaming et al., 2015). The performance of CNC leans on many aspects including cellulose sources, types of acid used for hydrolysis, reaction time, and temperature (Peng et al., 2011).

Brinchi et al. (2013) reported that different particle size, moisture content, crystallinity, porous structure, and surface area and molecular weight of CNC are caused from different origin and hydrolysis conditions of cellulose. For example, the CNC isolated from corncob has an average length (L) of 210.8 nm, a diameter (D) of 4.15 nm, and about 83.7% crystallinity and starts to degrade at around 185°C (Silvério et al., 2013), while CNC isolated from soy hulls has an average length (L) of 122.66 nm, diameter (D) of 2.77 nm, with crystallinity around 73.5%, and is degraded at around 200 °C. The CNC can be produced from any lignocellulosic materials by several approaches of chemical methods as shown in Table 6.3.

3.3 Bacterial Nanocellulose

BNC is another type of nanocellulose that was obtained via bacterial synthesis process (Mondal, 2018). Mostly, BNC is synthesized by the bacterium *Gluconacetobacter xylinus*. This bacterium is able to produce higher dense of lateral surface and gelatinous layer on the opposite side of nanofibrilar film and recorded a diameter of about 20–80 nm (Dima et al., 2017). The BNC can be considered as an alternative to CNF or CNC that was produced from plant sources as BNC contains higher cellulose purity compared with plant sources (Zhan, 2017). In fact, BNC also has high flexibility, high degree of crystallinity, good water absorption capacity, and good shape retention advantages (Azeredo et al., 2017; Lavoine et al., 2012). Despite of offering some advantageous, BNC however requires longer production times compared with CNF/CNC, mostly up to 21 days. For this reason, the axenic conditions for growing specific bacteria need to be considered, as well as the condition of the biosynthesis (Dima et al., 2017). In general, there were three main steps on BNC production by *G. xylinus*, which were (i) glucose residues polymerization in β-1-4 glucan, (ii) extracellular secretion of liner chains, and (iii) organization and crystallization of glucan chains via hydrogen bonds and forces of van der Waals (Jozala et al., 2016). The rigid process of BNC however makes it unfavorable to be produced in large scale, as compared with both CNF and CNC.

4 STARCH BIOPOLYMER

Conventional packaging materials (i.e., nonrenewable materials and nonbiodegradable) have transformed everyday life such household, medicine, food packaging, transportation, sports, electronics, construction, building, agriculture, and medical uses. Plastic usage has been escalating in numbers that have exceeded

TABLE 6.3
The Example of Isolation for Cellulose Nanocrystal (CNC) Production.

Methods	Source of Biomass	References
H_2SO_4 hydrolysis	*Acacia mangium*	Jasmani and Adnan (2017)
H_2SO_4 hydrolysis	Algae	Imai et al. (2003)
HCl hydrolysis	Arecanut husk fiber	Julie Chandra et al. (2016)
H_2SO_4 hydrolysis	Bacterial cellulose	Grunert and Winter (2002)
H_2SO_4 hydrolysis	Bamboo	Brito et al. (2012)
H_2SO_4 hydrolysis	Bamboo (*Pseudosasa amabilis*)	Liu et al. (2010)
$H_2C_2O_4$ hydrolysis	Banana fiber	Cherian et al. (2010)
TEMPO-mediated oxidation, formic acid hydrolysis	Banana pseudostem	Faradilla et al. (2017)
H_2SO_4 hydrolysis	Cassava bagasse	Teixeira et al. (2009)
H_2SO_4 hydrolysis	Coconut husk	Rosa et al. (2010)
H_2SO_4 hydrolysis	Colored cotton	de Morais Teixeira et al. (2010)
H_2SO_4 hydrolysis	Corncob	Liu et al. (2016)
H_2SO_4 hydrolysis	Cotton (cotton wool)	Morandi G et al. (2009)
HCl hydrolysis	Cotton linters	Braun et al. (2008)
H_2SO_4 hydrolysis	Cotton Whatmanfilter paper	Paralikar et al. (2008)
H_2SO_4 hydrolysis	Cotton (*Gossypium hirsutum*) linters	Morais et al. (2013)
TEMPO-mediated oxidation and H_2SO_4 hydrolysis	Cotton stalk	Soni et al. (2015)
H_2SO_4 hydrolysis	Cotton fiber	Pereda et al. (2014)
H_2SO_4, H_2SO_4/HCl, HCl hydrolysis	Curaua fiber	Corrêa et al. (2010)
H_2SO_4 hydrolysis	Eucalyptus kraft pulp	Tonoli et al. (2012)
HCl hydrolysis	Ginger fiber	Abral et al. (2020)
H_2SO_4 hydrolysis	Grass fibers	Pandey et al. (2009)
H_2SO_4 hydrolysis	Grass fibers (*Imperata brasiliensis*)	Benini et al. (2018)
H_2SO_4 hydrolysis	Groundnut shells	Bano and Negi (2017)
Steam explosion H_2SO_4 hydrolysis	*Hibiscus sabdariffa* fibers	Sonia and Priya Dasan (2013)
H2SO4 hydrolysis with high-temperature pretreatment	*Humulus japonicus* stem	Jiang et al. (2017)
H_2SO_4 hydrolysis	Industrial bioresidue	Oksman et al. (2010)
H_2SO_4 hydrolysis	Industrial bioresidue (sludge)	Herrera et al. (2012)
H_2SO_4 hydrolysis	Kraft pulp	He et al. (2019)
H_2SO_4 hydrolysis	Kenaf core wood	Chan et al. (2013)
H_2SO_4 hydrolysis	MCC	Bondeson et al. (2006)
H_2SO_4 hydrolysis	Mengkuang leaves	Sheltami et al. (2012)
H_2SO_4 hydrolysis	Mulberry	Li et al. (2009)
H_2SO_4 hydrolysis	Oil palm trunk	Lamaming et al. (2015)
H_2SO_4 hydrolysis	Oil palm empty fruit bunch (OPEFB)	Haafiz et al. (2014)
H_2SO_4 hydrolysis	*Phormium tenax* (harakeke) fiber	Fortunati et al. (2013)

Continued

TABLE 6.3
The Example of Isolation for Cellulose Nanocrystal (CNC) Production.—cont'd

Methods	Source of Biomass	References
H_2SO_4 hydrolysis	Potato peel waste	Chen et al. (2012)
H_2SO_4 hydrolysis	Flax fiber	Fortunati et al. (2013)
KOH hydrolysis	Ramie	Wahono et al. (2018)
H_2SO_4 hydrolysis	Ramie	Habibi and Vignon (2008)
H_2SO_4 hydrolysis	Ramie	Lu et al. (2006)
H_2SO_4 hydrolysis	Rice husk	Rosa et al. (2010)
H_2SO_4 hydrolysis	Rice straw	Lu and Hsieh (2012)
H_2SO_4 hydrolysis	Sesame husk	Purkait et al. (2011)
H_2SO_4 hydrolysis	Sisal fiber	Morán et al. (2008)
H_2SO_4 hydrolysis	Soy hulls	Flauzino Neto et al. (2013)
H_2SO_4 hydrolysis	Sugar palm fiber	Ilyas et al. (2018a, 2018b, 2018c)
HCl hydrolysis	*Zingiber officinale* tubers	Abral et al. (2020, 2019)
H_2SO_4 hydrolysis	Ramie	Syafri et al., 2018a, 2018b)
H_2SO_4 hydrolysis	Water hyacinth fiber	Asrofi et al. (2018c)
H_2SO_4 hydrolysis	Sugar palm frond	Sumaiyah et al. (2014)
H_2SO_4 hydrolysis	Sugarcane bagasse	Teixeira et al. (2011a, 2012b)
H_2SO_4 hydrolysis	Sago seed shells	Naduparambath et al. (2018)
H_2SO_4 hydrolysis	Tunicate	Favier et al. (1995)
HCl hydrolysis	Water hyacinth fiber	Asrofi et al. (2018a)
TEMPO oxidation followed by HCl hydrolysis	Wood pulp	Salajková et al. (2012)
H_2SO_4 hydrolysis	Wheat straw	Pereira et al. (2017)
HCl hydrolysis	*Valonia ventricosa*	Revol (1982)
HCl hydrolysis	Water hyacinth	Syafri et al. (2019b)
H_2SO_4 hydrolysis	Juncus plant stems	Kassab et al. (2019)
HCl hydrolysis	Juncus plant stems	Kassab et al. (2019)
Chemical (KOH)-ultrasonication	Ramie	Syafri et al. (2019a)

around 300 million tonnes for annual production. Based on this statement, about 50% of the plastics usage is used for single-use purposes (utilized for just a few moments). However, these plastics remain on the planet for at least several hundred years before it fully degraded, besides that, some of them are not degrade at all. Approximately, high quantity of plastic has been deposited into world's water basins, which is more than 8 million tons every year. Petroleum resources are widely utilized in fabricating these plastic polymers, in which it leads to concerns in terms of both environmental sustainability, human and wildlife health, and economic (landfill sites). This plastic especially plastic bags caused severe ecological problems such as accumulation and agglomeration of plastic waste in landfills and physical issues for flora and fauna including sea creatures, avian, and animals. This is caused from entanglement or ingestion in plastic and the leaching process of chemical to form by-product from plastic to river, ocean, and water catchment area. Later, it may cause the transfer of chemical by-product to living things including animals and humans, which can cause sickness. Therefore, over relies on petroleum resources can be lightened by replacing petroleum-based polymer with the bio-based polymer or bioplastic using annually renewable resources (plants and crops) or any biological resource (wildlife and microorganisms). Bioplastic is derived from renewable resources, which is produced from food/agro sources. They are also considered safe to be used in food applications especially for sustainable packaging material. Many scientist and researchers are currently working on bio-based polymer to derive new polymer from biological resources (physical and chemical methods or industrial biotechnology).

The main disadvantage of the synthetic plastic products is that they are lacking in term of biodegradation, which had led many scientists to conduct the experiment on the development and characteristic of biopolymers. One of the renowned biopolymeric material is starch, which had been studied as the raw material for production of the biodegradable materials because of its environmentally friendly nature, biodegradability, relatively low cost to produce from widely abundant raw materials, and natural renewable polysaccharide that can be obtained from a numerous types of plant and agricultural crops such as sugar palm, rice, tapioca, corn, potato, pea, etc. (Ilyas et al., 2019a; Medeiros et al., 2009). Besides that, starches also exhibit hydrophilic properties, which they can naturally exist in the nature in the form of partially and discrete crystalline microscopic granules being bonded by an extended micellar network of combined molecules. Starch is made up of both branched and linear polysaccharides recognized as amylopectin and amylose, respectively, and they vary according to their botanical origin. Figs. 6.2A and B dispatches the chemical structure of amylopectin and amylose, respectively. Amylose is a linear structure of α-1, 4 linked glucose units. Meanwhile, amylopectin is a highly branched structure of short α-1,4 chains linked by α-1,6 bonds. Usually, amylose molecules consist of 200–20,000 glucose units, which form a helix. This happened because of the bond angles between the glucose units. Amylopectin is comprising of short side chains of 30 glucose units and joined to every 20–30 glucose units along the chain. Amylopectin molecules could comprise round 2 million glucose units (Sanyang et al., 2016). The amylose linear structure causes it to closely look like the behavior of synthetic polymers. In the meantime, amylopectin branched structure aids to reduce the flexibility of the polymer chains. It also interferes with any tendency for them to become oriented closely enough to exhibit significant levels of hydrogen bonding. In most cases, native starch contains around 70%–85% amylopectin and 15%–30% amylose (Reddy et al., 2013). The processing of starch is affected by the existence of many intermolecular hydrogen bonds that resulted in higher starch softening temperature than its degradation temperature. Therefore, plasticizers (water, sorbitol, glycerol) are used to facilitate increasing of the free volume and hence decreasing the softening temperature as well as glass transition (Fishman et al., 2000). Thermoplastic starch (TPS) is formed when there is disruption of starch molecular structure, where heating of starch granules caused swelling and nonirreversible transition of amorphous regions in the presence of plasticizer, under specific condition (Sanyang et al., 2016).

5 NANOCELLULOSE REINFORCED STARCH BIOPOLYMER COMPOSITES

Over the last few decades, huge attention has been given to the usage of plant source-based stiff filler as reinforcement in starch-based thermoplastics as a way to enhance the mechanical performance and water sensitivity of starch biopolymer as well as reducing the usage

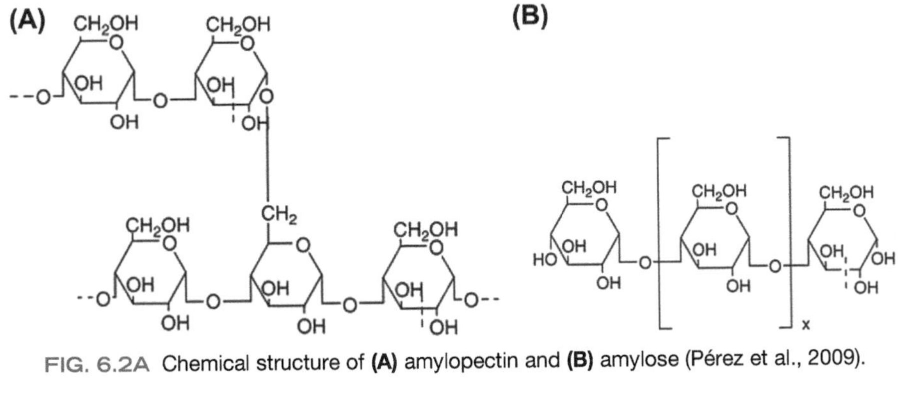

FIG. 6.2A Chemical structure of **(A)** amylopectin and **(B)** amylose (Pérez et al., 2009).

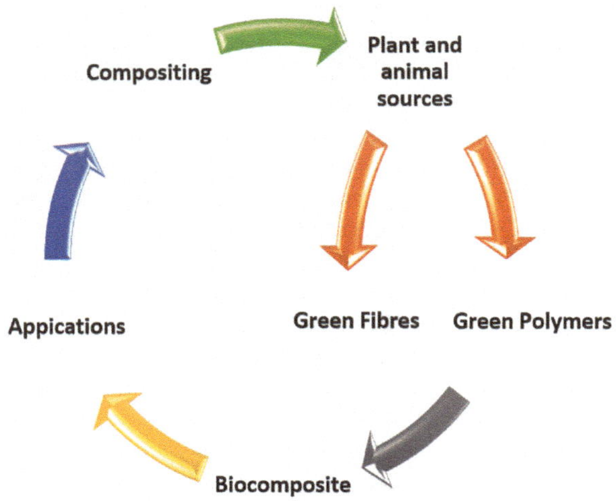

FIG. 6.2B Life cycle of biocomposites.

of non-degradable petroleum-based polymer and synthetic polymer (Halimatul et al., 2019a, 2019b; Jumaidin et al., 2019b). This phenomena happen because starch-based thermoplastics are brittle and have less water sensitivity and reduced mechanical properties. Hence, to improve both properties of water resistance and mechanical strength, fillers from plant sources have been reinforced to starch-based thermoplastic to produce green composites or biocomposites. Composite material is a material that is made from more than one constituent materials combined together to form a significant chemical and physical behaviors, and when combined, it resulted in a material with different characteristics compared with individual constituents (Atiqah et al., 2019; Ilyas and Sapuan, 2020; Mazani et al., 2019; Nurazzi et al., 2020). Composites have been implemented since prehistoric age. Unlike the modern composites nowadays, archeological evidence had proved that the earliest civilizations used the composites of straw and mud to produce strong brick (Medeiros et al., 2009). Natural composites are materials such as wood (plants) and blood vessels (animal and human). Wood consists of cellulose reinforced fibrils bound together by hemicellulose and lignin matrix (Asyraf et al., 2020; Jaafar et al., 2018). In early 1990s, modern nanocomposites advanced rapidly, in which researchers from the car manufacturer industry fabricated composites using reinforcement of nano-sized clay within polymer (i.e., clay-polymer nanocomposites), resulted significant improvements in stiffness, dimensional stability, and heat distortion temperature as compared with pure polymer. The biocomposites started to develop in late 1980s, where the biodegradable polymers (i.e., polycaprolactone, Mater Bi, BioPolm, Bioceta, and so on) have been examined for use as matrix polymer for the production of high-quality green composites (Takagi and Asano, 2008). Natural fiber composites are composites materials, in which at least the reinforcing fibers are originated from renewable resources such as plants, woods, or animals. Besides that, the utilization of biocomposites is beneficial not only for the environment and sustainability standpoint as they are renewable materials for new and green products but also for an economical point of view, where petroleum price keeps increasing due to depletion. Fig. 6.1 shows the design and life cycle assessment of biocomposites.

Nanocomposites not only enhanced physical properties (reduced flammability, improve barrier properties, etc.) and mechanical properties (elastic modulus, tensile strength, flexural, impact, etc.) but also they their optical transparency as compared with macrocomposite.

6 PREPARATION AND PROCESSING OF NANOCELLULOSE REINFORCED STARCH BIOPOLYMER COMPOSITE

There are numerous studies attempted to explain the challenges of nanocellulose as nanofiller in polymer matrix as they are difficult to disperse and distribute in nonpolar medium because of their polar surface. The vital phases in the preparation of nanocomposite is blending or mixing process. Therefore, nanocellulose has to be well dispersed and well distributed within the polymer matrix in order for the performance of nanocellulose to be fully optimized. This phenomenon can be dispatched in Fig. 6.3. Distribution and dispersion levels of nanocellulose reinforcement in polymer resin are important as the content of nanofiller is low, besides nanocellulose possesses high surface area, which has the tendency to accumulate and agglomerate rather than distribute and disperse in the polymer matrix (Medeiros et al., 2009; Azammi et al., 2020; Ilyas et al., 2017; Jumaidin et al., 2020). According to Siqueira et al. (2010), nanocelluloses are limited to either polar or aqueous environments. There are two methods that can be commonly applied to make films of polysaccharide nanocomposite as shown in Fig. 6.4 (Dufresne, 2009).

The first method is hugely dependent on the types of matrix polymer, and it is commonly used in starch-based nanocomposites preparation. For example, polymer emulsions, water-soluble polymers, and nonhydrosoluble polymers. The nanocellulose-based nanocomposite preparation implements solvent casting method, which can be seen in Fig. 6.5. Polymer starch-

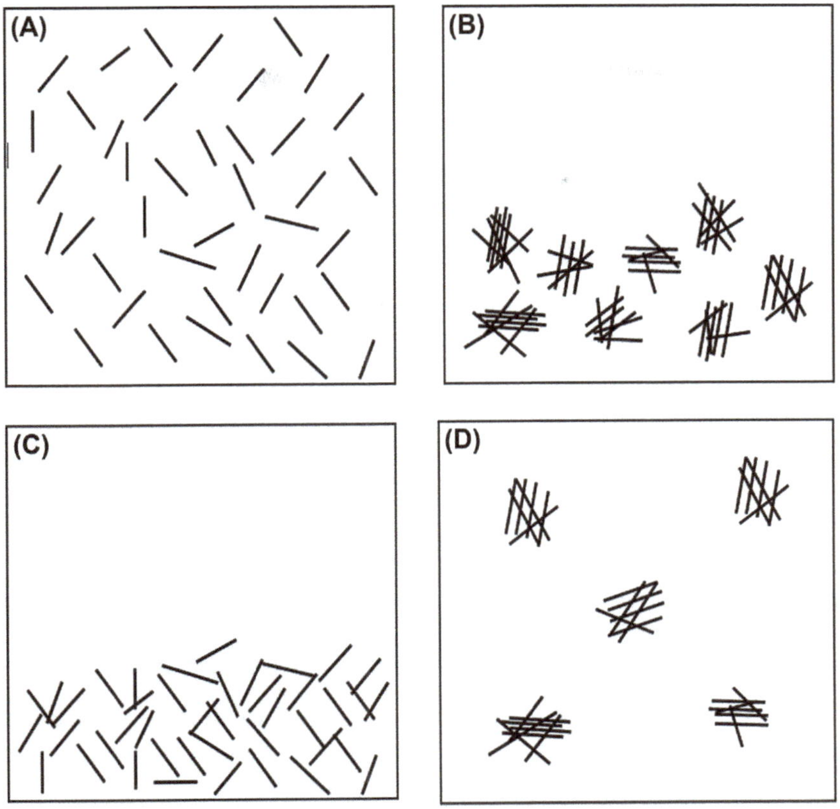

FIG. 6.3 Degree of distribution and dispersion of nanocellulose reinforcement in polymer matrix: **(A)** well dispersed and distributed; **(B)** less dispersed and distributed; **(C)** well dispersed but poorly distributed; and **(D)** well distributed but poorly dispersed.

based nanocomposite can be produced by mixing the polymer solution with nanocellulose suspension (i.e., BNC, CNC, and CNF). Besides that, these methods were done to preserve the dispersion of the nanocelluloses evenly in nanocomposite film. Table 6.4 reveals the comparisons between types of solvent casting, i.e., hydrosoluble, emulsion, and nonhydrosoluble.

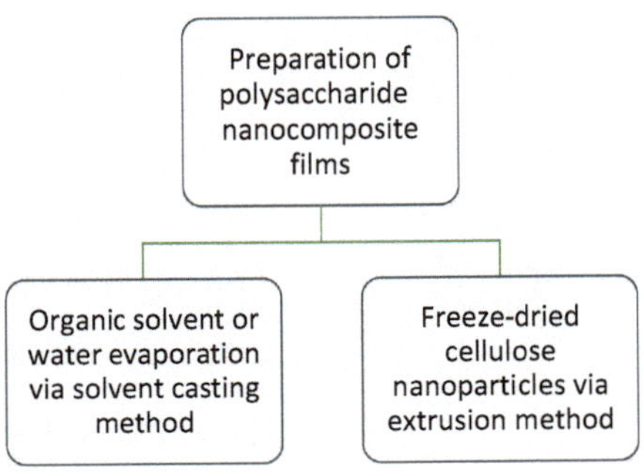

FIG. 6.4 Techniques to prepare polysaccharides nanocomposite films.

The second method applies a melting compounding method or an extrusion method to incorporate the nanocellulose with composite, forming nanocomposite (Siqueira et al., 2010). However, in this method, not many studies have been carried out due to poor dispersion and distribution of nanocellulose in polymeric resin. The preparation of nanocomposites using solution casting method shows higher mechanical properties compared with nanocomposite of the same polymer prepared by freeze-drying and extrusion methods. This phenomenon happened because of the strong linkage effect of the hydrogen bond in the solution casting method in which the formation of rigid nanocellulose networks is established. The advantage of the solution casting method is that it has a lower processing time compared with other method. On the other hand, the extrusion method has a limitation in nano-sized reinforcement because strong hydrogen bonds can only be established after the polysaccharide nanocellulose had dried.

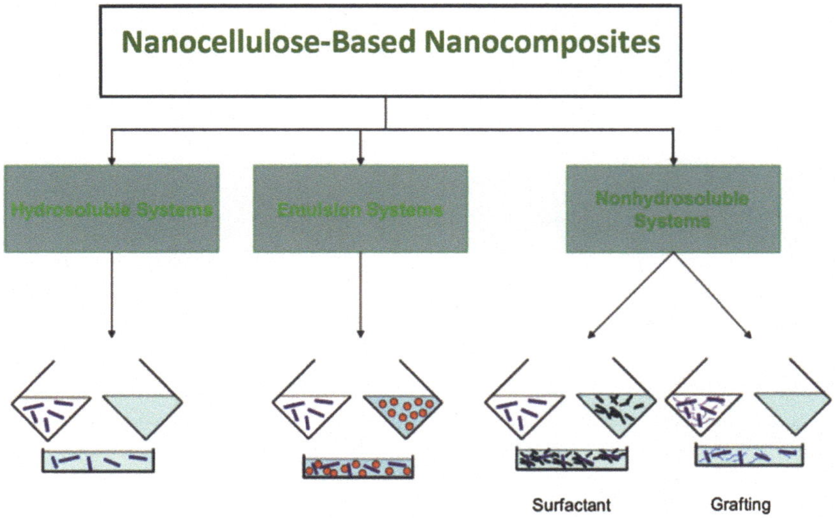

FIG. 6.5 Preparation of nanocellulose-based nanocomposites by solvent casting.

TABLE 6.4
Solvent Casting Techniques (Ilyas et al., 2018b, 2020, 2019b; Siqueira et al., 2010).

Types of Solvent Casting Systems	Hydrosoluble Systems	Nonhydrosoluble Systems	Emulsion Systems
Medium Preparation	Water	Surfactants is used along with organic media	Polymer in the form of latex
Processing	The nanoparticle suspension was added with the polymer previously dissolved in water, evaporating the liquid.	The nanoparticle suspension was mixed with the surfactant. Then, nanocomposite films were managed in hot toluene (110°C) by mixing solubilized polymer. The solvent was let to evaporate overnight at 110°C and kept under vacuum for 6 h and, finally, hot pressed at 150°C.	The nanoparticle aqueous suspension was added inside the polymer latex evenly, evaporating the liquid at room temperature (a temperature higher than T_g of poly (S-co-BuA), around 0°C).
Limitation	1. Limits the options of the matrix only for hydrosoluble polymers. 2. Highly sensitive to humidity.	–	–
To overcome	Store their films in vacuum oven	–	–
Polymers used	1. Polyvinyl acetate (PVA) 2. Hydroxypropylcellulose (HPC) 3. Starch	1. Polypropylene 2. Polylactic acid (PLA)	1. Synthetic latex (poly(S-coBuA)) 2. (β-hydroxyoctanoate) (PHO) 3. Polyvinyl chloride (PVC) 4. Waterborne epoxy 5. Natural rubber 6. Polyvinyl acetate (PVAc)

7 MECHANICAL, MORPHOLOGICAL, AND PHYSICAL PROPERTIES OF NANOCELLULOSE REINFORCED STARCH BIOPOLYMER

Recently, there are many studies that can be found in the journals and books discussing on the reinforcement of nanocellulose in starch biopolymeric matrices as well as their surface topography, thermal, physical mechanical, and water sensitivity behaviors of these bio-nanocomposites material. According to Siqueira et al. (2009), the well homogeneous dispersion and distribution of nanocellulose in the starch biopolymeric matrix is important as it will affect the mechanical properties (i.e., efficient load transfer from matrix to nanocellulose (Fu et al., 2008)) and the other performances of nanocomposites. Tables 6.5 and 6.6 demonstrate the tensile strength and Young's modulus of nanofibrillated cellulose and nanocrystalline cellulose reinforced starch biopolymer matrix, respectively. From Table 6.5, we can see that the highest tensile strength is the bio-nanocomposites made from the reinforcement of kenaf cellulose nanofibre reinforced corn starch with the value of 38.0 MPa. This research had been conducted by Babaee et al. (2015). In this study, they focused on the effect of chemical modification (acetic anhydride) of kenaf bast fiber (*Hibiscus cannabinus*) cellulose nanofibre on mechanical strength, water sensitivity, and biodegradability properties of kenaf CNF reinforced thermoplastic corn starch. The nanocomposites were fabricated using solution casting methods with water

TABLE 6.5
Starch-based Polymer, NFCs Nanocomposites, and Their Mechanical Properties.

Starch-Based Polymers	NFC Sources	NFC Preparation	Manufacturing Technique	Tensile Strength (MPa)	Young's Modulus (MPa)	Refernces
Cassava starch	Cassava bagasse	Hydrolyzed in 6.5 M H_2SO_4/ 40 min	Solution casting	4.8	84.3	Teixeira et al. (2009)
Mango puree	Wheat	–	Solution casting	8.76	322.05	Azeredo et al. (2009)
Maize starch	Wheat straw	High-pressurize homogenizer/ 15 min	Solution casting	6.75	220	Kaushik et al. (2010)
Maize starch	Cotton cellulose	Hydrolyzed in 6.5M sulfuric acid/75 min	Twin screw extruder and hot press	0.35	3.12	Teixeira et al. (2011a, 2012b)
Potato starch	Softwood wood flour	Supermasscolloider	Solution casting	17.5	1317.0	Hietala et al. (2013)
Potato starch	Rice straw	Ultrasonication	Solution casting	5.01	160	Nasri-Nasrabadi et al. (2014)
Maize starch	Kenaf	Supermasscolloider	Solution casting	2.35	53.6	Karimi et al. (2014)
Corn starch	Kenaf	Supermasscolloider	Solution casting	38.0	141.0	Babaee et al. (2015)
Corn starch	Bamboo fiber	–	Solution casting	11.2	12.4	Llanos and Tadini (2018)
Sugar palm starch	Sugar palm fiber	High-pressurize homogenizer, 500 bar	Solution casting	10.68	121.26	Ilyas et al. (2019c)
Yam bean starch	Water Hyacinth	Ultrasonic crusher treatment to	Solution casting	11.47	443	Asrofi et al. (2018b)

TABLE 6.6
Starch-based Polymer, CNC Nanocomposites, and Their Mechanical Properties.

Starch-based Polymers	NCCs Sources	Isolation Chemical/Time of NCCs	Manufacturing Technique	Tensile Strength (MPa)	Young's Modulus (MPa)	References
Pea starch	Hemp	64 wt% H_2SO_4/4 h	Solution casting	3.9–11.5	31.9–823.9	Cao et al. (2008a, 2008b)
Pea starch	Flax	64 wt% H_2SO_4/4 h	Solution casting	3.9–11.9	31.9–498.2	Cao et al. (2008a, 2008b)
Pea starch	Bamboo	50 wt% H_2SO_4/ 48 h	Solution casting	2.5–12	20.4–210.3	Liu et al. (2010)
Maize starch	Tunicin	55 wt% H_2SO_4/ 20 min	Solution casting	0.24–20	51–315	Anglès and Dufresne (2001)
Maize starch	Waxy maize starch	H_2SO_4/5 days	Solution casting	1–15	11–320	Angellier et al. (2006)
Maize starch	Tunicin	–	Solution casting	42	208–838	Mathew et al. (2008)
Wheat starch	Cottonseed linter	64 wt% H_2SO_4/4 h	Solution casting	2.5–7.8	36–301	Lu et al. (2005)
Plasticized starch	Cotton cellulose powders	H_2SO_4	Solution casting	–	–	Yang et al. (2014)
Wheat starch	Ramie	64 wt% H_2SO_4/4 h	Solution casting	2.8–6.9	56–480	Lu et al. (2006)
Potato starch	MCC	64 wt% H_2SO_4/2 h	Solution casting	13.7	460	Kvien et al. (2007)
Wheat starch	MCC	36.5 wt% HCl	Solution casting	3.15–10.98	–	Chang et al. (2010)
Potato starch	Cotton linter	64 wt% H_2SO_4/1 h	Solution casting	4.93	–	Noshirvani et al. (2016)
Potato starch	Potato peel waste	64 wt% H_2SO_4/ 90 min	Solution casting	–	460	Chen et al. (2012)
Maize starch	Sugarcane bagasse	64 wt% H_2SO_4/3 h	Solution casting	17.4	520	Slavutsky and Bertuzzi (2014)
Sugar palm starch	Sugar palm fiber	60 wt% H_2SO_4/ 45 min	Solution casting	11.5	178	Ilyas et al. (2018b)
Tapioca starch	Water hyacinth	HCl	Solution casting	5.8	403	Asrofi et al. (2018a)
Tapioca starch	Ramie fiber	30 w/v% H_2SO_4/ 60 min	Solution casting	12.48	479.8	Syafri et al. (2018b)
Bengkuang starch	Water hyacinth fiber	5M HCl/20h	Solution casting	–	–	Syafri et al. (2019b)

and glycerol added as the plasticizer. The results showed that the reinforcement of kenaf CNF significantly reduces the absorption of water and vapor permeability of the nanocomposites and enhanced the mechanical performance and degradation rate of the TPS.

The highest Young's modulus is the bionanocomposites made from the reinforcement of wheat cellulose nanofibre reinforced mango puree with the value of 322.05 MPa. This research had been conducted by Azeredo et al. (2009). Bionanocomposite have been produced by adding wheat CNF provided by FMC BioPolymer (Philadelphia, Pa., USA) in distinct loading around 36 g/100 g as reinforcement agent to mango puree-based edible films. For fabrication, the film-forming dispersions were executed to vacuum to eliminate air pocket, and it was casted and allowed to dry at relative humidity of 42% and 22°C for 16 h. The study showed that the incorporation of CNF significantly enhanced the mechanical, vapor permeability, and thermal properties of mango puree biopolymer films. Besides that, the incorporation of CNF had effectively increased the tensile properties and the glass transition temperature of films and significantly increases the Young's modulus of bionanocomposite, particularly at higher concentrations of CNF. This was due to the formation of a network form of nanofibrillar within the matrix polymer (Azeredo et al., 2009).

It can be seen from Table 6.5 that the lowest tensile properties are the bionanocomposites that are made from the reinforcement of cotton CNF reinforced maize starch with the value of 0.35 (tensile strength) and 3.12 MPa (Young's modulus). This research had been demonstrated by (Teixeira et al., 2011b). This study evaluated the reinforcement of cotton CNF as a reinforcement agent to corn starch biopolymer matrix to produce thermoplastic bionanocomposites. The bionanocomposites were compounded using twin screw extruder (140–160°C) and hot press (160°C for 5 min) with glycerol (30 wt%) added as the plasticizer. The result shows that the incorporation of CNF was observed to improve the mechanical properties of neat corn starch biopolymer in loading above 2.5 wt %, although some accumulation agglomeration could be observed. This is due to the fabrication process that had been used by authors, which is the twin screw extruder. The nanocellulose is not dispersed and distributed well in the polymer matrix, leads to CNF agglomeration, and, consequently, lowers the effectiveness of mechanical reinforcement of CNF.

The mechanical properties of various CNC reinforced starch-based nanocomposites can be observed through Table 6.6. From the table, we can observe that most researchers and scientists use sulfuric acid (H_2SO_4) hydrolysis method to isolate nanofibre from raw natural fiber. In hydrolysis method, natural fiber will let to undergo strong acidic environment. This method showed in high amount of negatively charged sulfate groups on the surface of CNC, consequently limiting the accumulation, agglomeration, and flocculation of CNC in polymer matrix. Furthermore, from Table 6.6, it can be observed that the bionanocomposites made from the reinforcement of tunicin CNC reinforced waxy maize starch biopolymer have the highest tensile strength and Young's modulus with the value of 42 MPa and 208–838 MPa, respectively. This experiment was conducted by Mathew et al. (2008). This research elaborated on the effect of nanofiller concentration (0–25 wt%) and the relative humidity levels (0%–98%) on the mechanical performance of the bionanocomposites films. The mixture was performed with boiled water autoclave reactor at 160°C for 5 min in a stirred preheated. After that, the mixture was placed under vacuum to remove air bubble, and the mixture was casted in a Teflon mold. The result shows that the bionanocomposites display good mechanical strength properties because of the strong interaction between tunicin CNC, waxy maize starch biopolymer matrix, water, and plasticizer and because of the capability of the CNC to form a three-dimensional rigid network. On the other hand, the mechanical properties of the bionanocomposites films increased proportionally with nanofiller concentration, displaying an effective transfer of stress between the biopolymer matrix and CNC nanofiller (Herrera et al., 2016; Mathew et al., 2008).

Besides that, from Table 6.6, it can be seen that the bionanocomposites made from the reinforcement of kenaf nanofibrillated cellulose reinforced corn starch polymer matrix has the tensile strength value of 20 MPa. This experiment was conducted by Anglès and Dufresne (2001). In their study, the mechanical properties of glycerol plasticized maize starch reinforced tunicin CNC bionanocomposites were determined in both the linear and the nonlinear. The mixture was conducted with boiled water autoclave reactor at 160°C for 5 min and stirred preheated. After that, the mixture was executed with vacuum operation to remove air bubble, and the mixture was casted in a Teflon mold. The results deduced that the incorporation of CNC to 25 wt% increased the mechanical properties of composite from 0.24 to 20 MPa compared with neat starch biopolymer. Anglès and Dufresne (2001) reported that the reinforcement effect of tunicin CNC was intensely depended on the capability of CNC filler to form a rigid network, caused from strong interactions between CNC (hydrogen bonds) and moisture content.

Moreover, from Table 6.6, it can be perceived that the lowest tensile strength is the bionanocomposites made from the reinforcement of cottonseed linter CNC reinforced what starch with the value of 2.5–7.8 MPa. This experiment was conducted by Lu et al. (2005). The films were produced by casting mixture of the aqueous suspension of the CNC and plasticizer (glycerol) aqueous dispersion in various blend ratios (0, 2.5, 5, 10, 15, 20, 25, and 30 wt%). The results depicted that the mechanical behaviors of films are in increasing trends proportionally with nanofiller load, due to no large agglomerates and good adhesion between the matrix and cellulose crystallites witnessed. Bionanocomposites displayed an increase in tensile strength and Young's modulus from 2.5 to 7.8 MPa and 36 to 301 MPa with increasing CNC content from 0 to 30 wt% comparing with the pure film. However, this value is still low compared with other bionanocomposites listed in Table 6.6. This is due to different starch as well as different sources of nanocellulose that would give different values of mechanical properties.

The reinforcement of the nanocellulose decreased the molecular mobility of starch biopolymer, hence making the bionanocomposite stiffer, less stretchable, and resistant to break as compared with the neat starch biopolymer film. Besides that, an experiment conducted by Ilyas et al. (2018b) revealed that the reinforcement of 0.5 wt% sugar palm nanocellulose had significantly increased the mechanical properties (i.e., tensile strength and Young' modulus by 11.47 and 178.83 MPa, respectively) of bionanocomposites film. According to Ilyas et al. (2018b, 2018f), nanocellulose had low density, abundant of hydroxyl group, high crystallinity, high mechanical strength, high aspect ratio, and high surface area, in which these high functionalies of nanocellulose make it desirable as a reinforcing and nucleating agent in starch biopolymers. Moreover, the improved of mechanical properties of bionanocomposites were also due to the ability of the starch biopolymer to mechanically interlock with the nanofibres. Incorporating of nanocellulose with a starch biopolymer would create new strong interfacial adhesion between nanocellulose and starch biopolymer upon processing the film.

Fig. 6.6 demonstrates the SEM micrograph of fracture surface of water hyacinth (*Eichhornia crassipes*) nanocellulose-filled bengkuang (*Pachyrhizus erosus*) starch biocomposites. Fig. 6.6 displayed well-dispersion and distribution of nanocellulose within the matrix and compact structures (Syafri et al., 2019b). According to Syafri et al. (2019b), these findings might be attributed to the kinetic energy produced from the ultrasonic bath, leading to improved interfacial bonding between the matrix and fiber. Besides that, ultrasonic treatment also resulted in well distribution and well dispersion of the nanocellulose in the starch-based matrix and consequently reduced the free OH bonding between the matrix and fibers.

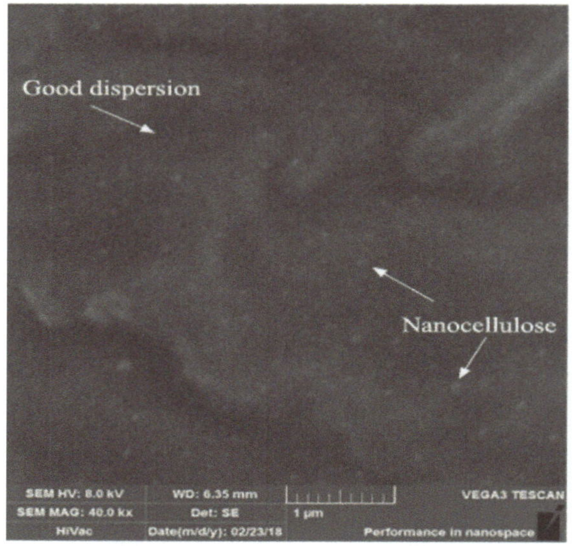

FIG. 6.6 Fracture surface of water hyacinth (*Eichhornia crassipes*) nanocellulose-filled bengkuang (*Pachyrhizus erosus*) starch biocomposites.

8 POTENTIAL APPLICATIONS

In this work, some broad reviews have already been described on the subject of nanocellulose reinforced starch-based films and their potential applications. Nanocellulose-based film is differentiated by its combination of nanocellulose with other functional materials such as macromolecular polymers, 0D, 1D, and 2D nanomaterials. This is due to enhance the properties of materials in term of mechanical, optical, and thermal properties with diverse functionalities, including their electrical, plasmonic, luminescent, and magnetic properties. This section emphases on the current progresses in nanocellulose reinforced starch-based bionanocomposites, along with their fundamentals and possible applications in emerging areas.

However, current focus seems to be in the field associated with reinforcement of cellulose with bio-based packaging, which is fast growing with a very high progress rate. This is due to the problems created by petroleum-based polymer that is nonrenewable and nonbiodegradable as they are being thrown away. Therefore, to overcome these problems, the nonbiodegradable packaging waste materials have to be reduced by shifting to eco-friendly packaging while maintaining the quality and food stability. Starch biopolymer had been considered as promising substitution polymers due to their "green" footprint, and there are huge potential products to be made from these materials, such as flushable liners, shopping bags, food and fruit packaging films, and medical delivery devices and system (Fishman et al., 2000; Ilyas et al., 2018e; Sapuan et al., 2018). However, starch-based biopolymer had several disadvantages such as low physical and mechanical performance compared with conventional polymers. These disadvantages can be reduced by the incorporation of nanocellulose as a reinforcing agent into the starch-based polymer matrix. Although the interest for designing cellulose for packaging applications grew fast in the recent past according to Ilyas et al. (2018b), it is worth to rectify (Fig. 6.7) that experiments regarding the packaging field have hardly increased since 2009 to 2018 from 25 journals to 172 journal. The increment of seven times from the year 2009–2018 shows that huge attention had been given to celluloses and packaging.

Nanocellulose fiber reinforced starch-based composites have many potential applications such as flexible optoelectronic and scaffolds for tissue regeneration. Table 6.7 shows starch-based polymer components, the manufacturing technique, and applications. From Table 6.7, we can observe that most of the potential applications of nanocellulose reinforced starch-based bionanocomposites are for food packaging materials. Besides that, these bionanocomposites can be used as transparent materials, air permeable, resistant, surface-sized paper, and for food packaging. Food packaging is become the most hot topic issue recently due to the problems caused by petroleum-based polymer, in which the alternatives to the conventional plastics could be replaced soon. Moreover, most of the researchers used casting technique to fabricate packaging film as a solution.

9 CONCLUSIONS

The dominancy of petroleum-based plastic in packaging industry has been seen even by today. This material has alarmed the global society in order to replace the material at various sector including agriculture, electronic, food, and packaging industry. Furthermore, this material has also been carried along with it issues concerning to the renewability and safe disposal of these materials as it would affect the environment and wildlife. Due to of the escalation of interest of nations over environmental problems of this material, awareness on polymer has changed toward the progress, production, and promoting the use of biopolymer. Biopolymer or bioplastic is coming from renewable resources, which are produced from agro/food sources and usually considered safe to be used in food applications. Additionally, starch has been considered as an excellent candidate to partially substitute synthetic polymer due to its abundantly available, highly

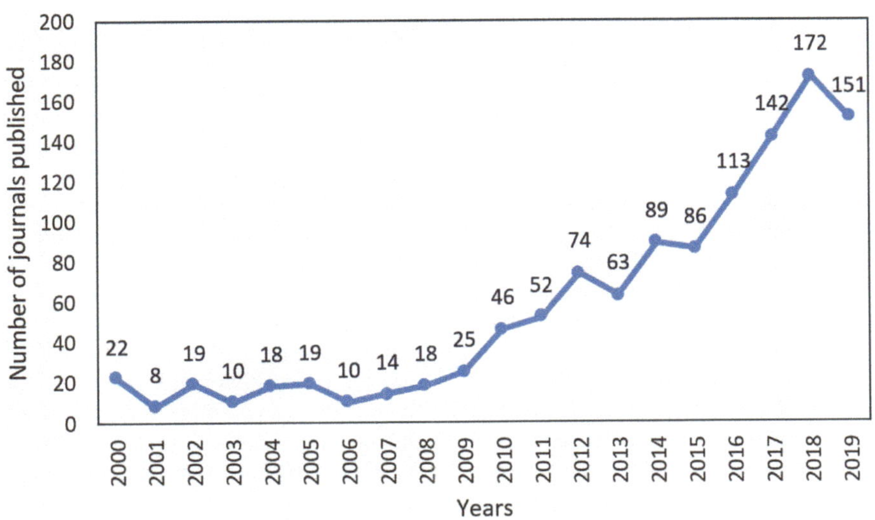

FIG. 6.7 Scopus database on the research output in form of research articles using celluloses and packaging as keywords.

TABLE 6.7
Starch-based Polymer Component Reinforced NanoCelluloses and the Manufacturing Technique and Applications.

Polymer Component	Manufacturing Techniques	Applications	References
Plasticized starch	Solution casting	Transparent materials	Nasseri and Mohammadi (2014)
Starch	Blending, solution casting	Air permeable, resistant, surface-sized paper, food packaging	Slavutsky and Bertuzzi (2014), Yang et al. (2014)
Starch	Solution casting	Food packaging	Liu et al. (2010)
Cassava starch	Solution casting	Food packaging	Teixeira et al. (2009)
Sugar palm starch	Solution casting	Food packaging	Ilyas et al. (2018)
Sugar palm starch	Solution casting	Food packaging	Ilyas et al. (2018b)
Sugar palm starch	Solution casting	Food packaging	Atikah et al. (2019), Ilyas et al. (2018)
Wheat starch	Solution casting	Food packaging	Lu et al. (2006)
Tuber native potato	Solution casting	Packaging	Montero et al. (2017)
Cereal corn starch	Solution casting	Packaging	Montero et al. (2017)
Legume pea starch	Solution casting	Packaging	Montero et al. (2017)
Yam bean	Solution casting	Packaging	Asrofi et al. (2018a)
Yam bean	Solution casting	Packaging	Asrofi et al. (2018b)
Cassava bagasse starch	Solution casting	Packaging	Teixeira et al. (2009)
Ramie starch	Solution casting	Packaging	Lu et al. (2006)
Potato	Solution casting	Packaging	Chen et al. (2012)
Cassava starch	Solution casting	Packaging	Syafri et al. (2018b)
Bengkuang starch	Solution casting	Packaging	Syafri et al. (2019b)

biodegradable, cheap, and renewable biopolymer. Nanocellulose can be defined as a material having a dimension of 100 nm (nm) or less with extremely high specific area, high porosity with excellent pore interconnectivity, low weight, and high biodegradability. In general, nanocellulose can be categorized into three main types based on their dimensions, functions, and method preparation. These types are CNF, CNCs, and BNC. There are various sources of cellulose fiber that can be applied to produce nanocellulose, such as sugar palm, water hyacinth, OPEFB, ramie, pineapple leaf, kenaf bast, roselle, cotton, arecanut husk (*A. catechu*) and sugarcane bagasse. These nanocellulose can be used to reinforce with starch biopolymer. There are two techniques that can be applied to make polysaccharides nanocomposite films, i.e., organic solvent or water evaporation by solvent casting method and extrusion method with freeze-dried nanocellulose. It is worth mentioning that the preparation of nanocomposites using solution casting method shows higher mechanical properties compared with nanocomposite of the same polymer ready by freeze-drying and extrusion methods. This is due to the strong linkage effect of the hydrogen bond in the solution casting method, in which the formation of rigid nanocellulose networks is established. The reinforcement of these nanocelluloses increased the mechanical, physical, and water sensitivity performance of starch biopolymer composites. To sum up, the reinforcement of nanocellulose and starch biopolymer has huge ability to partially substitute or replace the current synthetic polymer and can be utilized for different environmental, medical, packaging, industrial, agricultural mulch, and other low-cost and high-cost applications.

ACKNOWLEDGMENT

The authors would like to thank Universiti Putra Malaysia and Ministry of Education, Malaysia, for the financial support through the Universiti Putra Malaysia Grant scheme Hi-CoE (6369107) and Fundamental Research Grant Scheme FRGS/1/2017/TK05/UPM/01/1 (5540048).

REFERENCES

Abral, H., Ariksa, J., Mahardika, M., Handayani, D., Aminah, I., Sandrawati, N., Pratama, A.B., Fajri, N., Sapuan, S.M., Ilyas, R.A., 2020. Transparent and antimicrobial cellulose film from ginger nanofiber. Food Hydrocolloids 98, 105266. https://doi.org/10.1016/j.foodhyd.2019.105266.

Abral, H., Ariksa, J., Mahardika, M., Handayani, D., Aminah, I., Sandrawati, N., Sapuan, S.M., Ilyas, R.A., 2019. Highly transparent and antimicrobial PVA based bionanocomposites reinforced by ginger nanofiber. Polymer Testing 106186. https://doi.org/10.1016/j.polymertesting.2019.106186.

Ahmad, R., Mohd, S., Ibrahim, R., Abral, H., Roslim, M., Huzaifah, M., Mohd, A., Murat, A., Azammi, N., Adrinata, M., 2019. Cellulosic fibre hierarchy : a comprehensive approach from macro to nano scale. Integrative Medicine Research 1–14. https://doi.org/10.1016/j.jmrt.2019.04.011.

Aisyah, H.A., Paridah, M.T., Sapuan, S.M., Khalina, A., Berkalp, O.B., Lee, S.H., Lee, C.H., Nurazzi, N.M., Ramli, N., Wahab, M.S., Ilyas, R.A., 2019. Thermal properties of woven kenaf/carbon fibre-reinforced epoxy hybrid composite panels. International Journal of Polymer Science 2019, 1–8. https://doi.org/10.1155/2019/5258621.

Alemdar, A., Sain, M., 2008. Biocomposites from wheat straw nanofibers: morphology, thermal and mechanical properties. Composites Science and Technology 68, 557–565. https://doi.org/10.1016/j.compscitech.2007.05.044.

Angellier, H., Molina-Boisseau, S., Dole, P., Dufresne, A., 2006. Thermoplastic starch–waxy maize starch nanocrystals nanocomposites. Biomacromolecules 7, 531–539. https://doi.org/10.1021/bm050797s.

Anglès, M.N., Dufresne, A., 2001. Plasticized starch/tunicin whiskers nanocomposite materials. 2. Mechanical behavior. Macromolecules 34, 2921–2931. https://doi.org/10.1021/ma001555h.

Ariffin, H., Norrrahim, M.N.F., Yasim-Anuar, T.A.T., Nishida, H., Hassan, M.A., Ibrahim, N.A., Yunus, W.M.Z.W., 2017. Oil palm biomass cellulose-fabricated polylactic acid composites for packaging applications. Bionanocomposites for Packaging Applications 95–105. https://doi.org/10.1007/978-3-319-67319-6_5.

Asrofi, M., Abral, H., Kasim, A., Pratoto, A., Mahardika, M., Hafizulhaq, F., 2018a. Characterization of the sonicated yam bean starch bionanocomposites reinforced by nanocellulose water hyacinth fiber (WHF): the effect of various fiber loading. Journal of Engineering Science and Technology 13, 2700–2715.

Asrofi, M., Abral, H., Kasim, A., Pratoto, A., Mahardika, M., Hafizulhaq, F., 2018b. Mechanical properties of a water hyacinth nanofiber cellulose reinforced thermoplastic starch bionanocomposite: effect of ultrasonic vibration during processing. Fibers 6, 40. https://doi.org/10.3390/fib6020040.

Asrofi, M., Abral, H., Kasim, A., Pratoto, A., Mahardika, M., Park, J.W., Kim, H.J., 2018c. Isolation of nanocellulose from water hyacinth fiber (WHF) produced via digester-sonication and its characterization. Fibers and Polymers 19, 1618–1625. https://doi.org/10.1007/s12221-018-7953-1.

Asyraf, M.R.M., Ishak, M.R., Sapuan, S.M., Yidris, N., Ilyas, R.A., 2020. Woods and composites cantilever beam: a comprehensive review of experimental and numerical creep methodologies. Journal of Materials Research and Technology. https://doi.org/10.1016/j.jmrt.2020.01.013.

Atikah, M.S.N., Ilyas, R.A., Sapuan, S.M., Ishak, M.R., Zainudin, E.S., Ibrahim, R., Atiqah, A., Ansari, M.N.M., Jumaidin, R., 2019. Degradation and physical properties of sugar palm starch/sugar palm nanofibrillated cellulose bionanocomposite. Polimery 64, 27–36. https://doi.org/10.14314/polimery.2019.10.5.

Atiqah, A., Jawaid, M., Sapuan, S.M., Ishak, M.R., Ansari, M.N.M., Ilyas, R.A., 2019. Physical and thermal properties of treated sugar palm/glass fibre reinforced thermoplastic polyurethane hybrid composites. Journal of Materials Research and Technology 8, 3726–3732. https://doi.org/10.1016/j.jmrt.2019.06.032.

Azammi, A.M.N., Ilyas, R.A., Sapuan, S.M., Ibrahim, R., Atikah, M.S.N., Asrofi, M., Atiqah, A., 2020. Characterization studies of biopolymeric matrix and cellulose fibres based composites related to functionalized fibre-matrix interface. In: Interfaces in Particle and Fibre Reinforced Composites. Elsevier, London, pp. 29–93. https://doi.org/10.1016/B978-0-08-102665-6.00003-0.

Azeredo, H.M.C., Mattoso, L.H.C., Wood, D., Williams, T.G., Avena-Bustillos, R.J., McHugh, T.H., 2009. Nanocomposite edible films from mango puree reinforced with cellulose nanofibers. Journal of Food Science 74. https://doi.org/10.1111/j.1750-3841.2009.01186.x.

Azeredo, H.M.C., Rosa, M.F., Mattoso, L.H.C., 2017. Nanocellulose in bio-based food packaging applications. Industrial Crops and Products 97, 664–671. https://doi.org/10.1016/j.indcrop.2016.03.013.

Babaee, M., Jonoobi, M., Hamzeh, Y., Ashori, A., 2015. Biodegradability and mechanical properties of reinforced starch nanocomposites using cellulose nanofibers. Carbohydrate Polymers 132, 1–8. https://doi.org/10.1016/j.carbpol.2015.06.043.

Babu, R.P., O'Connor, K., Seeram, R., 2013. Current progress on bio-based polymers and their future trends. Progress in Biomaterials 2, 8. https://doi.org/10.1186/2194-0517-2-8.

Bano, S., Negi, Y.S., 2017. Studies on cellulose nanocrystals isolated from groundnut shells. Carbohydrate Polymers 157, 1041–1049. https://doi.org/10.1016/j.carbpol.2016.10.069.

de Benini, K.C.C.C., Voorwald, H.J.C., Cioffi, M.O.H., Rezende, M.C., Arantes, V., 2018. Preparation of nanocellulose from Imperata brasiliensis grass using Taguchi method. Carbohydrate Polymers 192, 337–346. https://doi.org/10.1016/j.carbpol.2018.03.055.

Bhattacharya, D., Germinario, L.T., Winter, W.T., 2008. Isolation, preparation and characterization of cellulose microfibers obtained from bagasse. Carbohydrate Polymers 73, 371–377. https://doi.org/10.1016/j.carbpol.2007.12.005.

Bondeson, D., Mathew, A., Oksman, K., 2006. Optimization of the isolation of nanocrystals from microcrystalline cellulose by acid hydrolysis. Cellulose 13, 171–180. https://doi.org/10.1007/s10570-006-9061-4.

Braun, B., Dorgan, J.R., Chandler, J.P., 2008. Cellulosic nanowhiskers. Theory and application of light scattering from polydisperse spheroids in the Rayleigh–Gans–Debye Regime. Biomacromolecules 9, 1255–1263. https://doi.org/10.1021/bm7013137.

Brinchi, L., Cotana, F., Fortunati, E., Kenny, J.M., 2013. Production of nanocrystalline cellulose from lignocellulosic biomass: technology and applications. Carbohydrate Polymers 94, 154–169. https://doi.org/10.1016/j.carbpol.2013.01.033.

Brito, B.S.L., Pereira, F.V., Putaux, J.-L., Jean, B., 2012. Preparation, morphology and structure of cellulose nanocrystals from bamboo fibers. Cellulose 19, 1527–1536. https://doi.org/10.1007/s10570-012-9738-9.

Camann, A., Dragsbeak, K., Krol, S., Sandgren, J., Song, D., 2010. Properties, recycling and alternatives to PE bags. Worcester Polytechnic Institute 1–132.

Cao, X., Chen, Y., Chang, P.R., Muir, A.D., Falk, G., 2008a. Starch-based nanocomposites reinforced with flax cellulose nanocrystals. Express Polymer Letters 2, 502–510. https://doi.org/10.3144/expresspolymlett.2008.60.

Cao, X., Chen, Y., Chang, P.R., Stumborg, M., Huneault, M.A., 2008b. Green composites reinforced with hemp nanocrystals in plasticized starch. Journal of Applied Polymer Science 109, 3804–3810. https://doi.org/10.1002/app.28418.

Chan, C.H., Chia, C.H., Zakaria, S., Ahmad, I., Dufresne, A., 2013. Production and characterisation of cellulose and nano- crystalline cellulose from kenaf core wood. BioResources 8, 785–794. https://doi.org/10.15376/biores.8.1.785-794.

Chang, P.R., Jian, R., Zheng, P., Yu, J., Ma, X., 2010. Preparation and properties of glycerol plasticized-starch (GPS)/cellulose nanoparticle (CN) composites. Carbohydrate Polymers 79, 301–305. https://doi.org/10.1016/j.carbpol.2009.08.007.

Chen, D., Lawton, D., Thompson, M.R., Liu, Q., 2012. Biocomposites reinforced with cellulose nanocrystals derived from potato peel waste. Carbohydrate Polymers 90, 709–716. https://doi.org/10.1016/j.carbpol.2012.06.002.

Chen, W., Abe, K., Uetani, K., Yu, H., Liu, Y., Yano, H., 2014. Individual cotton cellulose nanofibers: pretreatment and fibrillation technique. Cellulose 21, 1517–1528. https://doi.org/10.1007/s10570-014-0172-z.

Chen, W., Yu, H., Liu, Y., 2011a. Preparation of millimeter-long cellulose i nanofibers with diameters of 30–80 nm from bamboo fibers. Carbohydrate Polymers 86, 453–461. https://doi.org/10.1016/j.carbpol.2011.04.061.

Chen, W., Yu, H., Liu, Y., Chen, P., Zhang, M., Hai, Y., 2011b. Individualization of cellulose nanofibers from wood using high-intensity ultrasonication combined with chemical pretreatments. Carbohydrate Polymers 83, 1804–1811. https://doi.org/10.1016/j.carbpol.2010.10.040.

Chen, W., Yu, H., Liu, Y., Hai, Y., Zhang, M., Chen, P., 2011c. Isolation and characterization of cellulose nanofibers from four plant cellulose fibers using a chemical-ultrasonic process. Cellulose 18, 433–442. https://doi.org/10.1007/s10570-011-9497-z.

Cherian, B.M., Leão, A.L., de Souza, S.F., Thomas, S., Pothan, L.A., Kottaisamy, M., 2010. Isolation of nanocellulose from pineapple leaf fibres by steam explosion. Carbohydrate Polymers 81, 720–725. https://doi.org/10.1016/j.carbpol.2010.03.046.

Corre, Y.M., Bruzaud, S., Audic, J.L., Grohens, Y., 2012. Morphology and functional properties of commercial polyhydroxyalkanoates: a comprehensive and comparative study. Polymer Testing 31, 226–235. https://doi.org/10.1016/j.polymertesting.2011.11.002.

Corrêa, A.C., de Morais Teixeira, E., Pessan, L.A., Mattoso, L.H.C., 2010. Cellulose nanofibers from curaua fibers. Cellulose 17, 1183–1192. https://doi.org/10.1007/s10570-010-9453-3.

de Morais Teixeira, E., Corrêa, A.C., Manzoli, A., de Lima Leite, F., de Ribeiro Oliveira, C., Mattoso, L.H.C., 2010. Cellulose nanofibers from white and naturally colored cotton fibers. Cellulose 17, 595–606. https://doi.org/10.1007/s10570-010-9403-0.

De Oliveira Mori, C.L.S., Dos Passos, N.A., Oliveira, J.E., Mattoso, L.H.C., Mori, F.A., Carvalho, A.G., De Souza Fonseca, A., Tonoli, G.H.D., 2014. Electrospinning of zein/tannin bio-nanofibers. Industrial Crops and Products 52, 298–304. https://doi.org/10.1016/j.indcrop.2013.10.047.

Dima, S.O., Panaitescu, D.M., Orban, C., Ghiurea, M., Doncea, S.M., Fierascu, R.C., Nistor, C.L., Alexandrescu, E., Nicolae, C.A., Trica, B., Moraru, A., Oancea, F., 2017. Bacterial nanocellulose from side-streams of kombucha beverages production: preparation and physical-chemical properties. Polymers 9, 5–10. https://doi.org/10.3390/polym9080374.

Dufresne, A., 2009. Polymer nanocomposites from biological sources. Encyclopedia of Nanoscience and Nanotechnology X, 1–32.

Fahma, F., Iwamoto, S., Hori, N., Iwata, T., Takemura, A., 2010. Isolation, preparation, and characterization of nanofibers from oil palm empty-fruit-bunch (OPEFB). Cellulose 17, 977–985. https://doi.org/10.1007/s10570-010-9436-4.

Faradilla, R.H.F., Lee, G., Arns, J.Y., Roberts, J., Martens, P., Stenzel, M.H., Arcot, J., 2017. Characteristics of a free-standing film from banana pseudostem nanocellulose generated from TEMPO-mediated oxidation. Carbohydrate Polymers 174, 1156–1163. https://doi.org/10.1016/j.carbpol.2017.07.025.

Favier, V., Chanzy, H., Cavaille, J.Y., 1995. Polymer nanocomposites reinforced by cellulose whiskers. Macromolecules 28, 6365–6367. https://doi.org/10.1021/ma00122a053.

Fishman, M., Coffin, D., Konstance, R., Onwulata, C., 2000. Extrusion of pectin/starch blends plasticized with glycerol. Carbohydrate Polymers 41, 317–325. https://doi.org/10.1016/S0144-8617(99)00117-4.

Flauzino Neto, W.P., Silvério, H.A., Dantas, N.O., Pasquini, D., 2013. Extraction and characterization of cellulose nanocrystals from agro-industrial residue – soy hulls. Industrial Crops and Products 42, 480–488. https://doi.org/10.1016/j.indcrop.2012.06.041.

Fortunati, E., Puglia, D., Luzi, F., Santulli, C., Kenny, J.M., Torre, L., 2013. Binary PVA bio-nanocomposites containing cellulose nanocrystals extracted from different natural sources: Part I. Carbohydrate Polymers 97, 825–836. https://doi.org/10.1016/j.carbpol.2013.03.075.

Fu, S.-Y., Feng, X.-Q., Lauke, B., Mai, Y.-W., 2008. Effects of particle size, particle/matrix interface adhesion and particle loading on mechanical properties of particulate–polymer composites. Composites Part B: Engineering 39, 933–961. https://doi.org/10.1016/j.compositesb.2008.01.002.

Gadhave, R.V., Das, A., Mahanwar, P.A., Gadekar, P.T., 2018. Starch based bio-plastics: the future of sustainable packaging. Open Journal of Polymer Chemistry 08, 21–33. https://doi.org/10.4236/ojpchem.2018.82003.

Grunert, M., Winter, W.T., 2002. Nanocomposites of cellulose acetate butyrate reinforced with cellulose nanocrystals. Journal of Polymers and the Environment 10, 27–30.

Haafiz, M.K.M., Hassan, A., Zakaria, Z., Inuwa, I.M., 2014. Isolation and characterization of cellulose nanowhiskers from oil palm biomass microcrystalline cellulose. Carbohydrate Polymers 103, 119–125. https://doi.org/10.1016/j.carbpol.2013.11.055.

Habibi, Y., Vignon, M.R., 2008. Optimization of cellouronic acid synthesis by TEMPO-mediated oxidation of cellulose III from sugar beet pulp. Cellulose 15, 177–185. https://doi.org/10.1007/s10570-007-9179-z.

Halimatul, M.J., Sapuan, S.M., Jawaid, M., Ishak, M.R., Ilyas, R.A., 2019a. Water absorption and water solubility properties of sago starch biopolymer composite films filled with sugar palm particles. Polimery 64, 27–35. https://doi.org/10.14314/polimery.2019.9.4.

Halimatul, M.J., Sapuan, S.M., Jawaid, M., Ishak, M.R., Ilyas, R.A., 2019b. Effect of sago starch and plasticizer content on the properties of thermoplastic films: mechanical testing and cyclic soaking-drying. Polimery 64, 32–41. https://doi.org/10.14314/polimery.2019.6.5.

Hassan, M.A., Yee, L.N., Yee, P.L., Ariffin, H., Raha, A.R., Shirai, Y., Sudesh, K., 2013. Sustainable production of polyhydroxyalkanoates from renewable oil-palm biomass. Biomass and Bioenergy 50, 1–9. https://doi.org/10.1016/j.biombioe.2012.10.014.

He, Y., Boluk, Y., Pan, J., Ahniyaz, A., Deltin, T., Claesson, P.M., 2019. Corrosion protective properties of cellulose nanocrystals reinforced waterborne acrylate-based composite coating. Corrosion Science 155, 186–194. https://doi.org/10.1016/j.corsci.2019.04.038.

Hernandez, C.C., Ferreira, F.F., Rosa, D.S., 2018. X-ray powder diffraction and other analyses of cellulose nanocrystals obtained from corn straw by chemical treatments. Carbohydrate Polymers 193. https://doi.org/10.1016/j.carbpol.2018.03.085.

Herrera, M.A., Mathew, A.P., Oksman, K., 2012. Comparison of cellulose nanowhiskers extracted from industrial bio-residue and commercial microcrystalline cellulose. Materials Letters 71, 28–31. https://doi.org/10.1016/j.matlet.2011.12.011.

Herrera, N., Salaberria, A.M., Mathew, A.P., Oksman, K., 2016. Plasticized polylactic acid nanocomposite films with cellulose and chitin nanocrystals prepared using extrusion and compression molding with two cooling rates: effects on mechanical, thermal and optical properties. Composites Part A: Applied Science and Manufacturing 83, 89–97. https://doi.org/10.1016/j.compositesa.2015.05.024.

Hietala, M., Mathew, A.P., Oksman, K., 2013. Bionanocomposites of thermoplastic starch and cellulose nanofibers manufactured using twin-screw extrusion. European Polymer Journal 49, 950–956. https://doi.org/10.1016/j.eurpolymj.2012.10.016.

Ilyas, R.A., Sapuan, S.M., 2020. The preparation methods and processing of natural fibre bio-polymer composites. Current Organic Synthesis 16, 1068–1070. https://doi.org/10.2174/157017941608200120105616.

Ilyas, R.A., Sapuan, S.M., Atiqah, A., Ibrahim, R., Abral, H., Ishak, M.R., Zainudin, E.S., Nurazzi, N.M., Atikah, M.S.N., Ansari, M.N.M., Asyraf, M.R.M., Supian, A.B.M., Ya, H., 2019. Sugar palm (Arenga pinnata [Wurmb .] Merr) starch films containing sugar palm nanofibrillated cellulose as reinforcement: water barrier properties. Polymer Composites 1–9. https://doi.org/10.1002/pc.25379.

Ilyas, Ahmad, R., Sapuan, S.M., Ibrahim, R., Abral, H., Ishak, M.R., Zainudin, E.S., Asrofi, M., Atikah, M.S.N., Huzaifah, M.R.M., Radzi, A.M., Azammi, A.M.N., Shaharuzaman, M.A., Nurazzi, N.M., Syafri, E., Sari, N.H., Norrrahim, M.N.F., Jumaidin, R., 2019a. Sugar palm (Arenga pinnata (Wurmb.) Merr) cellulosic fibre hierarchy: a comprehensive approach from macro to nano scale. Journal of Materials Research and Technology 8, 2753–2766. https://doi.org/10.1016/j.jmrt.2019.04.011.

Ilyas, R.A., Sapuan, S.M., Ibrahim, R., Abral, H., Ishak, M.R., Zainudin, E.S., Atikah, M.S.N., Mohd Nurazzi, N., Atiqah, A., Ansari, M.N.M., Syafri, E., Asrofi, M., Sari, N.H., Jumaidin, R., 2019b. Effect of sugar palm nanofibrillated cellulose concentrations on morphological, mechanical and physical properties of biodegradable films based on agro-waste sugar palm (Arenga pinnata (Wurmb.) Merr) starch. Journal of Materials Research and Technology 8, 4819–4830. https://doi.org/10.1016/j.jmrt.2019.08.028.

Ilyas, R.A., Sapuan, S.M., Ibrahim, R., Abral, H., Ishak, M.R., Zainudin, E.S., Atiqah, A., Atikah, N., Syafri, E., Asrofi, M., Jumaidin, R., 2020. Thermal, biodegradability and water barrier properties of bio-nanocomposites based on plasticised sugar palm starch and nanofibrillated celluloses from sugar palm fibres. Journal of Biobased Materials and Bioenergy 14, 1–13. https://doi.org/10.1166/jbmb.2020.1951.

Ilyas, R.A., Sapuan, S.M., Ishak, M.R., 2018a. Isolation and characterization of nanocrystalline cellulose from sugar palm fibres (Arenga Pinnata). Carbohydrate Polymers 181,

1038–1051. https://doi.org/10.1016/j.carbpol.2017.11.045.

Ilyas, R.A., Sapuan, S.M., Ishak, M.R., Zainudin, E.S., 2017. Effect of delignification on the physical, thermal, chemical, and structural properties of sugar palm fibre. BioResources 12, 8734–8754. https://doi.org/10.15376/biores.12.4.8734-8754.

Ilyas, R.A., Sapuan, S.M., Ishak, M.R., Zainudin, E.S., 2019c. Sugar palm nanofibrillated cellulose (Arenga pinnata (Wurmb.) Merr): effect of cycles on their yield, physic-chemical, morphological and thermal behavior. International Journal of Biological Macromolecules 123, 379–388. https://doi.org/10.1016/j.ijbiomac.2018.11.124.

Ilyas, R.A., Sapuan, S.M., Ishak, M.R., Zainudin, E.S., 2018g. Water transport properties of bio-nanocomposites reinforced by sugar palm (Arenga Pinnata) nanofibrillated cellulose. Journal of Advanced Research in Fluid Mechanics and Thermal Sciences Journal 51, 234–246.

Ilyas, R.A., Sapuan, S.M., Ishak, M.R., Zainudin, E.S., 2018b. Development and characterization of sugar palm nanocrystalline cellulose reinforced sugar palm starch bionanocomposites. Carbohydrate Polymers 202, 186–202. https://doi.org/10.1016/j.carbpol.2018.09.002.

Ilyas, R.A., Sapuan, S.M., Ishak, M.R., Zainudin, E.S., 2018c. Sugar palm nanocrystalline cellulose reinforced sugar palm starch composite: degradation and water-barrier properties. In: IOP Conference Series: Materials Science and Engineering. https://doi.org/10.1088/1757-899X/368/1/012006.

Ilyas, R.A., Sapuan, S.M., Ishak, M.R., Zainudin, E.S., Atikah, M.S.N., 2018d. Nanocellulose reinforced starch polymer composites : a review of preparation, properties and application. In: Proceeding: 5th International Conference on Applied Sciences and Engineering (ICASEA, 2018). GLOBAL ACADEMIC EXCELLENCE (M) SDN BHD, Capthorne Hotel, Cameron Highlands, Malaysia, pp. 325–341.

Ilyas, R.A., Sapuan, S.M., Ishak, M.R., Zainudin, E.S., Atikah, M.S.N., 2018e. Characterization of sugar palm nanocellulose and its potential for reinforcement with a starch-based composite. In: Sugar Palm Biofibers, Biopolymers, and Biocomposites, first ed. CRC Press/Taylor & Francis Group, Boca Raton, FL, pp. 189–220. https://doi.org/10.1201/9780429443923-10.

Ilyas, R.A., Sapuan, S.M., Sanyang, M.L., Ishak, M.R., 2016. Nanocrystalline cellulose reinforced starch-based nanocomposite: a review. In: 5th Postgraduate Seminar on Natural Fiber Composites. Universiti Putra Malaysia, Serdang, Selangor, pp. 82–87.

Ilyas, R.A., Sapuan, S.M., Sanyang, M.L., Ishak, M.R., Zainudin, E.S., 2018f. Nanocrystalline cellulose as reinforcement for polymeric matrix nanocomposites and its potential applications: a review. Current Analytical Chemistry 14, 203–225. https://doi.org/10.2174/1573411013666171003155624.

Imai, T., Putaux, J., Sugiyama, J., 2003. Geometric phase analysis of lattice images from algal cellulose microfibrils. Polymer 44, 1871–1879. https://doi.org/10.1016/S0032-3861(02)00861-3.

Jaafar, C.N.A., Rizal, M.A.M., Zainol, I., 2018. Effect of kenaf alkalization treatment on morphological and mechanical properties of epoxy/silica/kenaf composite. International Journal of Engineering and Technology 7, 258–263. https://doi.org/10.14419/ijet.v7i4.35.22743.

Jasmani, L., Adnan, S., 2017. Preparation and characterization of nanocrystalline cellulose from Acacia mangium and its reinforcement potential. Carbohydrate Polymers 161, 166–171. https://doi.org/10.1016/j.carbpol.2016.12.061.

Jiang, F., Hsieh, Y.L., 2013. Chemically and mechanically isolated nanocellulose and their self-assembled structures. Carbohydrate Polymers 95, 32–40. https://doi.org/10.1016/j.carbpol.2013.02.022.

Jiang, Y., Zhou, J., Zhang, Q., Zhao, G., Heng, L., Chen, D., Liu, D., 2017. Preparation of cellulose nanocrystals from Humulus japonicus stem and the influence of high temperature pretreatment. Carbohydrate Polymers 164, 284–293. https://doi.org/10.1016/j.carbpol.2017.02.021.

Jozala, A.F., de Lencastre-Novaes, L.C., Lopes, A.M., de Carvalho Santos-Ebinuma, V., Mazzola, P.G., Pessoa- Jr., A., Grotto, D., Gerenutti, M., Chaud, M.V., 2016. Bacterial nanocellulose production and application: a 10-year overview. Applied Microbiology and Biotechnology 100, 2063–2072. https://doi.org/10.1007/s00253-015-7243-4.

Julie Chandra, C.S., George, N., Narayanankutty, S.K., 2016. Isolation and characterization of cellulose nanofibrils from arecanut husk fibre. Carbohydrate Polymers 142, 158–166. https://doi.org/10.1016/j.carbpol.2016.01.015.

Jumaidin, R., Ilyas, R.A., Saiful, M., Hussin, F., Mastura, M.T., 2019a. Water transport and physical properties of sugarcane bagasse fibre reinforced thermoplastic potato starch biocomposite. Journal of Advanced Research in Fluid Mechanics and Thermal Sciences 61, 273–281.

Jumaidin, R., Khiruddin, M.A.A., Asyul Sutan Saidi, Z., Salit, M.S., Ilyas, R.A., 2020. Effect of cogon grass fibre on the thermal, mechanical and biodegradation properties of thermoplastic cassava starch biocomposite. International Journal of Biological Macromolecules 146, 746–755. https://doi.org/10.1016/j.ijbiomac.2019.11.011.

Jumaidin, R., Saidi, Z.A.S., Ilyas, R.A., Ahmad, M.N., Wahid, M.K., Yaakob, M.Y., Maidin, N.A., Rahman, M.H.A., Osman, M.H., 2019b. Characteristics of cogon grass fibre reinforced thermoplastic cassava starch biocomposite: water absorption and physical properties. Journal of Advanced Research in Fluid Mechanics and Thermal Sciences 62 (62), 43–52.

Kaboorani, A., Riedl, B., 2015. Surface modification of cellulose nanocrystals (CNC) by a cationic surfactant. Industrial Crops and Products 65, 45–55. https://doi.org/10.1016/j.indcrop.2014.11.027.

Karimi, S., Tahir, P., Dufresne, A., Karimi, A., Abdulkhani, A., 2014. A comparative study on characteristics of nanocellulose reinforced thermoplastic starch biofilms prepared with different techniques. Nordic Pulp and Paper Research Journal 29, 41–45.

Kassab, Z., Syafri, E., Tamraoui, Y., Hannache, H., El Kacem Qaiss, A., El Achaby, M., 2019. Characteristics of sulfated and carboxylated cellulose nanocrystals extracted from Juncus plant stems. International Journal of Biological Macromolecules. https://doi.org/10.1016/j.ijbiomac.2019.11.023.

Kaushik, A., Singh, M., Verma, G., 2010. Green nanocomposites based on thermoplastic starch and steam exploded cellulose nanofibrils from wheat straw. Carbohydrate Polymers 82, 337–345. https://doi.org/10.1016/j.carbpol.2010.04.063.

Kaushik, M., Fraschini, C., Chauve, G., Putaux, J.-L., Moores, A., 2015. Transmission electron microscopy for the characterization of cellulose nanocrystals. In: The Transmission Electron Microscope - Theory and Applications. InTech, pp. 130–163. https://doi.org/10.5772/60985.

Khalil, H.P.S.A., Davoudpour, Y., Islam, M.N., Mustapha, A., Sudesh, K., Dungani, R., Jawaid, M., 2014. Production and modification of nanofibrillated cellulose using various mechanical processes: a review. Carbohydrate Polymers 99, 649–665. https://doi.org/10.1016/j.carbpol.2013.08.069.

Kian, L.K., Jawaid, M., Ariffin, H., Karim, Z., 2018. Isolation and characterization of nanocrystalline cellulose from roselle-derived microcrystalline cellulose. International Journal of Biological Macromolecules 114, 54–63. https://doi.org/10.1016/j.ijbiomac.2018.03.065.

Kim, J.H., Shim, B.S., Kim, H.S., Lee, Y.J., Min, S.K., Jang, D., Abas, Z., Kim, J., 2015. Review of nanocellulose for sustainable future materials. International Journal of Precision Engineering and Manufacturing - Green Technology 2, 197–213. https://doi.org/10.1007/s40684-015-0024-9.

Klemm, D., Schumann, D., Kramer, F., Hebler, N., Koth, D., Sultanova, B., 2009. Nanocellulose materials - different cellulose, different functionality. Macromolecular Symposia 280, 60–71. https://doi.org/10.1002/masy.200950608.

Kvien, I., Sugiyama, J., Votrubec, M., Oksman, K., 2007. Characterization of starch based nanocomposites. Journal of Materials Science 42, 8163–8171. https://doi.org/10.1007/s10853-007-1699-2.

Lamaming, J., Hashim, R., Sulaiman, O., Leh, C.P., Sugimoto, T., Nordin, N.A., 2015. Cellulose nanocrystals isolated from oil palm trunk. Carbohydrate Polymers 127, 202–208. https://doi.org/10.1016/j.carbpol.2015.03.043.

Lani, N.S., Ngadi, N., Johari, A., Jusoh, M., Lani, N.S., Ngadi, N., Johari, A., Jusoh, M, 2014. Isolation, characterization and application of nanoCellulose from oil palm empty fruit bunch. Journal of Nanomaterials 2014, 1–9. https://doi.org/10.1155/2014/702538.

Lavoine, N., Desloges, I., Dufresne, A., Bras, J., 2012. Microfibrillated cellulose - its barrier properties and applications in cellulosic materials: a review. Carbohydrate Polymers 90, 735–764. https://doi.org/10.1016/j.carbpol.2012.05.026.

Lee, H.V., Hamid, S.B.A., Zain, S.K., 2014. Conversion of lignocellulosic biomass to nanocellulose: structure and chemical process. Science World Journal 2014. https://doi.org/10.1155/2014/631013.

Li, J., Song, Z., Li, D., Shang, S., Guo, Y., 2014. Cotton cellulose nanofiber-reinforced high density polyethylene composites prepared with two different pretreatment methods. Industrial Crops and Products 59, 318–328. https://doi.org/10.1016/j.indcrop.2014.05.033.

Li, R., Fei, J., Cai, Y., Li, Y., Feng, J., Yao, J., 2009. Cellulose whiskers extracted from mulberry: a novel biomass production. Carbohydrate Polymers 76, 94–99. https://doi.org/10.1016/j.carbpol.2008.09.034.

Liu, C., Li, B., Du, H., Lv, D., Zhang, Y., Yu, G., Mu, X., Peng, H., 2016. Properties of nanocellulose isolated from corncob residue using sulfuric acid, formic acid, oxidative and mechanical methods. Carbohydrate Polymers 151, 716–724. https://doi.org/10.1016/j.carbpol.2016.06.025.

Liu, D., Zhong, T., Chang, P.R., Li, K., Wu, Q., 2010. Starch composites reinforced by bamboo cellulosic crystals. Bioresource Technology 101, 2529–2536. https://doi.org/10.1016/j.biortech.2009.11.058.

Llanos, J.H.R., Tadini, C.C., 2018. Preparation and characterization of bio-nanocomposite films based on cassava starch or chitosan, reinforced with montmorillonite or bamboo nanofibers. International Journal of Biological Macromolecules 107, 371–382. https://doi.org/10.1016/j.ijbiomac.2017.09.001.

Lu, P., Hsieh, Y., 2012. Preparation and characterization of cellulose nanocrystals from rice straw. Carbohydrate Polymers 87, 564–573. https://doi.org/10.1016/j.carbpol.2011.08.022.

Lu, Y., Weng, L., Cao, X., 2006. Morphological, thermal and mechanical properties of ramie crystallites—reinforced plasticized starch biocomposites. Carbohydrate Polymers 63, 198–204. https://doi.org/10.1016/j.carbpol.2005.08.027.

Lu, Y., Weng, L., Cao, X., 2005. Biocomposites of plasticized starch reinforced with cellulose crystallites from cottonseed linter. Macromolecular Bioscience 5, 1101–1107. https://doi.org/10.1002/mabi.200500094.

Luckachan, G.E., Pillai, C.K.S., 2011. Biodegradable polymers-A review on recent trends and emerging perspectives. Journal of Polymers and the Environment 19, 637–676. https://doi.org/10.1007/s10924-011-0317-1.

Mahardika, M., Abral, H., Kasim, A., Arief, S., Asrofi, M., 2018. Production of nanocellulose from pineapple leaf fibers via high-shear homogenization and ultrasonication. Fibers 6, 28. https://doi.org/10.3390/fib6020028.

Mathew, A.P., Thielemans, W., Dufresne, A., 2008. Mechanical properties of nanocomposites from sorbitol plasticized starch and tunicin whiskers. Journal of Applied Polymer Science 109, 4065–4074. https://doi.org/10.1002/app.28623.

Mazani, N., Sapuan, S.M., Sanyang, M.L., Atiqah, A., Ilyas, R.A., 2019. Design and fabrication of a shoe shelf from kenaf fiber reinforced unsaturated polyester composites. In: Lignocellulose for Future Bioeconomy. Elsevier, pp. 315–332. https://doi.org/10.1016/B978-0-12-816354-2.00017-7.

Medeiros, E., Dufresne, A., Orts, W., 2009. Starch-based nanocomposites. Starches 205–251. https://doi.org/10.1201/9781420080247-c9.

Mohammadi, M., Hassan, M.A., Shirai, Y., Man, H.C., Ariffin, H., Yee, L.N., Mumtaz, T., Chong, M.L., Phang, L.Y., 2012. Separation and purification of polyhydroxyalkanoates from newly isolated comamonas sp.

EB172 by simple digestion with sodium hydroxide. Separation Science and Technology 47, 534–541. https://doi.org/10.1080/01496395.2011.615788.

Mohammadkazemi, F., Doosthoseini, K., Ganjian, E., Azin, M., 2015. Manufacturing of bacterial nano-cellulose reinforced fiber–cement composites. Construction and Building Materials 101, 958–964. https://doi.org/10.1016/j.conbuildmat.2015.10.093.

Mondal, S., 2018. Review on nanocellulose polymer nanocomposites. Polymer - Plastics Technology and Engineering 57, 1377–1391. https://doi.org/10.1080/03602559.2017.1381253.

Montero, B., Rico, M., Rodríguez-Llamazares, S., Barral, L., Bouza, R., 2017. Effect of nanocellulose as a filler on biodegradable thermoplastic starch films from tuber, cereal and legume. Carbohydrate Polymers 157, 1094–1104. https://doi.org/10.1016/j.carbpol.2016.10.073.

Morais, J.P.S., Rosa, M.D.F., De Souza Filho, M., de sá, M., Nascimento, L.D., Do Nascimento, D.M., Cassales, A.R., 2013. Extraction and characterization of nanocellulose structures from raw cotton linter. Carbohydrate Polymers 91, 229–235. https://doi.org/10.1016/j.carbpol.2012.08.010.

Morán, J.I., Alvarez, V.A., Cyras, V.P., Vázquez, A., 2008. Extraction of cellulose and preparation of nanocellulose from sisal fibers. Cellulose 15, 149–159. https://doi.org/10.1007/s10570-007-9145-9.

Morandi, G., Heath, L., Thielemans, W., 2009. Cellulose nanocrystals grafted with polystyrene chains through surface-initiated atom transfer radical polymerization (SI-ATRP). Langmuir 25, 8280–8286.

Naduparambath, S., T.V, J., Shaniba, V., M.P, S., Balan, A.K., Purushothaman, E., 2018. Isolation and characterisation of cellulose nanocrystals from sago seed shells. Carbohydrate Polymers 180, 13–20. https://doi.org/10.1016/j.carbpol.2017.09.088.

Nasri-Nasrabadi, B., Behzad, T., Bagheri, R., 2014. Preparation and characterization of cellulose nanofiber reinforced thermoplastic starch composites. Fibers and Polymers 15, 347–354. https://doi.org/10.1007/s12221-014-0347-0.

Nasseri, R., Mohammadi, N., 2014. Starch-based nanocomposites: a comparative performance study of cellulose whiskers and starch nanoparticles. Carbohydrate Polymers 106, 432–439. https://doi.org/10.1016/j.carbpol.2014.01.029.

Norizan, M.N., Abdan, K., Ilyas, R.A., Biofibers, S.P., 2020. Effect of fiber orientation and fiber loading on the mechanical and thermal properties of sugar palm yarn fiber reinforced unsaturated polyester resin composites. Polimery 65, 34–43. https://doi.org/10.14314/polimery.2020.2.5.

Norrahim, M.N.F., Ariffin, H., Hassan, M.A., Ibrahim, N.A., Nishida, H., 2013. Performance evaluation and chemical recyclability of a polyethylene/poly(3- hydroxybutyrate-co-3-hydroxyvalerate) blend for sustainable packaging. RSC Advances 3, 24378–24388. https://doi.org/10.1039/c3ra43632b.

Norrahim, M.N.F., Ariffin, H., Yasim-Anuar, T.A.T., Ghaemi, F., Hassan, M.A., Ibrahim, N.A., Ngee, J.L.H., Yunus, W.M.Z.W., 2018a. Superheated steam pretreatment of cellulose affects its electrospinnability for microfibrillated cellulose production. Cellulose 25, 3853–3859. https://doi.org/10.1007/s10570-018-1859-3.

Norrahim, M.N.F., Ariffin, H., Yasim-Anuar, T.A.T., Hassan, M.A., Nishida, H., Tsukegi, T., 2018b. One-pot nanofibrillation of cellulose and nanocomposite production in a twin-screw extruder. IOP Conference Series: Materials Science and Engineering 368. https://doi.org/10.1088/1757-899X/368/1/012034.

Noshirvani, N., Ghanbarzadeh, B., Fasihi, H., Almasi, H., 2016. Starch-PVA nanocomposite film incorporated with cellulose nanocrystals and MMT: a comparative study. International Journal of Food Engineering 12, 37–48. https://doi.org/10.1515/ijfe-2015-0145.

Nurazzi, N.M., Khalina, A., Sapuan, S.M., Ilyas, R.A., 2019. Mechanical properties of sugar palm yarn/woven glass fiber reinforced unsaturated polyester composites : effect of fiber loadings and alkaline treatment. Polimery 64, 12–22. https://doi.org/10.14314/polimery.2019.10.3.

Nurazzi, N.M., Khalina, A., Sapuan, S.M., Ilyas, R.A., Rafiqah, S.A., Hanafee, Z.M., 2020. Thermal properties of treated sugar palm yarn/glass fiber reinforced unsaturated polyester hybrid composites. Journal of Materials Research and Technology 9, 1606–1618. https://doi.org/10.1016/j.jmrt.2019.11.086.

Oksman, K., Etang, J.A., Mathew, A.P., Jonoobi, M., 2010. Cellulose nanowhiskers separated from a bio-residue from wood bioethanol production. Biomass and Bioenergy 35, 146–152. https://doi.org/10.1016/j.biombioe.2010.08.021.

Pandey, J.K., Kim, C., Chu, W., Lee, C.S., Jang, D.-Y., Ahn, S., 2009. Evaluation of morphological architecture of cellulose chains in grass during conversion from macro to nano dimensions. E-Polymers 9, 1–15. https://doi.org/10.1515/epoly.2009.9.1.1221.

Paralikar, S.A., Simonsen, J., Lombardi, J., 2008. Poly(vinyl alcohol)/cellulose nanocrystal barrier membranes. Journal of Membrane Science 320, 248–258. https://doi.org/10.1016/j.memsci.2008.04.009.

Peng, Y., Gardner, D.J., Han, Y., 2011. Drying cellulose nanofibrils : in search of a suitable method. https://doi.org/10.1007/s10570-011-9630-z.

Pereda, M., Dufresne, A., Aranguren, M.I., Marcovich, N.E., 2014. Polyelectrolyte films based on chitosan/olive oil and reinforced with cellulose nanocrystals. Carbohydrate Polymers 101, 1018–1026. https://doi.org/10.1016/j.carbpol.2013.10.046.

Pereira, P.H.F., Waldron, K.W., Wilson, D.R., Cunha, A.P., de Brito, E.S., Rodrigues, T.H.S., Rosa, M.F., Azeredo, H.M.C., 2017. Wheat straw hemicelluloses added with cellulose nanocrystals and citric acid. Effect on film physical properties. Carbohydrate Polymers 164, 317–324. https://doi.org/10.1016/j.carbpol.2017.02.019.

Pérez, S., Baldwin, P.M., Gallant, D.J., 2009. Structural features of starch granules I. In: Starch, third ed. Elsevier Inc. https://doi.org/10.1016/B978-0-12-746275-2.00005-7.

Picheth, G.F., Pirich, C.L., Sierakowski, M.R., Woehl, M.A., Sakakibara, C.N., de Souza, C.F., Martin, A.A., da Silva, R., de Freitas, R.A., 2017. Bacterial cellulose in biomedical applications: a review. International Journal of Biological Macromolecules 104, 97–106. https://doi.org/10.1016/J.IJBIOMAC.2017.05.171.

Purkait, B.S., Ray, D., Sengupta, S., Kar, T., Mohanty, A., Misra, M., 2011. Isolation of cellulose nanoparticles from sesame husk. Industrial & Engineering Chemistry Research 50, 871–876. https://doi.org/10.1021/ie101797d.

Reddy, M.M., Vivekanandhan, S., Misra, M., Bhatia, S.K., Mohanty, A.K., 2013. Biobased plastics and bionanocomposites: current status and future opportunities. Progress in Polymer Science 38, 1653–1689. https://doi.org/10.1016/j.progpolymsci.2013.05.006.

Reddy, M.M., Misra, M., M, A., 2012. Bio-based materials in the new bio-economy. Chemical Engineering Progress 108, 37–42.

Revol, J.F., 1982. On the cross-sectional shape of cellulose crystallites in Valonia ventricosa. Carbohydrate Polymers 2, 123–134. https://doi.org/10.1016/0144-8617(82)90058-3.

Rohaizu, R., Wanrosli, W.D., 2017. Sono-assisted TEMPO oxidation of oil palm lignocellulosic biomass for isolation of nanocrystalline cellulose. Ultrasonics Sonochemistry 34, 631–639. https://doi.org/10.1016/j.ultsonch.2016.06.040.

Rosa, M.F.M., Medeiros, E.S., Malmonge, J.A.J., Gregorski, K.S., Wood, D.F., Mattoso, L.H.C., Glenn, G., Orts, W.J., Imam, S.H., 2010. Cellulose nanowhiskers from coconut husk fibers: effect of preparation conditions on their thermal and morphological behavior. Carbohydrate Polymers 81, 83–92. https://doi.org/10.1016/j.carbpol.2010.01.059.

Sabaruddin, F.A., Paridah, M.T., 2018. Effect of lignin on the thermal properties of nanocrystalline prepared from kenaf core. IOP Conference Series: Materials Science and Engineering 368, 012039. https://doi.org/10.1088/1757-899X/368/1/012039.

Salaberria, A.M., Fernandes, S.C.M., Diaz, R.H., Labidi, J., 2015. Processing of α-chitin nanofibers by dynamic high pressure homogenization: characterization and antifungal activity against *A. niger*. Carbohydrate Polymers 116, 286–291. https://doi.org/10.1016/j.carbpol.2014.04.047.

Salajková, M., Berglund, L.A., Zhou, Q., 2012. Hydrophobic cellulose nanocrystals modified with quaternary ammonium salts. Journal of Materials Chemistry 22, 19798. https://doi.org/10.1039/c2jm34355j.

Sanyang, M.L., Sapuan, S.M., Jawaid, M., Ishak, M.R., Sahari, J., 2016. Recent developments in sugar palm (*Arenga pinnata*) based biocomposites and their potential industrial applications: a review. Renewable and Sustainable Energy Reviews 54, 533–549. https://doi.org/10.1016/j.rser.2015.10.037.

Sapuan, S.M., Ilyas, R.A., 2017. Sugar palm: fibers, biopolymers and biocomposites. INTROPica 5–7.

Sapuan, S.M., Ilyas, R.A., Ishak, M.R., Leman, Z., Huzaifah, M.R.M., Ammar, I.M., Atikah, M.S.N., 2018. Development of sugar palm–based products: a community project. In: Sugar Palm Biofibers, Biopolymers, and Biocomposites, first ed. CRC Press/Taylor & Francis Group, Boca Raton, FL, pp. 245–266. https://doi.org/10.1201/9780429443923-12.

Sheltami, R.M., Abdullah, I., Ahmad, I., Dufresne, A., Kargarzadeh, H., 2012. Extraction of cellulose nanocrystals from mengkuang leaves (*Pandanus tectorius*). Carbohydrate Polymers 88, 772–779. https://doi.org/10.1016/j.carbpol.2012.01.062.

Shinoj, S., Visvanathan, R., Panigrahi, S., Kochubabu, M., 2011. Oil palm fiber (OPF) and its composites: a review. Industrial Crops and Products 33, 7–22. https://doi.org/10.1016/j.indcrop.2010.09.009.

Silvério, H.A., Flauzino Neto, W.P., Dantas, N.O., Pasquini, D., 2013. Extraction and characterization of cellulose nanocrystals from corncob for application as reinforcing agent in nanocomposites. Industrial Crops and Products 44, 427–436. https://doi.org/10.1016/j.indcrop.2012.10.014.

Siqueira, G., Bras, J., Dufresne, A., 2010. Cellulosic bionanocomposites: a review of preparation, properties and applications. Polymers 2, 728–765. https://doi.org/10.3390/polym2040728.

Siqueira, G., Bras, J., Dufresne, A., 2009. Cellulose whiskers versus microfibrils: influence of the nature of the nanoparticle and its surface functionalization on the thermal and mechanical properties of nanocomposites. Biomacromolecules 10, 425–432. https://doi.org/10.1021/bm801193.

Slavutsky, A.M., Bertuzzi, M.A., 2014. Water barrier properties of starch films reinforced with cellulose nanocrystals obtained from sugarcane bagasse. Carbohydrate Polymers 110, 53–61. https://doi.org/10.1016/j.carbpol.2014.03.049.

Soni, B., Hassan, E.B., Mahmoud, B., 2015. Chemical isolation and characterization of different cellulose nanofibers from cotton stalks. Carbohydrate Polymers 134, 581–589. https://doi.org/10.1016/j.carbpol.2015.08.031.

Sonia, A., Priya Dasan, K., 2013. Chemical, morphology and thermal evaluation of cellulose microfibers obtained from Hibiscus sabdariffa. Carbohydrate Polymers 92, 668–674. https://doi.org/10.1016/j.carbpol.2012.09.015.

Sreekala, M.S., Kumaran, M.G., Thomas, S., 1997. Oil palm fibers: morphology, chemical composition, surface modification, and mechanical properties. Journal of Applied Polymer Science 66, 821–835. https://doi.org/10.1002/(sici)1097-4628(19971031)66:5<821::aid-app2>3.3.co;2-l.

Sumaiyah, B.,W., Karsono, M.P.,N., S, G., 2014. Preparation and characterization of nanocrystalline cellulose from sugar palm bunch. International Journal of PharmTech Research 6, 814–820.

Syafri, E., Kasim, A., Abral, H., Asben, A., 2019a. Cellulose nanofibers isolation and characterization from ramie using a chemical-ultrasonic treatment. Journal of Natural Fibers 16, 1145–1155. https://doi.org/10.1080/15440478.2018.1455073.

Syafri, E., Kasim, A., Abral, H., Asben, A., 2018a. Cellulose nanofibers isolation and characterization from ramie using a chemical-ultrasonic treatment. Journal of Natural Fibers 00, 1–11. https://doi.org/10.1080/15440478.2018.1455073.

Syafri, E., Kasim, A., Abral, H., Sudirman, Sulungbudi, G.T., Sanjay, M.R., Sari, N.H., 2018b. Synthesis and characterization of cellulose nanofibers (CNF) ramie reinforced cassava starch hybrid composites. International Journal of

Biological Macromolecules 120, 578–586. https://doi.org/10.1016/j.ijbiomac.2018.08.134.

Syafri, E., Sudirman, M., Yulianti, E., Deswita, Asrofi, M., Abral, H., Sapuan, S.M., Ilyas, R.A., Fudholi, A., 2019b. Effect of sonication time on the thermal stability, moisture absorption, and biodegradation of water hyacinth (*Eichhornia crassipes*) nanocellulose-filled bengkuang (Pachyrhizus erosus) starch biocomposites. Journal of Materials Research and Technology 8, 6223–6231. https://doi.org/10.1016/j.jmrt.2019.10.016.

Takagi, H., Asano, A., 2008. Effects of processing conditions on flexural properties of cellulose nanofiber reinforced "green" composites. Composites Part A: Applied Science and Manufacturing 39, 685–689. https://doi.org/10.1016/j.compositesa.2007.08.019.

de Teixeira, E., Bondancia, T.J., Teodoro, K.B.R., Corrêa, A.C., Marconcini, J.M., Mattoso, L.H.C., 2011a. Sugarcane bagasse whiskers: extraction and characterizations. Industrial Crops and Products 33, 63–66. https://doi.org/10.1016/j.indcrop.2010.08.009.

de Teixeira, E., Pasquini, D., Curvelo, A.A.S.S., Corradini, E., Belgacem, M.N., Dufresne, A., 2009. Cassava bagasse cellulose nanofibrils reinforced thermoplastic cassava starch. Carbohydrate Polymers 78, 422–431. https://doi.org/10.1016/j.carbpol.2009.04.034.

Teixeira, E.D.M., Lotti, C., Corrêa, A.C., Teodoro, K.B.R., Marconcini, J.M., Mattoso, L.H.C., 2011b. Thermoplastic corn starch reinforced with cotton cellulose nanofibers. Journal of Applied Polymer Science 120, 2428–2433. https://doi.org/10.1002/app.33447.

Tian, C., Yi, J., Wu, Y., Wu, Q., Qing, Y., Wang, L., 2016. Preparation of highly charged cellulose nanofibrils using high-pressure homogenization coupled with strong acid hydrolysis pretreatments. Carbohydrate Polymers 136, 485–492. https://doi.org/10.1016/j.carbpol.2015.09.055.

Tibolla, H., Pelissari, F.M., Menegalli, F.C., 2014. Cellulose nanofibers produced from banana peel by chemical and enzymatic treatment. Lebensmittel-Wissenschaft und -Technologie- Food Science and Technology 59, 1311–1318. https://doi.org/10.1016/j.lwt.2014.04.011.

Tonoli, G.H.D., Teixeira, E.M., Corrêa, A.C., Marconcini, J.M., Caixeta, L.A., Pereira-Da-Silva, M.A., Mattoso, L.H.C., 2012. Cellulose micro/nanofibres from Eucalyptus kraft pulp: preparation and properties. Carbohydrate Polymers 89, 80–88. https://doi.org/10.1016/j.carbpol.2012.02.052.

Verhoogt, H., Ramsay, B.A., Favis, B.D., 1994. Polymer blends containing poly (3-hydroxyalkanoate)s. Polymer Review. Polymer 35, 5155–5169.

Wahono, S., Irwan, A., Syafri, E., Asrofi, M., 2018. Preparation and characterization of ramie cellulose nanofibers/CaCO3Unsaturated polyester resin composites. ARPN Journal of Engineering and Applied Sciences 13, 746–751. https://doi.org/10.1039/c7nr02736b.

Xiao, S., Gao, R., Lu, Y., Li, J., Sun, Q., 2015. Fabrication and characterization of nanofibrillated cellulose and its aerogels from natural pine needles. Carbohydrate Polymers 119, 202–209. https://doi.org/10.1016/j.carbpol.2014.11.041.

Xu, X., Liu, F., Jiang, L., Zhu, J.Y., Haagenson, D., Wiesenborn, D.P., 2013. Cellulose nanocrystals vs. Cellulose nanofibrils: a comparative study on their microstructures and effects as polymer reinforcing agents. ACS Applied Materials and Interfaces 5, 2999–3009. https://doi.org/10.1021/am302624t.

Yang, S., Tang, Y., Wang, J., Kong, F., Zhang, J., 2014. Surface treatment of cellulosic paper with starch-based composites reinforced with nanocrystalline cellulose. Industrial & Engineering Chemistry Research 53, 13980–13988. https://doi.org/10.1021/ie502125s.

Zhan, T., 2017. Improved Bacterial Nanocellulose Production by Co-cultivation, pp. 1–56.

Zhao, Y., Moser, C., Lindström, M.E., Henriksson, G., Li, J., 2017. Cellulose nanofibers from softwood, hardwood, and tunicate: preparation-structure-film performance interrelation. ACS Applied Materials and Interfaces 9, 13508–13519. https://doi.org/10.1021/acsami.7b01738.

Zhu, J., Wei, S., Patil, R., Rutman, D., Kucknoor, A.S., Wang, A., Guo, Z., 2011. Ionic liquid assisted electrospinning of quantum dots/elastomer composite nanofibers. Polymer 52, 1954–1962. https://doi.org/10.1016/j.polymer.2011.02.051.

Zimmermann, T., Bordeanu, N., Strub, E., 2010. Properties of nanofibrillated cellulose from different raw materials and its reinforcement potential. Carbohydrate Polymers 79, 1086–1093. https://doi.org/10.1016/j.carbpol.2009.10.045.

CHAPTER 7

Mechanical Testing of Sugar Palm Fiber Reinforced Sugar Palm Biopolymer Composites

R.A. ILYAS • S.M. SAPUAN • ABUDUKEREMU KADIER • SANTHANA KRISHNAN • M.S.N. ATIKAH • RUSHDAN IBRAHIM • A. NAZRIN • R. SYAFIQ • S. MISRI • M.R.M. HUZAIFAH • M.D. HAZROL

1 INTRODUCTION

Increased awareness of environmental issues and policies regarding plastic packaging has led to an increase in the importance of using polymers obtained from renewable sources that are generally biodegradable (Mazani et al., 2019). A renewable resource is a resource that can be used repeatedly and replaced naturally. Biopolymers have attracted a great deal of attention due to their environmental benefits and awareness toward the depleted petroleum resources (Sanyang et al., 2018). A wide range of naturally occurring biopolymers gained from renewable resources are accessible for material applications. Some of these such as bacteria and plants (chitin, starch, and cellulose) are actively used in commercial products today, while many others remain underutilized (Ilyas et al., 2018b). Biopolymers can be classified as synthetic or natural based on their origins, such as agro polymers (starch or cellulose), animal polymer-based (chitin), microbial (exopolysaccharides and polyhydroxyalkanoate) polymers, chemically synthesized from agro-based resource monomer (polylactic acid), and chemically synthesized from conventionally synthesized monomers (Abral et al., 2019; Ilyas et al., 2016). Among these biopolymers, starch has been investigated as potential alternatives to conventional plastic packaging. This is due to its wide availability, renewable, affordable, and biodegradability (Ortega-Toro et al., 2017).

Nevertheless, starch biopolymers have been described to have low water barrier resistance and poor mechanical properties. These disadvantages have greatly limited their wide-ranging applications, especially for plastic packaging purposes. In an effort to resolve these problems, many studies have been conducted by scientists and engineers to enhance the water sensitivity and improve the mechanical properties of starch biopolymer without losing their biodegradation properties (Hazrati et al., 2019; Ilyas et al., 2018g; Nazrin et al., 2018). One of the ways to enhance both mechanical and water barrier properties is by reinforcing starch biopolymer with natural fiber. Sahari et al. (2012a) used plasticized sugar palm starch (SPS) as biomatrix combined with raw sugar palm fiber (SPF) (different loading) as the reinforcing material. Sahari et al. (2014a,b) applied SPS plasticized with 30 wt% glycerol as biometric and using fibrous material obtained from SPF as reinforcement. Development of SPS derived green composite reinforced with sugar palm cellulose (SPC) was performed by Sanyang et al. (2016b). Ilyas et al. (2018c,f) prepared biodegradable polymer from thermoplastic SPS and sugar palm nanocrystalline cellulose (SPNCC) to enhance the poor tensile properties of thermoplastic SPS. Recently, Atikah et al. (2019) applied SPS plasticized with 30 wt% of combined glycerol and sorbitol (at 1:1 glycerol to sorbitol ratio) as biometric and using nanofibrous material, which is nanofibrillated cellulose obtained from SPF as reinforcement.

As nowadays, fully biodegradable composites are being developed by more and more scientists, which are called "green" composites or biocomposites that make up of natural biopolymer matrix and natural fibers (Jumaidin et al., 2019). Besides that, rising awareness among the world population to safeguard our environment has promoted research in agricultural waste or residue. Agricultural crop remnants such as SPF, wheat straw, soy hull, arecanut husk fiber, pineapple leaf fiber, ramie fiber, oil palm fiber, curauna fiber, banana fiber,

Advanced Processing, Properties, and Applications of Starch and Other Bio-Based Polymers. https://doi.org/10.1016/B978-0-12-819661-8.00007-X

sugarcane bagasse, kenaf fiber, sugar beet fiber, and mengkuang leaves are being manufactured in billions of tons around the globe. They are abundant in nature, available at low cost, and are also renewable sources of biomass (Abral et al., 2020; Faruk et al., 2012; Ilyas et al., 2018b).

Tropical countries like Malaysia, Philippine, Cambodia, Brunei, and Indonesia are habitats to many natural fibers types, for instance, sugar palm (*Arenga pinnata* [Wurmb] Merr.) fiber. *A. pinnata* (Wurmb) Merr. is a plant, normally survive in wild, found in many places, such as in Malaysia. One such example is in Langkawi. Bukit Nau in Langkawi is taken from the word enau in Malay, which is sugar palm (Ilyas et al., 2018f,g; Sapuan et al., 2018, 2017a). The trees can be found in abundance in the past in that area. In Perak, plants are found in abundance in Bruas and Parit and the people of Perak refer the fruits of this tree as buah kanto, which is normally used in local delicacies such as ABC and cendol. In Sabah, sugar palm are planted into plantation such as found in Balung, Tawau. In Pahang, the plants are found in wild, but people made money but of these trees to make palm sugar, or locally known as gula kabung. In Negeri Sembilan it is known as gula anau. At UPM, researchers have worked with this plant for more than 15 years. One of the reasons for exploring the research in biopolymers and natural fibers from sugar palm is to utilize unwanted sugar palm trunks from being wasted (Fig. 7.1) (Sapuan and Ilyas, 2017; Sapuan et al., 2017b). To date, very few studies have been done on the importance of SPF and its polymer composites. This chapter presents the reviews on recent advances in research on SPFs and sugar palm biopolymers from sugar palm trees and their composites. This chapter also seeks to discover other prospective applications

FIG. 7.1 Sugar palm free is not utilized and is regarded as waste.

of sugar palm biopolymer and its fiber composites to continue developing sugar palm as a new crop in the near future. The main aim of this chapter is to review the mechanical testing of palm sugar-fortified palm oil biopolymer composites.

2 SUGAR PALM FIBERS

2.1 Morphological, Physical, and Chemical Analysis of Macro-, Micro-, and Nano-Sized Sugar Palm Fibers

One of the vital products of sugar palm plant is its fiber, which is also known as ijuk fiber, aren, black, and gomoti. This versatile fibers can be utilized to make a wide range of products such as brushes, ropes, mats, brooms, shelter for aquatic life breeding, and filters. SPF is known for its resistance and resilience to seawater. SPF has high mechanical property compared with others fibers. The characterization on tensile properties of SPF was conducted by Ishak et al. (2011). In this experiment, the fiber was collected from different heights of sugar palm tree (1, 3, 5, 7, 9, 11, 13, and 15 m). Then, this fiber was tested for single fiber tensile test. The results revealed that the fibers collected from upper part (palm frond) revealed higher mechanical properties (tensile strength, modulus, elongation at break, and toughness), compared with fibers collected at the bottom part of sugar palm tree. According to Ishak et al. (2012), the differences in tensile mechanical properties of these different heights were because of the differences in their chemical compositions. They also added that aging process affects the chemical composition of the fiber, specifically the ones at the bottom part of the tree.

In another study conducted by Sahari et al. (2012a,b) on the tensile properties of SPF, comparison of fibers collected from different parts of sugar palm tree, frond, trunk, and bunch fibers, was reported. The results (Table 7.2) displayed that frond SPF has the highest tensile properties (tensile strength, tensile modulus, and elongation at break) recorded followed by bunch fiber, ijuk fiber, and trunk fiber. These findings were strongly influenced by their cellulose contents, in which cellulose provides strength and stability to the cell walls of the fibers (Sahari et al., 2012a).

For micro-sized SPF, the characterization on physical properties of SPF with treated and untreated was performed by Ilyas et al. (2017). The main objective of the study was to identify the effects of the delignification and mercerization treatments to the physical, thermal, chemical, and structural properties of SPF. For delignification process, there were three different treatment durations, 5, 6, and 7 h, and the treated fibers

TABLE 7.1
Physical Properties of Sugar Palm Fibers, Bleached Fibers, Alkali-Treated Fibers, SPNCCs, and Other Materials.

Samples	Diameter (μm)	Density (g/ cm^{-3})	Moisture Content (wt%)	X_c (%)	Surface Area (m^2/g)	Pore Volume (cm^3/g)	DP	M_w (g/ mol)	References
Sugar palm fibers	212.01 ± 2.17	1.50	8.36 ± 0.0984	55.8	7.58	0.0607	–	–	Ilyas et al. (2017)
SPBF05	122.95 ± 0.05	1.36	8.36	64.9	–	–	3417.9	554,215	Ilyas et al. (2017)
SPBF06	104.45 ± 0.02	1.33	6.13	64.8	–	–	3164.5	513,137	Ilyas et al. (2017)
SPBF07	94.49 ± 0.03	1.30	6.27	65.9	10.35	0.0678	2963.3	480,513	Ilyas et al. (2017)
SPC05	11.84 ± 2.48	1.36	6.25	73.4	–	–	1165.7	189,024	Ilyas et al. (2017)
SPC06	10.68 ± 2.27	1.32	4.19	74.2	–	–	1047.9	169,920	Ilyas et al. (2017)
SPC07	8.81 ± 1.65	1.28	4.31	76.0	13.18	0.1950	946.4	153,458	Ilyas et al. (2017)
Sugar palm NCC	9 ± 1.96	1.05 ± 0.0023	17.901 ± 0.0623	85.9	14.47	0.2260	142.86	23,164.7	Ilyas et al. (2018a)
Sugar palm NFC	5.5 ± 0.99	1.1 ± 0.0026	12.855 ± 0.8912	81.2	14.01	0.2109	289.79	46,989	Rushdan Ahmad Ilyas et al. (2019)

Results expressed as mean ± standard deviation.

TABLE 7.2
Mechanical Properties of Fiber From Various Parts of Sugar Palm Tree (Sahari et al., 2012a,b).

Fibers	Sugar Palm Frond	Sugar Palm Bunch	Sugar Palm Trunk	Ijuk
Tensile strength (MPa)	421.4	365.1	198.3	276.6
Tensile modulus (GPa)	10.4	8.6	3.1	5.9
Elongation at break (%)	9.8	12.5	29.7	22.3

were labeled as sugar palm bleached fiber 05 (SPBF05), sugar palm bleached fiber 06 (SPBF06), and sugar palm bleached fiber 07 (SPBF07), respectively. Meanwhile for mercerization process, the SPC obtained from the fiber was treated for 5, 6, and 7 h and denoted as SPC05, SPC06, and SPC07, respectively. From the result obtained, the average diameter of bleached fibers decreased after chemical treatment, which was from 212.01 ± 2.17 to 121.80 ± 10.57 μm. The measurement was conducted after the partial elimination of lignin and hemicelluloses. After mercerization treatment, the size of the SPF was reduced from 121.80 ± 10.57 to 8.81 μm. The result showed that treated SPF, particularly SPBF07 and SPC07, possessed certain advantages, such as a higher crystallinity, higher purity of cellulose, and better thermal stability, compared with the other fibers. The raw SPF had a slightly higher density and moisture content than the treated fibers (Table 7.1). The chemical treatments performed on the fibers also affected the characterization of the separation of micro-sized fibers from the fibers bundle into individual micro-sized fibers.

Fig. 7.2A shows the sugar palm tree. Fig. 7.2B and C shows the view from the outer to the inner part that showed SPFs consisting of a middle lamella (1.98 ± 0.15 μm); a primary (10.38 ± 0.57 μm), a secondary, and a tertiary cell wall, build up around an opening; and the lumen (3.72 ± 0.15 μm). From their work, it can be observed that the bleaching and alkali treatments have affected the structure of the fibers surfaces (Fig. 7.2) and chemical composition (Table 7.4) of the treated fibers. The color of SPFs changed from black (Fig. 7.2D) to light brown after the bleaching treatment (delignification) (Fig. 7.2E) and turned white after alkali treatment (mercerization) (Fig. 7.2F). Besides that, Fig. 7.2G showed that SPFs with diameter size of approximately 212.01 ± 2.17 μm, in the original form, were reinforced together by cement components known as middle lamella, which were partly removed after the bleaching treatment (Fig. 7.2H). Fig. 7.2G shows the longitudinal section of SPF surface topography that is rough, with pore-like spots, which are also known as tyloses, appeared in almost regular intervals, whereby these similar spots were also observed at the surface of coir fibers (Ticoalu et al., 2013). The chemical composition of the raw SPFs comprises of 43.88% cellulose, 7.24% hemicellulose, 33.24% lignin, 2.73% extractive, and 1.01% ash. However, after delignification, mercerization, and beating treatments, the cellulose content increased by 88.79%, whereas the lignin was slightly reduced by about 0.04%. These are due to the elimination of hemicellulose and lignin component in fiber. According to Alemdar and Sain (2008), the modifications of chemical compositions of SPF after all treatments lead to better crystalline degree of cellulose; therefore, it has improved the strength and thermal properties of the fibers.

For nano-sized of the SPF, Ilyas et al. (2018a,b,c,d,e,f,g) had carried out experiments on isolation and characterization of nanocrystalline cellulose from SPFs (*A. pinnata*) (Fig. 7.3). Initially, cellulose was extracted out from SPF (*A. pinnata*) by performing delignification and mercerization treatments (Fig. 7.2). Consequently, SPNCCs were isolated from the extracted cellulose with 60 wt% concentrated sulfuric acid (H_2SO_4). The findings revealed that lignin and hemicellulose were eliminated from the extracted cellulose via delignification and mercerization process, respectively. The crystallinity of the SPF was increased from 55.8% in raw SPFs to 85.9% in SPNCCs. The isolated SPNCCs's length and diameter are 130 ± 30 and 9 ± 1.96 nm, respectively (Table 7.1). Another study was conducted by Ilyas et al. (2018d,g) on the effect of hydrolysis treatment on the size of SPNCC (Fig. 7.4). In this experiment, SPCs were mixed with 60 wt% aqueous sulfuric acid solution at different hydrolysis time of 30, 45, and 60 min (indicated as SPS/SPNCCs-30, SPS/SPNCCs-45, and SPS/SPNCCs-60) having the ratio of the SPC to liquor as 5:100 (wt%). Results showed that the average length of the SPNCCs-30, SPNCCs-45, and SPNCCs-60 was approximately 175 ± 37.01, 130 ± 30.23, and 110 ± 33.69 nm, respectively. The diameter of the SPNCCs-30, SPNCCs-45, and SPNCCs-60 was found to be approximately 13 ± 1.73, 9 ± 1.96, and 7.5 ± 1.35 nm, respectively (Table 7.3). These were attributed by the

FIG. 7.2 Photographs of **(A)** the sugar palm tree, **(D)** raw sugar palm fibers, **(E)** bleached fibers, and **(F)** alkali-treated fibers; FESEM micrographs of sugar plant fibers: **(G)** cross section, **(B)** longitudinal section, **(C)** primary and secondary cell wall, and middle lamella, **(H)** alkali-treated fibers, and **(I)** bleached fibers (Ilyas et al., 2018c,f).

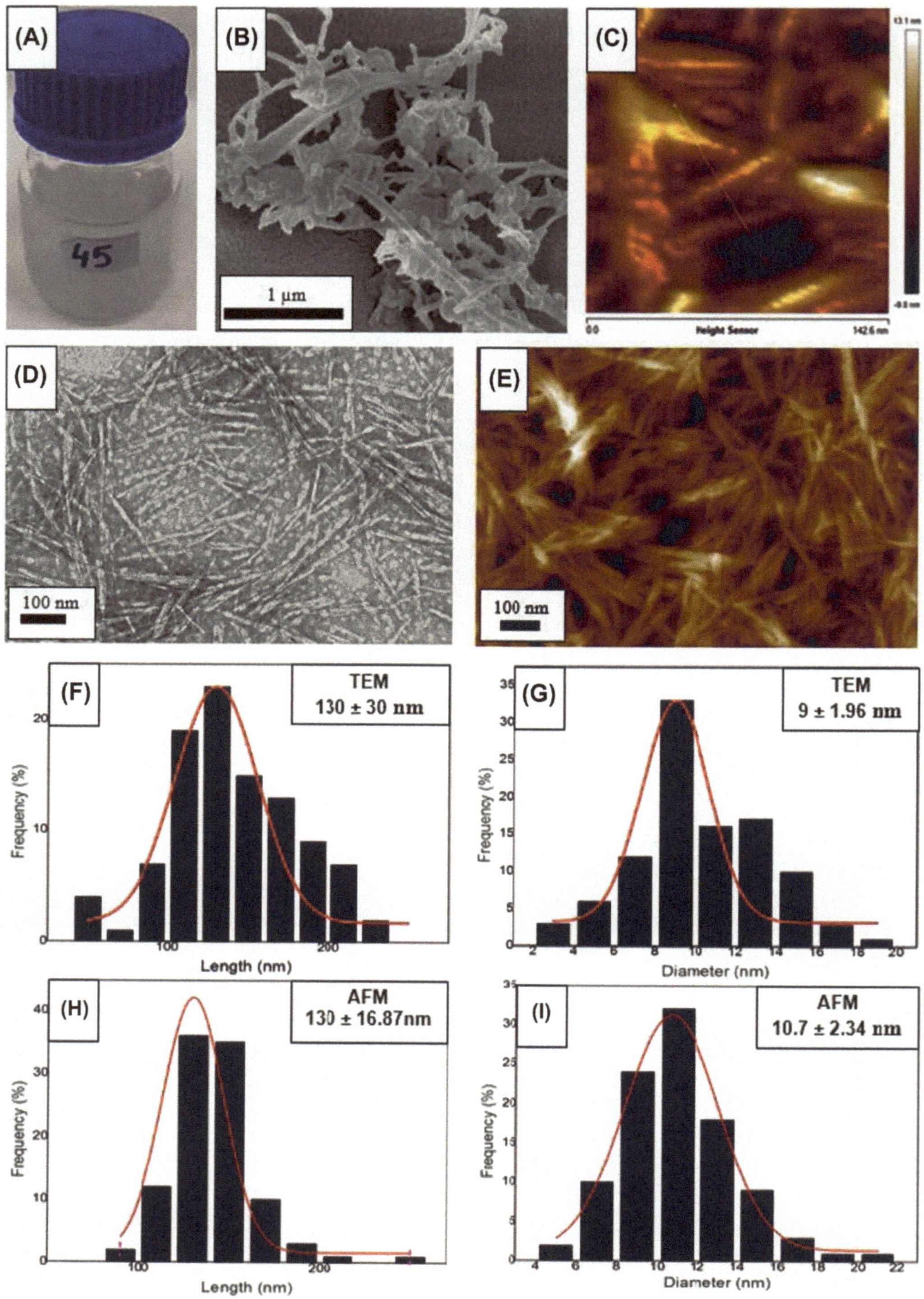

FIG. 7.3 **(A)** Aqueous suspension (2 wt%), **(B)** FESEM micrograph, **(C)** height measurement of NCCs using atomic force microscopy (AFM), **(D)** transmission electron (TEM) micrograph, **(E)** atomic force microscopy (AFM), and their length **(F and H)** and diameter **(G and I)** histograms of nanocrystalline cellulose extracted from sugar palm fibers.

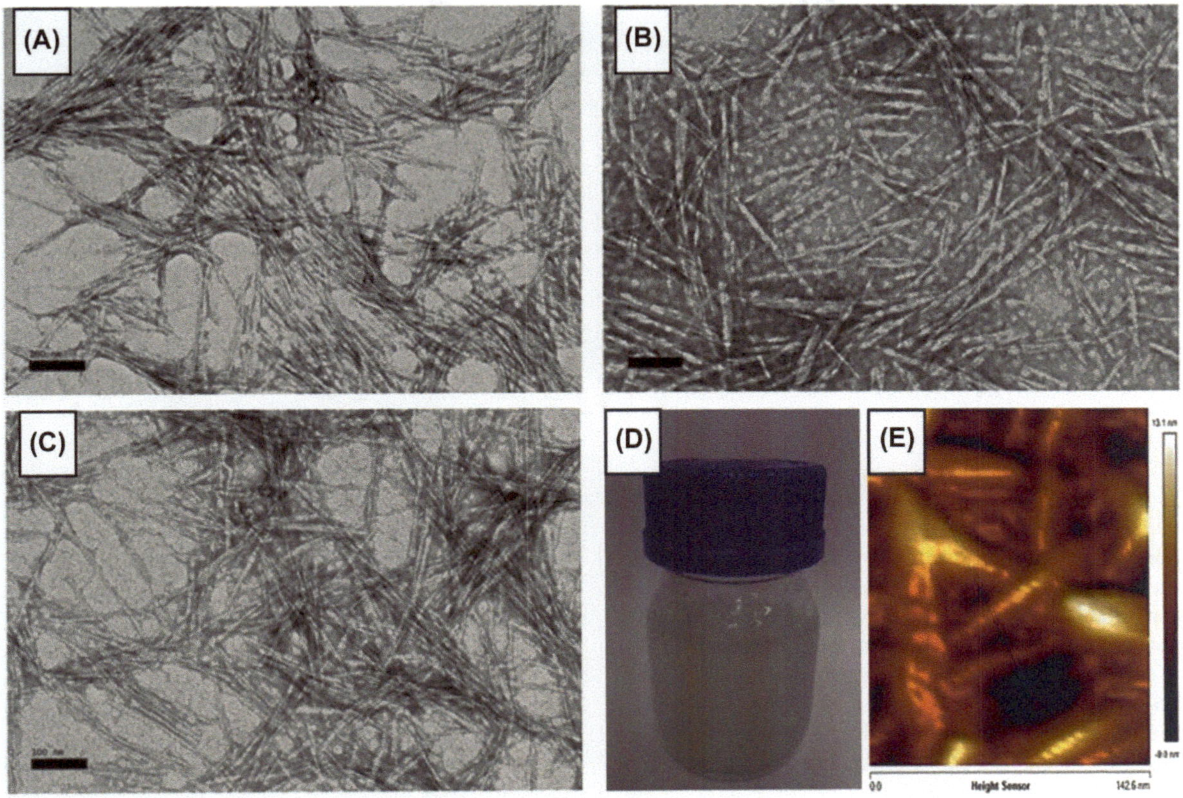

FIG. 7.4 TEM micrographs of the **(A)** SPNCCs-30, **(B)** SPNCCs-45, and **(C)** SPNCCs-60; **(D)** Aqueous suspension (2 wt%) of SPNCCs; and **(E)** AFM image of SPNCCs-45 (Ilyas et al., 2018a,b,c,d,e,f,g).

TABLE 7.3
Dimension Size of SPNFCs-5, SPNFCs-10, SPNFCs-15, SPNCCs-30, SPNCCs-45, and SPNCCs-60 Analyzed by TEM, FESEM, and AFM Microscopies.

Fibers	TEM (Diameter Size Range) (nm)	AFM (Diameter Size Range) (nm)	FESEM (Diameter Range) (nm)	X_c%	References
Sugar palm cellulose (SPC)	–	–	11,870	55.8	R.A. Ilyas et al. (2019a)
SPNFCs-5	21.37 ± 6.91	65 ± 6.19	90 ± 17.98	75.73	R.A. Ilyas et al. (2019a)
SPNFCs-10	11.54 ± 2.77	27.5 ± 7.35	50 ± 25.04	75.38	R.A. Ilyas et al. (2019a)
SPNFCs-15	5.5 ± 0.99	9 ± 1.89	35 ± 8.61	81.19	R.A. Ilyas et al. (2019a)
SPNCCs-30	13 ± 1.73	19.0 ± 3.49	37.5 ± 5.55	80.97	R.A. Ilyas et al. (2018d)
SPNCCs-45	9 ± 1.96	10.7 ± 2.34	35.0 ± 8.68	85.90	R.A. Ilyas et al. (2018d)
SPNCCs-60	7.5 ± 1.35	8.5 ± 1.16	25.0 ± 0.80	83.50	R.A. Ilyas et al. (2018d)

TABLE 7.4
Chemical Composition of Sugar Palm Fibers at Different Stages of Treatment.

Samples	Cellulose (%)	Hemicellulose (%)	Holocellulose (%)	Lignin (%)	Extractive (%)	Ash (%)	References
Sugar palm fibers	43.88	7.24	51.12	33.24	2.73	1.01	R.A. Ilyas et al. (2017)
Bleached fibers	56.67	19.8	76.47	0.27	0.23	2.16	R.A. Ilyas et al. (2017)
Alkali-treated fibers	82.33	3.97	86.3	0.06	–	0.72	R.A. Ilyas et al. (2017)
Sugar palm refined fibers	88.79	0.04	3.85	0.04	–	0.74	Rushdan Ahmad Ilyas et al. (2019)

longer duration of hydrolysis treatment on the fibers, which eliminated the amorphous region from nanofibres. Moreover, extended hydrolysis period could irritate the structure of SPNCCs. Ilyas et al. (2018d,g) also conclude that the longer the reaction times of hydrolysis, the shorter the length and diameter of the nanofibres.

Recently, Ilyas et al. (2019a,b,c) performed a study on the isolation of sugar palm nanofibrillated cellulose (SPNFCs) from SPFs produced by a chemomechanical approach (Fig. 7.5). Nanofibrillated cellulose with diameter of 21.37 ± 6.91, 11.54 ± 2.77, and 5.5 ± 0.99 nm were obtained through high pressurized homogenization process from three different passes, named SPNFCs-5, SPNFCs-10, and SPNFCs-15 (Table 7.3). Initially, beating pretreatment was conducted on the SPC to prevent the high pressurized homogenizer from clogging problem due to its very small orifice size. The results showed that the highest yield was attained by SPNFCs-15 yield of 92.52%, followed by SPNFCs-10 and SPNFCs-5 with 69.91% and 78.12%, respectively. The SPNFCs exhibited higher crystallinity (81.19%) and thermal stability compared with the raw and treated fibers (Ilyas et al., 2019a,b,c).

3 SUGAR PALM STARCH

Commercially used industrial starches are commonly coming from tubers (potatoes, sweet potatoes, etc.), cereals (rice, wheat, etc.), roots (cassava, yam, etc.), and legumes (green pea, bean, etc.). However, these precious materials are a main staple source of food for several poor countries, in which the consumption of such biomaterial matrix in composite has come to severe accusation and criticism. Therefore, in order to overcome these problems, numerous researches have been conducted to shift the uses of food sources to a nonfood source as polymeric matrix to develop biopolymer. One of them is SPS, which it located in the core of the plant stem. It was reported that the average yield of each sugar palm tree is 50–100 kg (Sapuan et al., 2018). Sugar palm are mostly cultivated to gain fibers and sugar-rich sap from its flower. However, Sanyang et al. (2016) reported that not all the sugar palm produced sap and that the nonproductive plants sometime yield up to half the trees in the cultivated area. The extraction process of SPS is displayed in Fig. 7.6. Starch is usually reaped from these unproductive plants trunk, which is then the next step following the step of producing SPS (Sanyang et al., 2016). After that, the collected white powdered starch was dried in an air circulating oven at a temperature of 120°C for 24 h (Sahari et al., 2013c).

Sahari et al. (2013c) studied the properties of SPS to discover their potential as a novel alternative polymer. It was reported that SPS has great potential to be polymer substitute due to its superior amylose (37.60%), compared with the other starches such as maize (26%–28%), potato (20%–25%), tapioca (17%), wheat (26%–27%), and sago (24%–27%) (Table 7.5). Amylopectin is defined as branched polysaccharide component in starch that makes up hundreds of short chains formed of α-D-glucopyranosyl residue with (1 → 4) linkages. These are intertwined by (1 → 6)-α-linkages and 5%–6% that occurred at the branch points. SPS has high amylopectin with high molecular weight (10^7–10^9); however, it has low intrinsic viscosity (120–190 mL/g) due to it highly branched molecular structure. Besides that, SPS has low fat and protein content of 0.27% and 0.10% (w/w), respectively. In

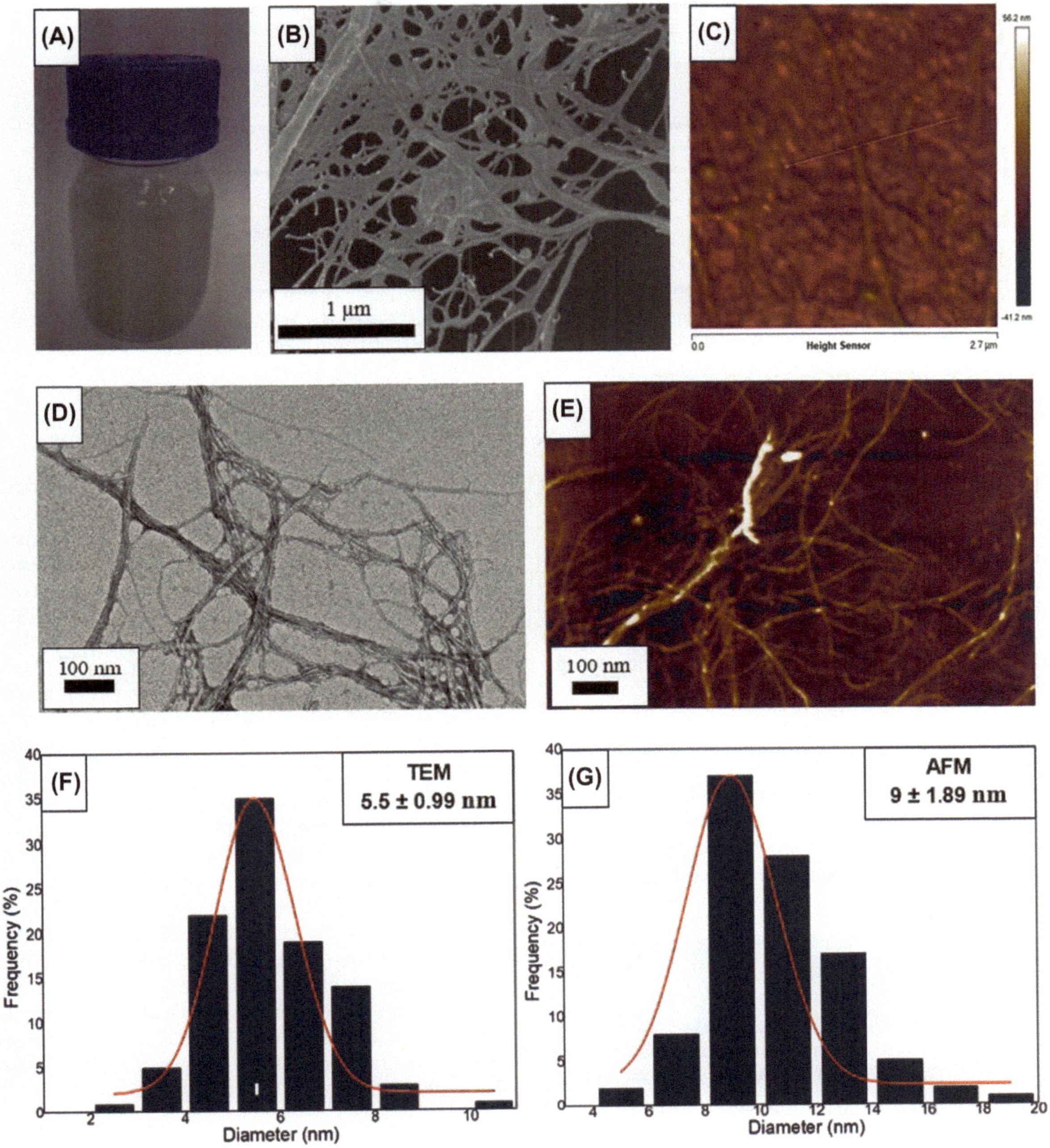

FIG. 7.5 **(A)** 2 wt% of nanocrystalline cellulose suspension, **(B)** field emission scanning electron microscopy (FESEM) micrograph, **(C)** height dimension of NFCs by means of atomic force microscopy (AFM) nanograph, **(D)** transmission electron microscopy (TEM) nanograph, **(E)** atomic force microscopy (AFM) nanograph, and **(F and G)** their diameter histograms based on TEM and AFM nanograph.

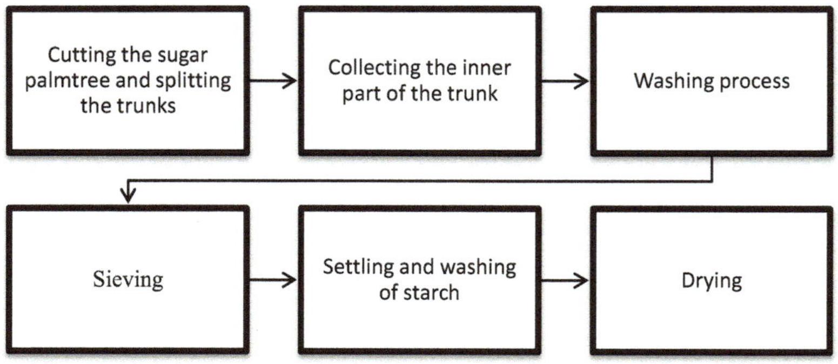

FIG. 7.6 Extraction of sugar palm starch. Modified from Sahari, J., Sapuan, S.M., Zainudin, E.S., Maleque, M.A., (2014a). Physico-chemical and thermal properties of starch derived from sugar palm tree (*Arenga pinnata*). Asian Journal of Chemistry 26 (4), 955–959. https://doi.org/10.14233/ajchem.2014.15652 and Sahari, J., Salit, M.S., Zainudin, E.S., Maleque, M. A, (2014b). Degradation characteristics of SPF/SPS biocomposites fabrication of SPF/SPS biocomposites. Fibres and Textiles in Eastern Europe 22 (5107), 96–98.

terms of density, SPS (1.54 g/cm^3) has high density compared with other biopolymer as displayed in Table 7.5. SPS also has high moisture content of 15% under normal atmospheric condition. This is due to hydroxyl functional groups that are present, as showed by the strong peak at 3200–3500 cm^{-1} from FTIR graph analysis. Besides, SPS has lower ash content than potato (0.4%), however similar ash content with wheat, tapioca, and sago (0.2%).

4 SUGAR PALM FIBER-SUGAR PALM STARCH BIOPOLYMER COMPOSITES

It is a well-known fact that the development of biopolymers producing environmentally friendly materials as synthetic polymers substitute for broad applications.

TABLE 7.5
The Chemical Composition of Commercial Starches and SPS (Sanyang et al., 2016).

Starch	Density	Ash (%)	Amylose (%)	Water Content (%)
Wheat	1.44	0.2	26–27	13
Tapioca	1.446 –1.461	0.2	17	13
Maize	1.5	0.1	26–28	12–13
Potato	1.54 –1.55	0.4	20–25	18–19
Sago	–	0.2	24–27	10–20
SPS	1.54	0.2	37.60	15

In past decades, SPF had been reinforced with various polymers such as epoxy, unsaturated polyester, polyester resin, phenol formaldehyde, high impact polystyrene, polyurethane, phenolic, SPS, vinyl ester, polyurethane, sago starch, and cassava starch (Table 7.6). To date, different researchers have studied various types of SPS-based biopolymers, such as Ilyas et al. (2018a,b,c,d,e,f,g), Sahari et al. (2013a, 2012a,b), Sanyang et al. (2016a, 2016b, 2017), Jumaidin et al. (2016, 2019, 2017), Atikah et al. (2019), Adawiyah et al. (2013), Jatmiko et al. (2017), Jatmiko et al. (2016), Poeloengasih et al. (2016), Apriyana et al. (2016), and Mansor et al. (2015). Another potential product from sugar palm plant is biopolymer; since starch is obtained from its trunk, it can be used in the manufacturing of biodegradable polymer. The presence of the starch is confirmed by many researchers (Ayana et al., 2014; Baratter et al., 2017; Belibi et al., 2014; Bootklad and Kaewtatip, 2013; Cao et al., 2007; Edhirej et al., 2017b; González et al., 2015; Halimatul et al., 2019b; Ilyas et al., 2018a,b,c,d,e,f,g; Ilyas et al., 2018d; Liao and Wu, 2008; Lu et al., 2006; Montero et al., 2017; Tang et al., 2008; Teixeira et al., 2012; Thakur et al., 2017).

5 MACROSIZE SUGAR PALM FIBER-SUGAR PALM STARCH BIOPOLYMER COMPOSITES

Sahari et al. (2013b) studied the influence of fiber content on the mechanical properties, water absorption behavior, and thermal properties of SPF reinforced plasticized SPS biocomposites. The composites were

TABLE 7.6
Sugar Palm Fiber Reinforced Various Polymer Composites.

	Fiber	Matrix	Authors
Macrosize	Sugar palm fiber	Epoxy	Leman et al. (2008)
		Unsaturated polyester	Sahari et al. (2011)
			Misri et al. (2010)
			Ishak et al. (2013a,b)
			Norizan et al. (2019)
		Polyester resin	Ticoalu et al. (2010)
			Oumer and Bachtiar (2014)
		Phenol formaldehyde	Ishak et al. (2013a,b)
		High impact polystyrene	Bachtiar et al. (2011)
		Polyurethane	Mohammed et al. (2018)
		Phenolic	Agrebi et al. (2018)
		Sugar palm starch	Sahari et al. (2013c)
		Vinyl ester	Razali et al. (2018)
			Huzaifah et al. (2018)
			Ammar et al. (2018)
			Nadlene et al. (Razali et al., 2018)
		Polyurethane	Atiqah et al. (2018, 2019)
			Radzi et al. (2017, 2019)
		Sago starch	Halimatul et al. (2019b, 2019a)
		Cassava starch	Edhirej et al. (2017a)
Microsize	Sugar palm cellulose	Sugar palm starch	Sanyang et al. (2016b)
Nanosize	Sugar palm nanocrystalline cellulose	Sugar palm starch	Ilyas et al. (2018c,f)
	Sugar palm nanofibrillated cellulose	Sugar palm starch	Ilyas et al. (2019a,b,c)
			Hazrol et al. (2019)

fabricated with various loading (i.e., 10%, 20%, and 30% by weight percent and coded as SPS, SPF10, SPF20, and SPF30, respectively) by using glycerol as the starch plasticizer. SPS of 70 wt% and glycerol of 30 wt% were combined using mechanical stirrer and later hot pressed at 130°C for 30 min under the load of 10 tonne. Fig. 7.7 displays image of scanning electron microscopy micrographs of SPF reinforced SPS (SPF/SPS) biocomposites containing different loading of SPF. From the figure, the relative homogeneous nature of these biocomposites can be observed. Smooth fracture surfaces with homogeneous SPS biopolymer matrix and good adhesion of the SPF to the thermoplastic SPS matrix play an important part in enhancing the mechanical performance of the biocomposites.

Besides that, from the experiment conducted by Sahari et al. (2013b), it was found that the mechanical properties of SPS biopolymer matrix improved with the reinforcement of SPF. The tensile strength and modulus of SPF/SPS biocomposites showed an increasing trend with increasing SPF loading. The addition of the SPF resulted in reduction in elongation from 8.03% to 3.32% (Fig. 7.8). According to Sahari et al. (2013b), this phenomenon might be attributed to the outstanding intrinsic adhesion of the fiber-matrix interface, produced by the chemical similarity of starch and the cellulose fiber. However, elongation data showed decreasing pattern when fiber loading was increased, due to a normal effect of the increasing fibers wt%, having low strain compared with rubbery SPS materials. Besides that, the flexural strength and modulus of SPF/SPS biocomposites revealed increasing pattern with increasing fiber loading. Parallel trend also was displayed for impact strength of biocomposites, in which, the impact strength increased with fiber loading. As conclusion to their work, mechanical properties of plasticized SPS are highly affected by the incorporation of SPF.

Sahari et al. (2012) investigated the effect of water absorption on the mechanical properties of SPF reinforced SPS (SPF/SPS) biocomposites. The experiment was conducted due to concern of moisture absorption in biocomposites for further improvement potential application of biocomposite. Therefore, the effect of water absorption on the mechanical properties of palm fiber

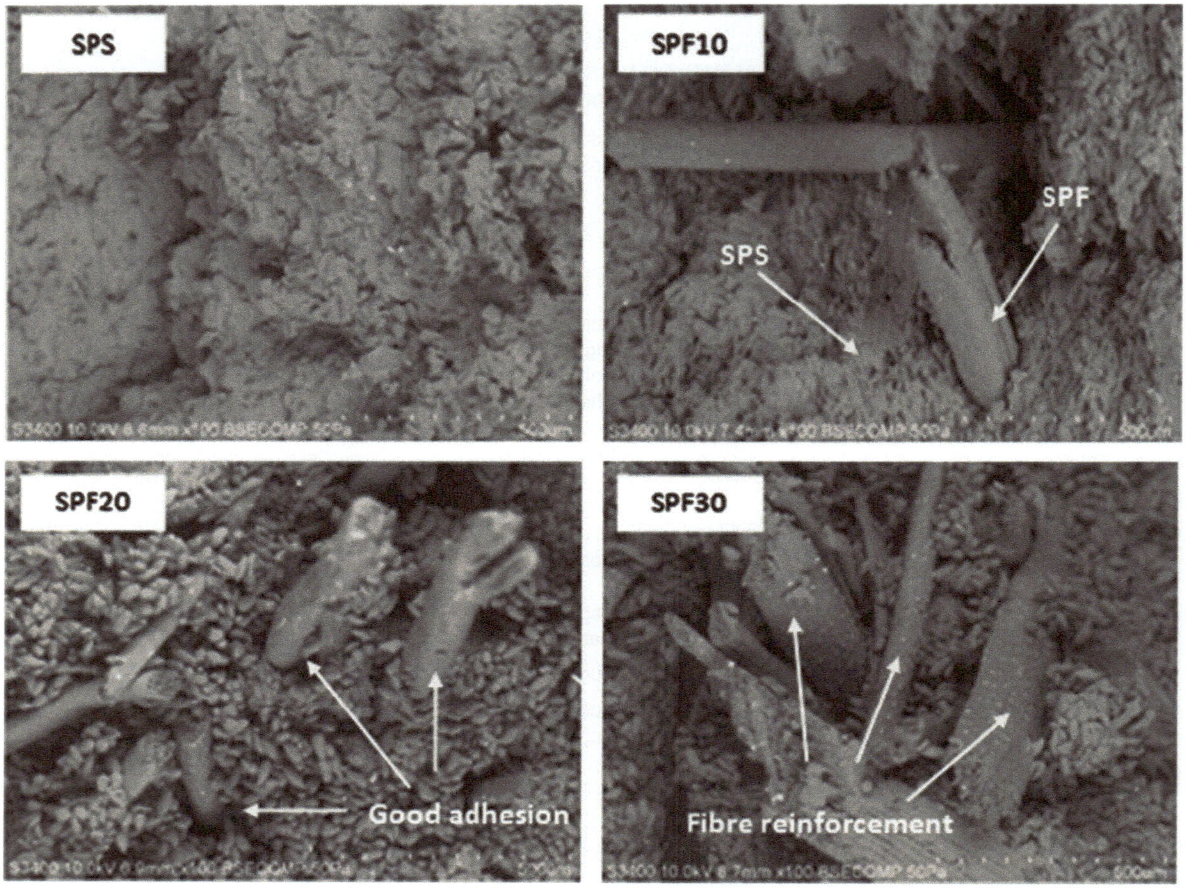

FIG. 7.7 Scanning Electron Microscopy of sugar palm fiber reinforced sugar palm starch (SPF/SPS) biocomposites (Sahari et al., 2013b).

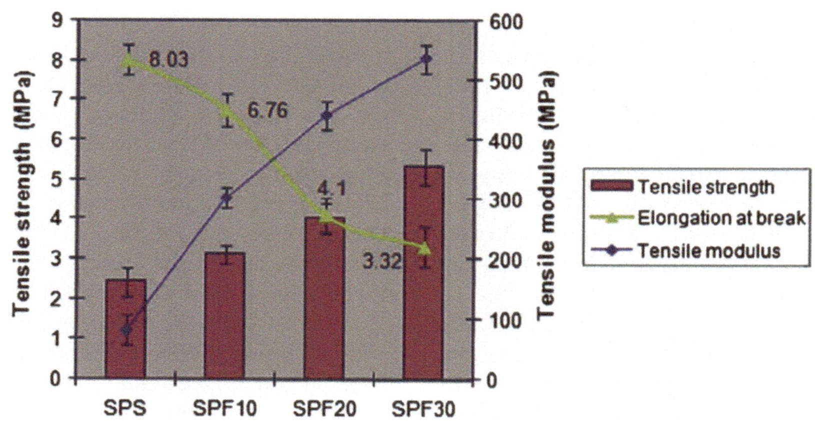

FIG. 7.8 Effect of fiber loading on tensile properties of SPF/SPS biocomposites (Sahari et al., 2013b).

reinforced sugar palm starch (SPF/SPS) biocomposite has been studied. For the fabrication of SPF/SPS biocomposites, SPS and glycerol with ratio of 70:30 wt% were mixed together using mechanical stirrer for 30 min and later was cured by hot pressing in a Carver hydraulic hot press at 130°C for 30 min under the load of 10 tonnes. The loading of SPFs was varied to 0, 10, 20, and 30 wt% coded as SPS, SPF10, SPF20, and SPF30, respectively. The samples were stored in humidity chamber for 72 h at RH = 75%. The result of SEM micrograph shows that there was good compatibility between SPF and SPS biopolymer. However, there are significant differences that can be observed through the SEM micrograph between both sample of (A) before being exposed to 75% RH and (B) after being exposed to 75% RH. SEM micrograph reveals the presence of SPF as reinforcement in biocomposite without any crack. However, after the biocomposite being stored at RH = 75% for 72 h, it can be observed that large cracks appear on the surface of the matrix (SPS). This phenomenon was due to water absorption by starch biopolymer, which resulted in its failure, hence decreasing the mechanical properties of biocomposites.

In terms of mechanical properties of tensile strength, as the SPFs loading increases from 0 to 30 wt%, the tensile strength of biocomposite also increases from 2.4 to 5.31 MPa, respectively. The improvement of tensile strength was due to the effectiveness of SPF in acting as good reinforcement with SPS biopolymer. However, after the biocomposite was exposed to 75% RH for 72 h, the tensile strength was dropped drastically from 0.42 to 1.73 MPa, respectively. Same phenomena were also witnessed for the impact strength of SPF/SPS biocomposite, where the impact strength increases with the increase of fiber loading and decreases after being exposed to 75% RH for 72 h. This was due to the presence of high hydroxyl group presented in starch biopolymer, which tends to show low moisture resistance and leads to the degradation of fiber-matrix interface region. Subsequently, this may result in reduction of dimensional variation, poor interfacial bonding of SPS, and poor mechanical properties of SPF/SPS biocomposites.

Another studies conducted by Sahari et al. (2014a,b) on degradation characteristics of SPF/SPS biocomposites, by incorporating SPS into SPS with glycerol (ratio of 70:30) using hot press at 130°C for 30 min at a load of 10 tonnes. Later samples were treated in weathering chamber to determine the environmental effect on the biocomposites. Fig. 7.9 shows the micrograph of the degradation of the SPS and SPF/SPS biocomposite after being buried in compost soil for 72 h. It can be observed through the microscope that most of the starch had degrade in both samples; however, starch in SPF/SPS had degraded less compared with neat SPS (Fig. 7.9).

The tensile test was then performed after the weathering test. The results show that the tensile strength of the SPS and SPF/SPS biocomposites after being exposed to the environmental effect were reduced by 78.09% (from 2.42 to 0.53 MPa) and 53.67% (from 5.31 to 2.46 MPa), respectively. This phenomenon might be due to the degradation of both SPF and SPS as the presence of oxygen and UV causes oxidative degradation of the polymer and change the morphology of the polymeric material by means of chemical cross-linking or chain scission, which resulted in low mechanical properties of SPS and biocomposites (Sahari et al., 2014a,b) Fig. 7.10.

6 MICROSIZE SUGAR PALM FIBER-SUGAR PALM STARCH BIOPOLYMER COMPOSITES

Surface treatment on natural fiber is compulsory in order to improve interfacial properties of natural fiber and polymer composite. Sanyang et al. (2016b) conducted investigation on effect of sugar palm-derived cellulose reinforcement on the mechanical and water barrier properties of SPS biocomposite films. SPS reinforced SPC composite films (SPS-C) were fabricated using different SPC contents (1–10 wt%) using a solution casting method. Fig. 7.11 displayed the surface morphological of SPS-C biocomposite. It can be seen that the neat SPS films displayed a continuous and smooth surface with no trace of starch crack or granular. Meanwhile, SPS-C10 presented rougher and random distribution of SPC surface with some of the SPC fiber overlapping each other.

Mechanical properties of SPS reinforced SPC composite film such as tensile strength, elongation, and tensile modulus are displayed in Fig. 7.12. The mechanical properties of the composite films showed increasing tensile strength and modulus, while the elongation at break decreased with the increasing of SPC loading. Incorporation of 1–10 wt% SPC into SPS biopolymer decreased the elongation at break for the biocomposite films (from 40.99% to 32.8%) and increased the tensile modulus and tensile strength values of the biocomposite films (from 31.38 to 92.33 MPa and from

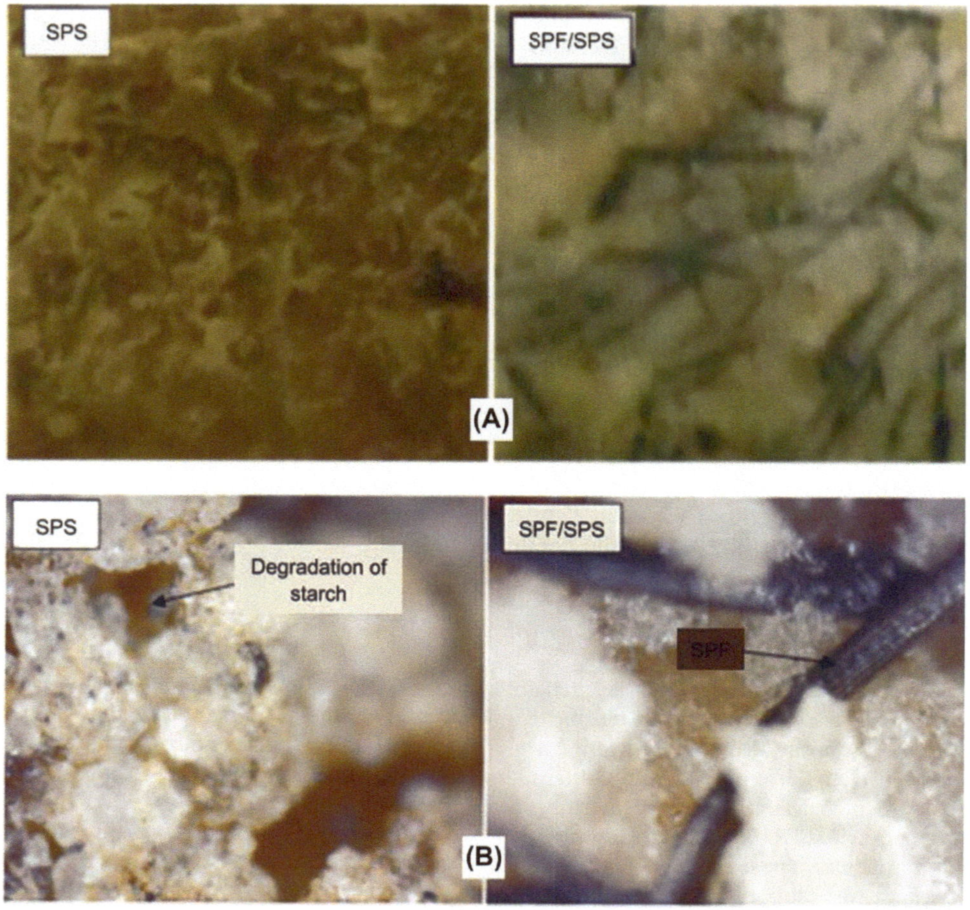

FIG. 7.9 Surface morphology of SPS and SPF/SPS biocomposites: **(A)** before and **(B)** after being buried (Sahari et al., 2014a,b).

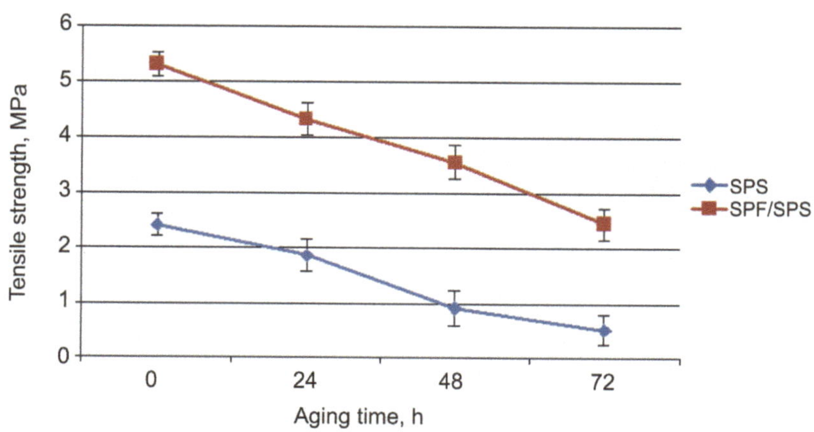

FIG. 7.10 Tensile strength of SPS and SPF/SPS biocomposites (Sahari et al., 2014a,b).

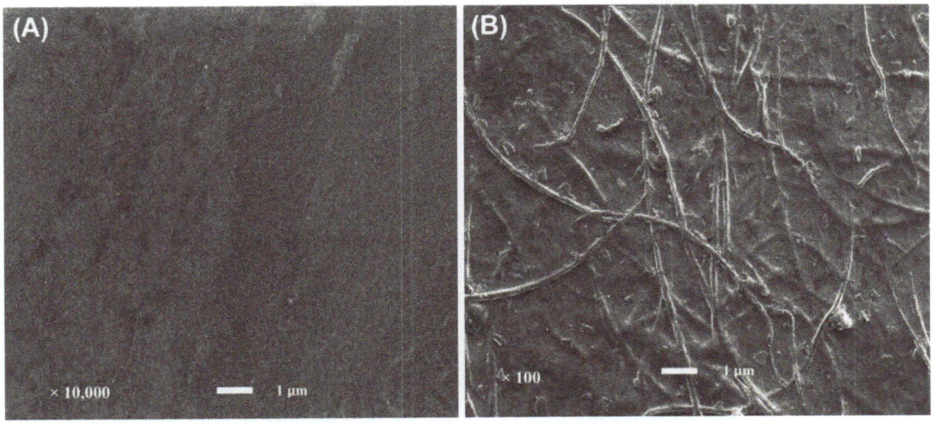

FIG. 7.11 FESEM micrographs of **(A)** neat SPS film and **(B)** SPS-C10 film (Sanyang et al., 2016b).

10.5 to 19.68 MPa, respectively). This might be attributed to the high compatibility between the SPS biopolymer and SPC (Sanyang et al., 2016b).

7 NANOSIZE SUGAR PALM FIBER-SUGAR PALM STARCH BIOPOLYMER COMPOSITES

Ilyas et al. (2019a,b,c) conducted studies on effect of SPNFC concentrations on morphological, mechanical, and physical properties of biodegradable films based on agrowaste sugar palm (*A. pinnata* [Wurmb] Merr.) starch.

Bionanocomposites were prepared by mixing SPS and sorbitol/glycerol with different SPNFC loading (0−1.0 wt%) using solution casting method. The FESEM micrograph of the SPS/SPNFCs nanocomposites displayed a continuous and smooth surface with no trace of starch cracks and agglomerations of SPNFCs (Fig. 7.13A). Moreover, Fig. 7.13C shows high dispersion of SPNFCs within SPS biopolymer indicating a good indication of strong interfacial adhesion between the two components of the SPS/SPNFCs nanocomposite film.

In term of mechanical properties of SPNFCs reinforced SPS (Tables 7.7 and 7.8), the results showed that as the SPNFC concentration increased from 0.1 to 1.0 wt%, and the tensile strength and tensile modulus of SPS/SPNFCs nanocomposite films were increased from 6.80 to 10.68 MPa and 59.07 to 121.26 MPa, respectively. These might be due to the compatible interaction between the SPNFCs and SPS polymer matrices, which facilitated sufficient interfacial adhesion because of their chemical similarities (Ilyas et al., 2018a,b,c,d,f,g). However, the elongation at break for the nanocomposite films decreased from 38.1% to 25.38%, as the SPNFCs concentration was increased from 0 to 1.0 wt%. This was due to the addition of SPNFCs concentration, which restricts the molecular mobility and ductility of the SPS matrix, making the composite materials stiffer. Ilyas et al. (2019a,b,c) concluded that there were three main factors that had possibly affected the mechanical performances of the nanocomposite material: (1) the processing method, (2) the dimension and morphology of the nanofiller, and (3) the micro/nanostructure of the matrix and matrix/filler interface.

The FESEM result shows that the neat SPS films displayed an even surface with no trace of starch cracks or granules (Fig. 7.14), whereas the addition of 0.1 wt.% SPNCCs to a neat SPS film (SPS/SPNCCs-0.1) showed even, smooth, random, and well-dispersed SPNCCs inside the matrix of SPS, deprived of deformation cracks and pores with no visible clusters or agglomerations of SPNCCs. High distribution and dispersion of SPNCC indicated a strong interfacial adhesion between the SPNCCs and SPS. This strong nanoadhesion could be attributed into high tensile strength.

The SPS/SPNCCs with 1.0 wt% had undergone an increment in both the tensile strength and Young's modulus when compared with the neat SPS film, from 4.80 to 11.47 MPa and 53.97 to 178.83 MPa, respectively. Meanwhile, the elongation at break (%) indicates a decrease in value (from 38.1% to 24.01%) because of the reduction in the molecular mobility of the SPS biomatrix, making film become stiffer.

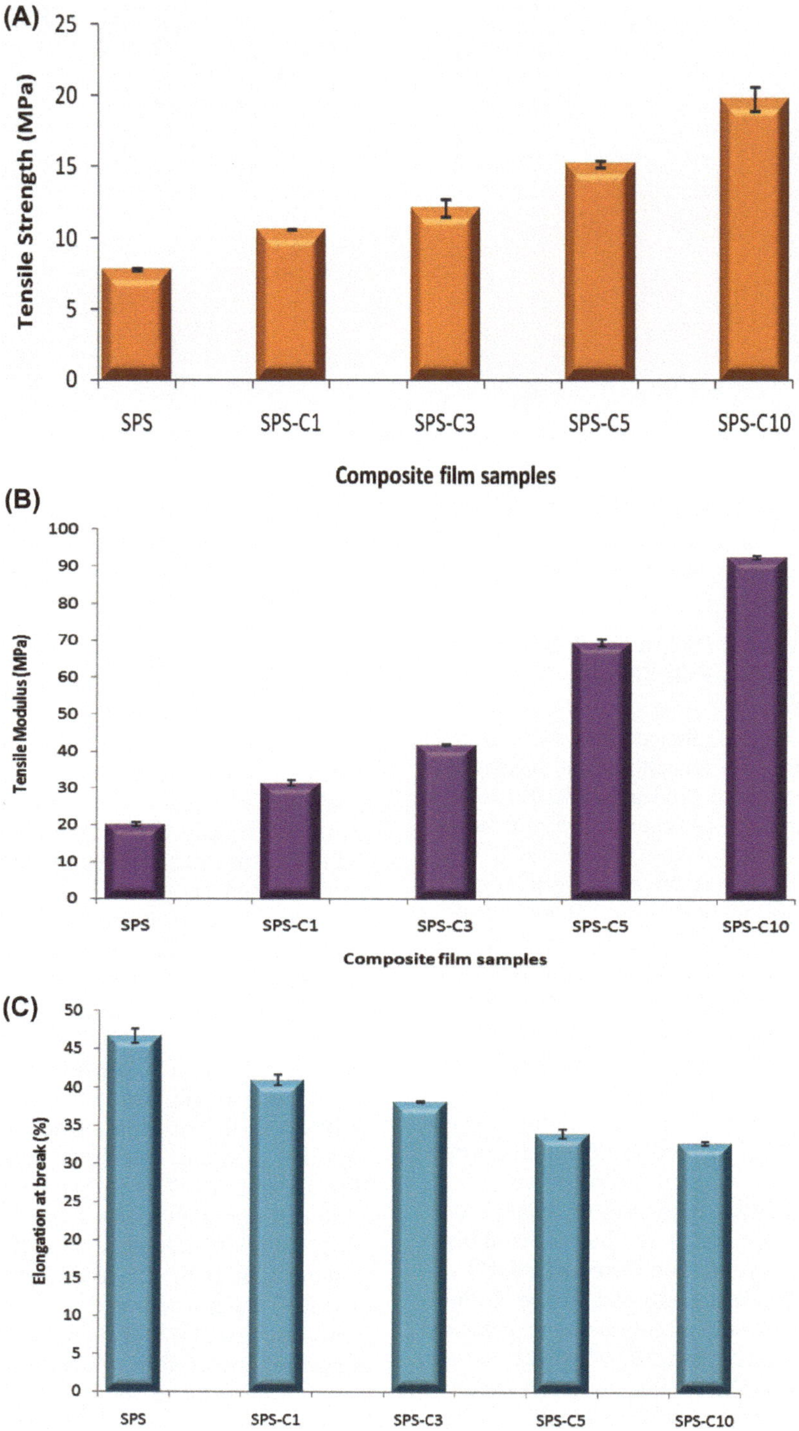

FIG. 7.12 Effect of SPC loading on the **(A)** tensile strength, **(B)** tensile modulus, and **(C)** elongation at break of SPS-C composite films compared with neat SPS films (Sanyang et al., 2016b).

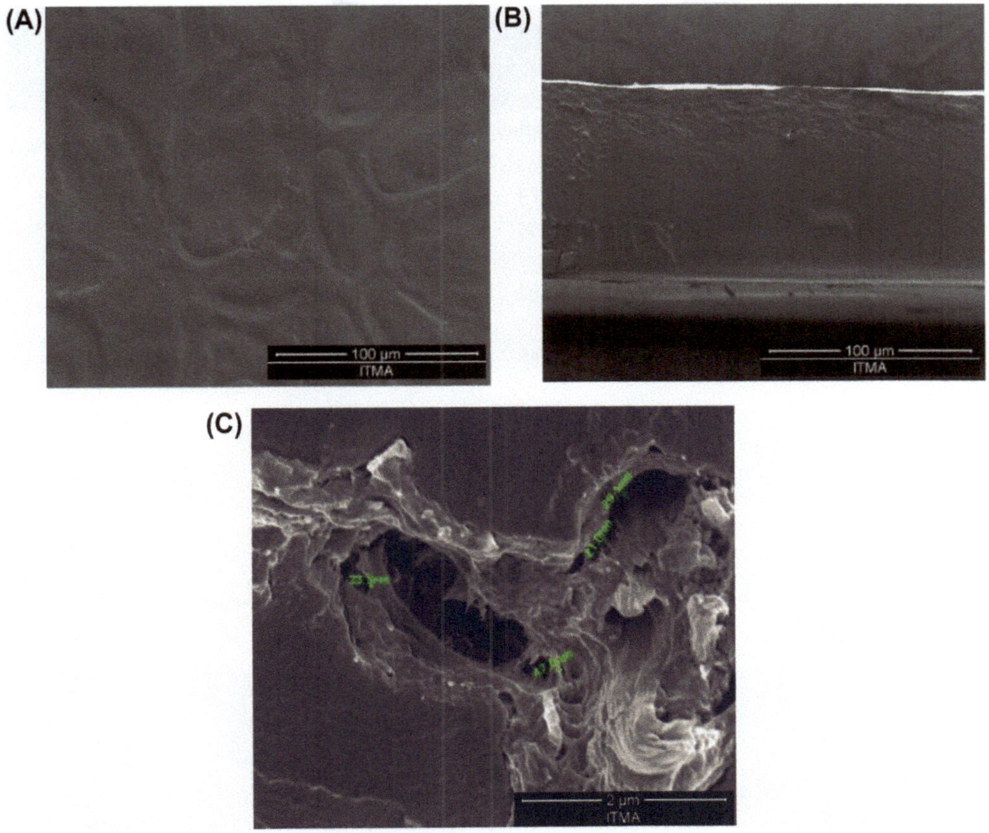

FIG. 7.13 Surface morphology of SPF/SPNFCs bionanocomposite: **(A)** surface, **(B)** cross-sectional (100 μm), and **(C)** cross-sectional (2 μm) (Ilyas et al., 2018a,b,c,d,f,g).

TABLE 7.7
Tensile Properties of SPS and SPS/SPNFCs Nanocomposite Films in Different Concentrations.

Samples	Tensile Strength (MPa)	Tensile Modulus (MPa)	Elongation at Break (%)
SPS	4.80 ± 0.41^{a}	53.97 ± 8.74^{a}	38.10 ± 1.16^{f}
SPS/SPNFCs-0.1	6.80 ± 0.25^{b}	$59.07 \pm 2.10^{a,b}$	37.70 ± 0.66^{f}
SPS/SPNFCs-0.2	$7.30 \pm 0.11^{b,c}$	63.63 ± 2.09^{b}	36.21 ± 0.75^{e}
SPS/SPNFCs-0.3	$7.55 \pm 0.07^{c,d}$	72.46 ± 3.51^{c}	34.24 ± 0.55^{d}
SPS/SPNFCs-0.4	$8.10 \pm 0.23^{d,e}$	84.46 ± 4.21^{d}	30.74 ± 0.67^{c}
SPS/SPNFCs-0.5	$8.53 \pm 0.13^{e,f}$	99.52 ± 7.20^{e}	27.54 ± 0.76^{b}
SPS/SPNFCs-1.0	10.68 ± 0.67^{f}	121.26 ± 5.69^{f}	25.38 ± 0.50^{a}

Values with different letters in the same column are significantly different ($P < .05$).

Other related studies on mechanical properties of nano-sized SPF reinforced SPS can be found in Ilyas et al. (2018c,f). Ilyas et al. (2018c,f) also studied the potential of hydrolysis treatment for surface modification of SPF used to reinforce SPS composites. The experiment was conducted to investigate the effect of different concentrations (0–1.0 wt%) of SPNCC reinforced SPS on morphological, mechanical, and physical properties of the bionanocomposites film.

TABLE 7.8
Effect of SPNCCs Loading on the Tensile Strength, Tensile Modulus, and Elongation at Break (%) of SPS-SPNCCs Nanocomposite Films Compared With Neat SPS Films (Ilyas et al., 2018c).

Samples	Tensile Strength (MPa)	Tensile Modulus (GPa)	Elongation at Break (%)	WVP(10^{-10} × g/s^{-1}.m^{-1}.Pa^{-1})
SPS	4.80 ± 0.41	53.97 ± 8.74	38.10 ± 1.16	9.58 ± 0.01
SPS/SPNCCs-0.1	6.60 ± 0.47	98.10 ± 9.54	35.52 ± 0.43	8.91096 ± 0.02
SPS/SPNCCs-0.2	7.19 ± 0.25	107.98 ± 10.52	32.06 ± 0.93	8.66335 ± 0.03
SPS/SPNCCs-0.3	8.15 ± 0.43	122.93 ± 10.45	28.30 ± 1.24	8.46507 ± 0.02
SPS/SPNCCs-0.4	8.60 ± 0.48	133.94 ± 10.82	26.08 ± 0.32	8.26768 ± 0.03
SPS/SPNCCs-0.5	11.47 ± 0.25	178.83 ± 4.48	24.42 ± 0.96	8.17125 ± 0.01
SPS/SPNCCs-1.0	7.78 ± 0.46	117.19 ± 8.45	24.02 ± 1.15	7.81457 ± 0.02

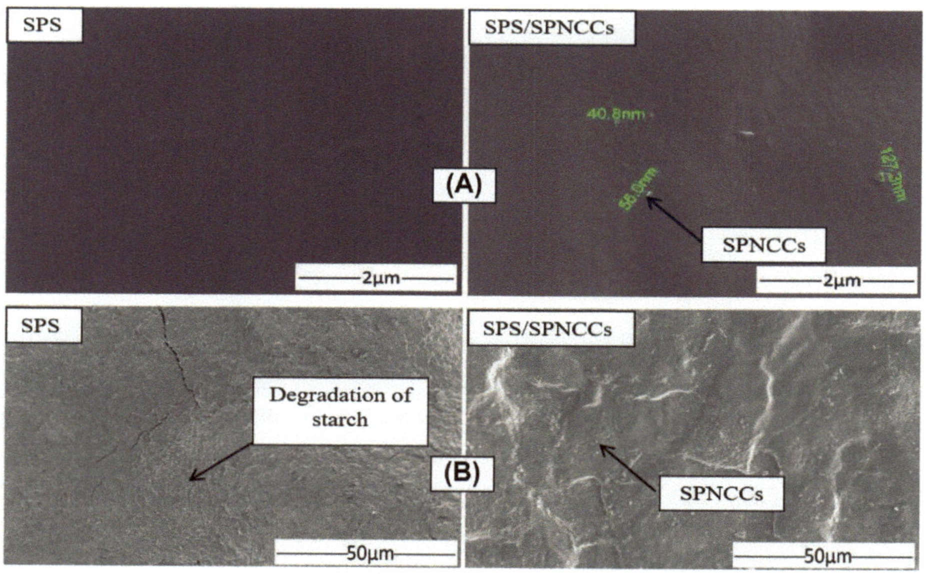

FIG. 7.14 Surface morphology of SPS and SPF/SPNCCs bionanocomposites: **(A)** before and **(B)** after being buried.

8 CONCLUSIONS

SPF has outstanding potential to be used as reinforcement in SPS biopolymer composites. Previously, this fiber had been modified (delignification, mercerization, hydrolysis, defilation, and high pressurize homogenization processes, etc.) to improve its interfacial properties in polymer composite. These modifications not only effected the properties of the sugar palm but also reduced the size of sugar palm itself from macrosize to nanosize. Reducing the size of the SPF was known to be a good reinforcing agent because of its high surface area, high crystallinity, high aspect ratio, abundant hydroxyl groups, large specific surface area, low weight, high strength, biodegradability, and stiffness as well as great thermal resistance and good mechanical properties. To date, numerous successful SPF reinforced SPS biocomposite products are being developed. Since sugar palm is not well known and very little information is available about it, more research needs to be conducted to show its importance and to promote its use for the benefit of the public.

ACKNOWLEDGMENTS

The authors would like to thank Universiti Putra Malaysia and Ministry of Education, Malaysia, for the financial support through the Graduate Research Fellowship (GRF) scholarship, Universiti Putra Malaysia Grant scheme Hi-CoE (6369107) and Fundamental Research Grant Scheme FRGS/1/2017/TK05/UPM/01/1 (5540048).

REFERENCES

Abral, H., Ariksa, J., Mahardika, M., Handayani, D., Aminah, I., Sandrawati, N., et al., 2020. Transparent and antimicrobial cellulose film from ginger nanofiber. Food Hydrocolloids 98, 105266. https://doi.org/10.1016/j.foodhyd.2019.105266.

Abral, H., Basri, A., Muhammad, F., Fernando, Y., Hafizulhaq, F., Mahardika, M., et al., 2019. A simple method for improving the properties of the sago starch films prepared by using ultrasonication treatment. Food Hydrocolloids 93, 276–283. https://doi.org/10.1016/j.foodhyd.2019.02.012.

Adawiyah, D.R., Sasaki, T., Kohyama, K., 2013. Characterization of arenga starch in comparison with sago starch. Carbohydrate Polymers 92 (2), 2306–2313. https://doi.org/10.1016/j.carbpol.2012.12.014.

Agrebi, F., Ghorbel, N., Rashid, B., Kallel, A., Jawaid, M., 2018. Influence of treatments on the dielectric properties of sugar palm fiber reinforced phenolic composites. Journal of Molecular Liquids 263, 342–348. https://doi.org/10.1016/j.molliq.2018.04.130.

Alemdar, A., Sain, M., 2008. Isolation and characterization of nanofibers from agricultural residues - wheat straw and soy hulls. Bioresource Technology 99 (6), 1664–1671. https://doi.org/10.1016/j.biortech.2007.04.029.

Ammar, I.M., Huzaifah, M.R.M., Sapuan, S.M., Ishak, M.R., Leman, Z.B., 2018. Development of sugar palm fiber reinforced vinyl ester composites. In: Natural Fiber Reinforced Vinyl Ester and Vinyl Polymer Composites. Elsevier, pp. 211–224. https://doi.org/10.1016/B978-0-08-102160-6.00011-1.

Apriyana, W., Poeloengasih, C.D., Hernawan, Hayati, S.N., Pranoto, Y., 2016. Mechanical and microstructural properties of sugar palm (*Arenga pinnata* Merr.) starch film: effect of aging. AIP Conference Proceedings 1755. https://doi.org/10.1063/1.4958576.

Atikah, M.S.N., Ilyas, R.A., Sapuan, S.M., Ishak, M.R., Zainudin, E.S., Ibrahim, R., et al., 2019. Degradation and physical properties of sugar palm starch/sugar palm nanofibrillated cellulose bionanocomposite. Polimery 64 (10), 27–36. https://doi.org/10.14314/polimery.2019.10.5.

Atiqah, A., Jawaid, M., Ishak, M.R., Sapuan, S.M., 2018. Effect of alkali and silane treatments on mechanical and interfacial bonding strength of sugar palm fibers with thermoplastic polyurethane. Journal of Natural Fibers 15 (2), 251–261. https://doi.org/10.1080/15440478.2017.1325427.

Atiqah, A., Jawaid, M., Sapuan, S.M., Ishak, M.R., Ansari, M.N.M., Ilyas, R.A., 2019. Physical and thermal properties of treated sugar palm/glass fiber reinforced thermoplastic polyurethane hybrid composites. Journal of Materials Research and Technology. https://doi.org/10.1016/j.jmrt.2019.06.032. July.

Ayana, B., Suin, S., Khatua, B.B., 2014. Highly exfoliated eco-friendly thermoplastic starch (TPS)/poly (lactic acid)(PLA)/clay nanocomposites using unmodified nanoclay. Carbohydrate Polymers 110, 430–439. https://doi.org/10.1016/j.carbpol.2014.04.024.

Bachtiar, D., Salit, M.S., Zainudin, E., Abdan, K., Dahlan, K.Z.H.M., 2011. Effects of alkaline treatment and a compatibilizing agent on tensile properties of sugar palm fiberinforced high impact polystyrene composites. BioResources 6 (4), 4815–4823.

Baratter, M., Weschenlfelder, E.F., Stoffel, F., Zeni, M., Piemolini-barreto, L.T., 2017. Analysis and evaluation of cassava starch-based biodegradable trays as an alternative packaging to fresh strawberry (*Fragaria ananassa* cv San Andreas). To cite this article American Journal of Polymer Science and Technology 3 (4), 76–81. https://doi.org/10.11648/j.ajpst.20170304.14.

Belibi, P.C., Daou, T.J., Ndjaka, J.M.B., Nsom, B., Michelin, L., Durand, B., 2014. A comparative study of some properties of cassava and tree cassava starch films. Physics Procedia 55, 220–226. https://doi.org/10.1016/j.phpro.2014.07.032.

Bootklad, M., Kaewtatip, K., 2013. Biodegradation of thermoplastic starch/eggshell powder composites. Carbohydrate Polymers 97 (2), 315–320. https://doi.org/10.1016/j.carbpol.2013.05.030.

Cao, X., Chen, Y., Chang, P.R., Huneault, M.A., 2007. Preparation and properties of plasticized starch/multiwalled carbon nanotubes composites. Journal of Applied Polymer Science 106 (2), 1431–1437. https://doi.org/10.1002/app.26799.

Edhirej, A., Sapuan, S.M., Jawaid, M., Zahari, N.I., 2017a. Cassava/sugar palm fiber reinforced cassava starch hybrid composites: physical, thermal and structural properties. International Journal of Biological Macromolecules 101, 75–83. https://doi.org/10.1016/j.ijbiomac.2017.03.045.

Edhirej, A., Sapuan, S.M., Jawaid, M., Zahari, N.I., 2017b. Preparation and characterization of cassava bagasse reinforced thermoplastic cassava starch. Fibers and Polymers 18 (1), 162–171. https://doi.org/10.1007/s12221-017-6251-7.

Faruk, O., Bledzki, A.K., Fink, H.-P., Sain, M., 2012. Biocomposites reinforced with natural fibers: 2000–2010. Progress in Polymer Science 37 (11), 1552–1596. https://doi.org/10.1016/j.progpolymsci.2012.04.003.

González, K., Retegi, A., González, A., Eceiza, A., Gabilondo, N., 2015. Starch and cellulose nanocrystals together into thermoplastic starch bionanocomposites. Carbohydrate Polymers 117. https://doi.org/10.1016/j.carbpol.2014.09.055.

Halimatul, M.J., Sapuan, S.M., Jawaid, M., Ishak, M.R., Ilyas, R.A., 2019a. Effect of sago starch and plasticizer content on the properties of thermoplastic films: mechanical testing and cyclic soaking-drying. Polimery 64 (6), 32–41. https://doi.org/10.14314/polimery.2019.6.5.

Halimatul, M.J., Sapuan, S.M., Jawaid, M., Ishak, M.R., Ilyas, R.A., 2019b. Water absorption and water solubility properties of sago starch biopolymer composite films filled with sugar palm particles. Polimery 64 (9), 27–35. https://doi.org/10.14314/polimery.2019.9.4.

Hazrati, K.Z., Sapuan, S.M., Ilyas, R.A., 2019. Biobased food packaging using natural fiber: a review. In: Prosiding Seminar Enau Kebangsaan 2019. Institute of Tropical Forest and Forest Products (INTROP), Universiti Putra Malaysia, Bahau, Negeri Sembilan, Malaysia, pp. 140–142.

Hazrol, M.D., Sapuan, S.M., Zuhri, M.Y.M., Ilyas, R.A., 2019. Electrical properties of sugar palm nanocellulose fiber reinforced sugar palm starch biopolymer composite. In: Prosiding Seminar Enau Kebangsaan 2019. Institute of Tropical Forest and Forest Products (INTROP), Universiti Putra Malaysia, Bahau, Negeri Sembilan, Malaysia, pp. 57–62.

Huzaifah, M.R.M., Sapuan, S.M., Leman, Z., Ishak, M.R., Ilyas, R.A., 2018. Effect of soil burial on water absorption of sugar palm fiber reinforced vinyl ester composites. In: 6th Postgraduate Seminar on Natural Fiber Reinforced Polymer Composites 2018, pp. 52–54 (Serdang, Selangor).

Ilyas, R.A., Sapuan, S.M., Sanyang, M.L., Ishak, M.R., 2016. Nanocrystalline cellulose reinforced starch-based nanocomposite: a review. In: 5th Postgraduate Seminar on Natural Fiber Composites. Serdang, Selangor: Universiti Putra Malaysia, pp. 82–87. Retrieved from. https://www.researchgate.net/publication/315675302_Nanocrystalline_cellulose_reinforced_starch-based_nanocomposites_A_Review.

Ilyas, R.A., Sapuan, S.M., Ishak, M.R., Zainudin, E.S., 2017. Effect of delignification on the physical, thermal, chemical, and structural properties of sugar palm fiber. BioResources 12 (4), 8734–8754. https://doi.org/10.15376/biores.12.4.8734-8754.

Ilyas, R.A., Sapuan, S.M., Ishak, M.R., 2018a. Isolation and characterization of nanocrystalline cellulose from sugar palm fibers (*Arenga Pinnata*). Carbohydrate Polymers 181, 1038–1051. https://doi.org/10.1016/j.carbpol.2017.11.045.

Ilyas, R.A., Sapuan, S.M., Sanyang, M.L., Ishak, M.R., Zainudin, E.S., 2018b. Nanocrystalline cellulose as reinforcement for polymeric matrix nanocomposites and its potential applications: a review. Current Analytical Chemistry 14 (3), 203–225. https://doi.org/10.2174/1573411013666171003155624.

Ilyas, R.A., Sapuan, S.M., Ishak, M.R., Zainudin, E.S., 2018c. Development and characterization of sugar palm nanocrystalline cellulose reinforced sugar palm starch bionanocomposites. Carbohydrate Polymers 202, 186–202. https://doi.org/10.1016/j.carbpol.2018.09.002.

Ilyas, R.A., Sapuan, S.M., Ishak, M.R., Zainudin, E.S., 2018d. Sugar palm nanocrystalline cellulose reinforced sugar palm starch composite: degradation and water-barrier properties. In: IOP Conference Series: Materials Science and Engineering, vol. 368. https://doi.org/10.1088/1757-899X/368/1/012006.

Ilyas, R.A., Sapuan, S.M., Ishak, M.R., Zainudin, E.S., 2018e. Water transport properties of bio-nanocomposites reinforced by sugar palm (*Arenga pinnata*) nanofibrillated cellulose. Journal of Advanced Research in Fluid Mechanics and Thermal Sciences Journal 51 (2), 234–246.

Ilyas, R.A., Sapuan, S.M., Ishak, M.R., Zainudin, E.S., Atikah, M.S.N., 2018f. Characterization of sugar palm nanocellulose and its potential for reinforcement with a starch-based composite. In: Sugar Palm Biofibers, Biopolymers, and Biocomposites, first ed. CRC Press/Taylor & Francis Group, Boca Raton, FL, pp. 189–220. https://doi.org/10.1201/9780429443923-10.

Ilyas, R.A., Sapuan, S.M., Ishak, M.R., Zainudin, E.S., Atikah, M.S.N., 2018g. Nanocellulose reinforced starch polymer composites: a review of preparation, properties and application. In: Proceeding: 5th International Conference on Applied Sciences and Engineering (ICASEA, 2018). GLOBAL ACADEMIC EXCELLENCE (M) SDN BHD, Capthorne Hotel, Cameron Highlands, Malaysia, pp. 325–341.

Ilyas, R.A., Sapuan, S.M., Ishak, M.R., Zainudin, E.S., 2019a. Sugar palm nanofibrillated cellulose (*Arenga pinnata* (Wurmb.) Merr): effect of cycles on their yield, physic-chemical, morphological and thermal behavior. International Journal of Biological Macromolecules 123, 379–388. https://doi.org/10.1016/j.ijbiomac.2018.11.124.

Ilyas, R.A., Sapuan, S.M., Ibrahim, R., Abral, H., Ishak, M.R., Zainudin, E.S., et al., 2019b. Sugar palm (*Arenga pinnata* (Wurmb.) Merr) cellulosic fiber hierarchy: a comprehensive approach from macro to nano scale. Journal of Materials Research and Technology 8 (3), 2753–2766. https://doi.org/10.1016/j.jmrt.2019.04.011.

Ilyas, R.A., Sapuan, S.M., Ibrahim, R., Abral, H., Ishak, M.R., Zainudin, E.S., et al., 2019c. Effect of sugar palm nanofibrillated cellulose concentrations on morphological, mechanical and physical properties of biodegradable films based on agro waste sugar palm (Arenga pinnata (Wurmb.) Merr) starch. Journal of Materials Research and Technology 8 (5), 4819–4830.

Ishak, M.R., Leman, Z., Sapuan, S.M., Rahman, M.Z.A., Anwar, U.M.K., 2011. Effects of impregnation time on physical and tensile properties of impregnated sugar palm (*Arenga pinnata*) fibers. Key Engineering Materials 471, 1147–1152.

Ishak, M.R., Sapuan, S.M., Leman, Z., Rahman, M.Z.A., Anwar, U.M.K., 2012. Characterization of sugar palm (*Arenga pinnata*) fibers. Journal of Thermal Analysis and Calorimetry 109 (2), 981–989. https://doi.org/10.1007/s10973-011-1785-1.

Ishak, M.R., Leman, Z., Sapuan, S.M., Rahman, M.Z.A., Anwar, U.M.K., 2013a. Impregnation modification of sugar palm fibers with phenol formaldehyde and unsaturated polyester. Fibers and Polymers 14 (2), 250–257. https://doi.org/10.1007/s12221-013-0250-0.

Ishak, M.R., Sapuan, S.M., Leman, Z., Rahman, M.Z.A., Anwar, U.M.K., Siregar, J.P., 2013b. Sugar palm (*Arenga pinnata*): its fiber, polymers and composites. Carbohydrate Polymers 91 (2), 699–710. https://doi.org/10.1016/j.carbpol.2012.07.073.

Jatmiko, T.H., Poeloengasih, C.D., Prasetyo, D.J., Hernawan, 2017. Modelling of moisture adsorption for sugar palm (*Arenga pinnata*) starch film. AIP Conference Proceedings 020007. https://doi.org/10.1063/1.4978080.

Jatmiko, T.H., Poeloengasih, C.D., Prasetyo, D.J., Rosyida, V.T., 2016. Effect of plasticizer on moisture sorption isotherm of sugar palm (*Arenga pinnata*) starch film. AIP Conference Proceedings 080004. https://doi.org/10.1063/1.4941651.

Jumaidin, R., Sapuan, S.M., Jawaid, M., Ishak, M.R., Sahari, J., 2016. Effect of seaweed on physical properties of thermoplastic sugar palm starch/agar composites. Journal of Mechanical Engineering and Sciences 10 (3), 2214–2225. https://doi.org/10.15282/jmes.10.3.2016.1.0207.

Jumaidin, R., Sapuan, S.M., Ilyas, R.A., 2019. Physio-mechanical properties of thermoplastic starch composites: a review. In: Prosiding Seminar Enau Kebangsaan 2019. Institute of Tropical Forest and Forest Products (INTROP), Universiti Putra Malaysia, Bahau, Negeri Sembilan, Malaysia, pp. 104–108.

Jumaidin, R., Sapuan, S.M., Jawaid, M., Ishak, M.R., Sahari, J., 2017. Thermal, mechanical, and physical properties of seaweed/sugar palm fiber reinforced thermoplastic sugar palm starch/agar hybrid composites. International Journal of Biological Macromolecules 97, 606–615. https://doi.org/10.1016/j.ijbiomac.2017.01.079.

Leman, Z., Sapuan, S.M., Azwan, M., Ahmad, M.M.H.M., Maleque, M.A., 2008. The effect of environmental treatments on fiber surface properties and tensile strength of sugar palm fiber-reinforced epoxy composites. Polymer - Plastics Technology and Engineering 47 (6), 606–612. https://doi.org/10.1080/03602550802059451.

Liao, H.-T., Wu, C.-S., 2008. New biodegradable blends prepared from polylactide, titanium tetraisopropylate, and starch. Journal of Applied Polymer Science 108 (4), 2280–2289. https://doi.org/10.1002/app.27901.

Lu, Y., Weng, L., Cao, X., 2006. Morphological, thermal and mechanical properties of ramie crystallites—reinforced plasticized starch biocomposites. Carbohydrate Polymers 63 (2), 198–204. https://doi.org/10.1016/j.carbpol.2005.08.027.

Mansor, M.R., Sapuan, S.M., Salim, M.A., Akop, M.Z., Tahir, M.M., 2015. Modeling of kenaf reinforced sugar palm starch biocomposites mechanical behaviour using Halpin-Tsai model. Recent Advances in Environment, Ecosystems and Development 94–99.

Mazani, N., Sapuan, S.M., Sanyang, M.L., Atiqah, A., Ilyas, R.A., 2019. Design and fabrication of a shoe shelf from kenaf fiber reinforced unsaturated polyester composites. In: Lignocellulose for Future Bioeconomy. Elsevier, pp. 315–332. https://doi.org/10.1016/B978-0-12-816354-2.00017-7.

Misri, S., Leman, Z., Sapuan, S.M., Ishak, M.R., 2010. Mechanical properties and fabrication of small boat using woven glass/sugar palm fibers reinforced unsaturated polyester hybrid composite. IOP Conference Series: Materials Science and Engineering 11, 012015. https://doi.org/10.1088/1757-899x/11/1/012015.

Mohammed, A.A., Bachtiar, D., Rejab, M.R.M., Siregar, J.P., 2018. Effect of microwave treatment on tensile properties of sugar palm fiber reinforced thermoplastic polyurethane composites. Defence Technology 14 (4), 287–290. https://doi.org/10.1016/j.dt.2018.05.008.

Montero, B., Rico, M., Rodríguez-Llamazares, S., Barral, L., Bouza, R., 2017. Effect of nanocellulose as a filler on biodegradable thermoplastic starch films from tuber, cereal and legume. Carbohydrate Polymers 157, 1094–1104. https://doi.org/10.1016/j.carbpol.2016.10.073.

Nazrin, A., Sapuan, S.M., Ilyas, R.A., 2018. Thermoplastic starch blended poly(lactic) acid for Food packaging application: mechanical properties. In: 6th Postgraduate Seminar on Natural Fiber Reinforced Polymer Composites 2018, pp. 79–84 (Serdang).

Norizan, M.N., Abdan, K., Ilyas, R.A., 2019. Effect of water absorption on treated sugar palm yarn fiber/glass fiber hybrid composites. In: Prosiding Seminar Enau Kebangsaan 2019. Institute of Tropical Forest and Forest Products (INTROP), Universiti Putra Malaysia, Bahau, Negeri Sembilan, Malaysia, pp. 78–81.

Ortega-Toro, R., Bonilla, J., Talens, P., Chiralt, A., 2017. Future of starch-based materials in food packaging. In: Starch-Based Materials in Food Packaging: Processing, Characterization and Applications. https://doi.org/10.1016/B978-0-12-809439-6.00009-1.

Oumer, A.N., Bachtiar, D., 2014. Modeling and experimental validation of tensile properties of sugar palm fiber reinforced high impact polystyrene composites. Fibers and Polymers 15 (2), 334–339. https://doi.org/10.1007/s12221-014-0334-5.

Poeloengasih, C.D., Pranoto, Y., Hayati, S.N., Hernawan, Rosyida, V.T., Prasetyo, D.J., et al., 2016. A physicochemical study of sugar palm (*Arenga pinnata*) starch films plasticized by glycerol and sorbitol. AIP Conference Proceedings 080003. https://doi.org/10.1063/1.4941650.

Radzi, A.M., Sapuan, S.M., Jawaid, M., Mansor, M.R., 2017. Influence of fiber contents on mechanical and thermal properties of roselle fiber reinforced polyurethane composites. Fibers and Polymers 18 (7), 1353–1358. https://doi.org/10.1007/s12221-017-7311-8.

Radzi, A.M., Sapuan, S.M., Jawaid, M., Mansor, M.R., 2019. Effect of alkaline treatment on mechanical , physical and thermal properties of roselle/sugar palm fiber reinforced thermoplastic polyurethane hybrid composites. Fiber and Polymers 20 (4), 847–855. https://doi.org/10.1007/s12221-019-1061-8.

Razali, N., Sapuan, S.M., Razali, N., 2018. Mechanical properties and morphological analysis of roselle/sugar palm fiber reinforced vinyl ester hybrid composites. In: Natural Fiber Reinforced Vinyl Ester and Vinyl Polymer Composites.

Elsevier, pp. 169–180. https://doi.org/10.1016/B978-0-08-102160-6.00008-1.

Sahari, J., Sapuan, S.M., Ismarrubie, Z.N., Rahman, M.Z.A., 2011. Investigation on bending strength and stiffness of sugar palm fiber from different parts reinforced unsaturated polyester composites. Key Engineering Materials 471–472, 502–506. https://doi.org/10.4028/www.scientific.net/KEM.471-472.502. October 2015.

Sahari, J., Sapuan, S.M., Ismarrubie, Z.N., Rahman, M.Z., 2012a. Physical and chemical properties of different morphological parts of sugar palm fibers. Fibers and Textiles in Eastern Europe 91 (2), 21–24.

Sahari, J., Sapuan, S.M., Zainudin, E.S., Maleque, M.A., 2012b. Effect of water absorption on mechanical properties of sugar palm fiber reinforced sugar palm starch (spf/sps) biocomposites. Journal of Biobased Materials and Bioenergy 6, 1–5. https://doi.org/10.1166/jbmb.2012.1267.

Sahari, J., Sapuan, S.M., Zainudin, E.S., Maleque, M.A., 2013a. Mechanical and thermal properties of environmentally friendly composites derived from sugar palm tree. Materials and Design 49 (2), 285–289. https://doi.org/10.1016/j.matdes.2013.01.048.

Sahari, J., Sapuan, S.M., Zainudin, E.S., Maleque, M.A., 2013b. Mechanical and thermal properties of environmentally friendly composites derived from sugar palm tree. Materials and Design 49, 285–289. https://doi.org/10.1016/j.matdes.2013.01.048.

Sahari, J., Sapuan, S.M., Zainudin, E.S., Maleque, M.A., 2013c. Thermo-mechanical behaviors of thermoplastic starch derived from sugar palm tree (*Arenga pinnata*). Carbohydrate Polymers 92 (2), 1711–1716. https://doi.org/10.1016/j.carbpol.2012.11.031.

Sahari, J., Sapuan, S.M., Zainudin, E.S., Maleque, M.A., 2014a. Physico-chemical and thermal properties of starch derived from sugar palm tree (*Arenga pinnata*). Asian Journal of Chemistry 26 (4), 955–959. https://doi.org/10.14233/ajchem.2014.15652.

Sahari, J., Salit, M.S., Zainudin, E.S., Maleque, M.A., 2014b. Degradation characteristics of SPF/SPS biocomposites fabrication of SPF/SPS biocomposites. Fibers and Textiles in Eastern Europe 22 (5107), 96–98.

Sanyang, M.L., Sapuan, S.M., Jawaid, M., Ishak, M.R., Sahari, J., 2016. Recent developments in sugar palm (*Arenga pinnata*) based biocomposites and their potential industrial applications: a review. Renewable and Sustainable Energy Reviews 54, 533–549. https://doi.org/10.1016/j.rser.2015.10.037.

Sanyang, M.L., Sapuan, S.M., Jawaid, M., Ishak, M.R., Sahari, J., 2016a. Development and characterization of sugar palm starch and poly(lactic acid) bilayer films. Carbohydrate Polymers 146, 36–45. https://doi.org/10.1016/j.carbpol.2016.03.051.

Sanyang, M.L., Sapuan, S.M., Jawaid, M., Ishak, M.R., Sahari, J., 2016b. Effect of sugar palm-derived cellulose reinforcement on the mechanical and water barrier properties of sugar palm starch biocomposite films. BioResources 11 (2), 4134–4145. https://doi.org/10.15376/biores.11.2.4134-4145.

Sanyang, M.L., Ilyas, R.A., Sapuan, S.M., Jumaidin, R., 2018. Sugar palm starch-based composites for packaging applications. In: Bionanocomposites for Packaging Applications. Springer International Publishing, Cham, pp. 125–147. https://doi.org/10.1007/978-3-319-67319-6_7.

Sanyang, M.L., Muniandy, Y., Sapuan, S.M., Sahari, J., 2017. Tea tree (Melaleuca alternifolia) fiber as novel reinforcement material for sugar palm biopolymer based composite films. BioResources 12 (2), 3751–3765. https://doi.org/10.15376/biores.12.2.3751-3765.

Sapuan, S.M., Ilyas, R.A., 2017. Sugar Palm: Fibers, Biopolymers and Biocomposites. INTROPica, pp. 5–7.

Sapuan, S.M., Ilyas, R.A., Ishak, M.R., Leman, Z., Huzaifah, M.R.M., Ammar, I.M., Atikah, M.S.N., 2018. Development of sugar palm–based products: a community project. In: Sugar Palm Biofibers, Biopolymers, and Biocomposites, first ed. CRC Press/Taylor & Francis Group, Boca Raton, FL, pp. 245–266. https://doi.org/10.1201/9780429443923-12.

Sapuan, S.M., Ishak, M.R., Leman, Z., Huzaifah, M.R.M., Ilyas, R.A., Ammar, I.M., et al., 2017a. In: Sapuan, S.M. (Ed.), Pokok Enau: Potensi Dan Pembangunan Produk, first ed. Penerbit Universiti Putra Malaysia, Serdang, Selangor.

Sapuan, S.M., Ishak, M.R., Leman, Z., Ilyas, R.A., Huzaifah, M.R.M., 2017b. Development of Products from Sugar Palm Trees (*Arenga pinnata* Wurb. Merr): A Community Project. INTROPica, pp. 12–13. Retrieved from. http://www.introp.upm.edu.my/upload/dokumen/2018091416164313.5_Prof._Sapuan.pdf.

Tang, X., Alavi, S., Herald, T.J., 2008. Effects of plasticizers on the structure and properties of starch-clay nanocomposite films. Carbohydrate Polymers 74 (3), 552–558. https://doi.org/10.1016/j.carbpol.2008.04.022.

Teixeira, E. de M., Curvelo, A.A.S., Corrêa, A.C., Marconcini, J.M., Glenn, G.M., Mattoso, L.H.C., 2012. Properties of thermoplastic starch from cassava bagasse and cassava starch and their blends with poly (lactic acid). Industrial Crops and Products 37 (1), 61–68. https://doi.org/10.1016/j.indcrop.2011.11.036.

Thakur, R., Pristijono, P., Golding, J.B., Stathopoulos, C.E., Scarlett, C.J., Bowyer, M., et al., 2017. Amylose-lipid complex as a measure of variations in physical, mechanical and barrier attributes of rice starch- ι -carrageenan biodegradable edible film. Food Packaging and Shelf Life 14, 108–115. https://doi.org/10.1016/j.fpsl.2017.10.002.

Ticoalu, A., Aravinthan, T., Cardona, F., 2010. Experimental investigation into gomuti fibers/polyester composites. In: 21st Australasian Conference on the Mechanics of Structures and Materials (ACMSM 21). CRC Press/Balkema, Melbourne, Australia, pp. 451–456.

Ticoalu, A., Aravinthan, T., Cardona, F., 2013. A review on the characteristics of gomuti fiber and its composites with thermoset resins. Journal of Reinforced Plastics and Composites 32 (2), 124–136. https://doi.org/10.1177/0731684412463109.

CHAPTER 8

Properties and Characterization of PLA, PHA, and Other Types of Biopolymer Composites

R.A. ILYAS • S.M. SAPUAN • ABUDUKEREMU KADIER • MOHD SAHAID KALIL • RUSHDAN IBRAHIM • M.S.N. ATIKAH • N. MOHD NURAZZI • A. NAZRIN • C.H. LEE • MOHD NOR FAIZ NORRRAHIM • NASMI HERLINA SARI • EDI SYAFRI • HAIRUL ABRAL • LATIFAH JASMANI • M.I.J. IBRAHIM

1 INTRODUCTION

The production of petroleum-based plastic had increased tremendously, reaching about 350 million tons annually (Garside, 2019). According to Rochman et al. (2013), the earth will accumulate with about 33 billion tons of plastic wastes by 2050, if current rates of consumption continue. Besides that, inappropriate usage and disposal of plastics waste would lead to substantial pollution of both terrestrial and marine ecosystems. The world produces about 350 million tons of plastic waste annually, and surprisingly, only 9% of this waste has been recycled (UNEP, 2018). Moreover, it had been estimated that each year at least 8 million tons of plastic waste go to the rivers and oceans, and some of them often decompose into small microplastics that end up stopping in our food chain (UNEP, 2018). Because of the environmental challenges aiming to reduce this environmental impact, many researchers have formulate eco-friendly and biodegradable composites polymer to replace conventional petroleum-based polymer.

Recently, many countries have banned petroleum-based plastics because of the huge volume of plastic waste that harmfully affects the ecosystem, wildlife, and environment (Aisyah et al., 2019; Asyraf et al., 2020; Atikah et al., 2019; Norizan et al., 2020; Nurazzi et al., 2019a). These nondegradable plastic are responsible for the "white pollution" worldwide. The "white pollution," including plastic bags, plastic bottles, plastic silverware, and other materials that are made from the plastic, kills the wildlife, marine life, and avifauna and degrades the quality and features of the environment on the Mother Nature (Thiagamani et al., 2019). Remarkably, Malaysia is the foremost country in Southeast Asia region to take courageous act to confront "white pollution." The Government of Malaysia has broadcasted that the government will ban single-use plastic by year of 2030 (UNEP, 2018). Although Malaysia is a bit behind when it comes to enacting against single-use plastics, nevertheless according to New Strait Times, Federal Territories of Malaysia has announced that from March 2019, a pollution charge of 20 cent imposed for a single plastic bag. Therefore, customers will either have to pay 20 cent for a reusable bag or bring their own bags. Besides Malaysia, others countries such as Kenya, China, Rwanda, Uganda, Ireland, South Africa, Morocco, Taiwan, India, France, and Canada have already forbid and eliminated completely the use of single-use plastic and plastic bags.

Therefore, in order to cater this problems, biodegradable polymers were introduced. Biodegradable polymers are one of the potential solutions to the problems associated with discarded wastes (Abral et al., 2020a,b; 2019a; Atiqah et al., 2019; Ilyas et al., 2018, 2017; Nurazzi et al., 2019b). This is due to their fast degradation by the action of naturally occurring microorganisms in the environment (Ilyas et al., 2018a). Biodegradable polymer can be produced via (1) bio-based (i.e., polyhydroxyalkanoates [PHA], starch, protein, polylactic acid [PLA], chitin, chitosan, polybutylene succinate [PBS], and cellulosics) and (2) fossil- or petroleum-based (i.e., polycaprolactone [PCL], poly(vinyl alcohol), and poly(butylene adipate-co-terephthalate) materials.

Generally, polymer is made up of long chain of polymer with many repeated subunits. Both man-made

Advanced Processing, Properties, and Applications of Starch and Other Bio-Based Polymers. https://doi.org/10.1016/B978-0-12-819661-8.00008-1

(synthetic) and natural polymers play important and ubiquitous roles in daily life. This is because of their various unique properties (Ilyas et al., 2018a). They are made by the process of condensation and addition polymerization reactions. They can be categorized either as thermosetting or thermoplastic polymers. "Bio-based plastic" can be defined as a plastic that are made from natural resources and it is renewable, whereas "biodegradable plastic" is referred on how the plastics behave at end of its life. The illustrations of bioplastic is displayed in Fig. 8.1.

Biodegradable plastic is a material that is decomposed naturally when introduced in the environment by the action of living organisms, usually microorganisms. It is eco-friendly compared with conventional plastics. Besides that, this plastic is commonly produced via microorganism, renewable raw materials, petrochemicals, or combination of all three. There are a lot of materials that can be used to make biodegradable and bio-based plastics such as plants, starches (cassava, corn, potato, sugar palm, yam bean, pea, wheat tapioca, bengkuang), peels from citrus fruits, and corn oil (Atikah et al., 2019; Halimatul et al., 2019a,b; Jumaidin et al., 2019a–c; Syafri et al., 2019). Biodegradable and bio-based plastic offer a material that is made from natural resources, therefore the risk of breaking down these material are much fewer compared with conventional petroleum-based plastic. Fig. 8.2 shows some of the list of the advantages and disadvantages of utilizing biodegradable plastic.

Biodegradable plastic is often reflected as a savior product. Shifting to these materials would decrease the levels of greenhouse gas emission, reduce carbon dioxide levels, reduce energy of manufacturing, reduce amount of waste produced, and create new plastic industries. However, there is also the potential for biodegradable plastic to create more pollution. This is because biodegradable plastic is difficult to be decomposed in ocean water like it would during composting in soil. This plastic will either float on the ocean or river surface like other conventional plastic or more worse it create microplastic that is harmful to marine life, sea life, or ocean life. Therefore, shifting to this biodegradable plastic is not a final solution that will resolve our pollution problems. More research must be done to develop total biodegradable plastic that can be degraded fully in every condition. The advantages and disadvantages of biodegradable plastic provide opportunity to reduce human dependence in fossil fuels especially in reliance on crude oil or petroleum. There are several types of biodegradable and bio-based polymer such as PHA, PLA, starch, protein, chitin, chitosan, and PBS (Azammi et al., 2020). Therefore, this chapter focuses on providing an overview of recent advancements on biodegradable bio-based polymer for various application.

2 POLYHYDROXYALKANOATES

PHA are a kind of degradable plastics. It has good environmental effects compared with petroleum-based

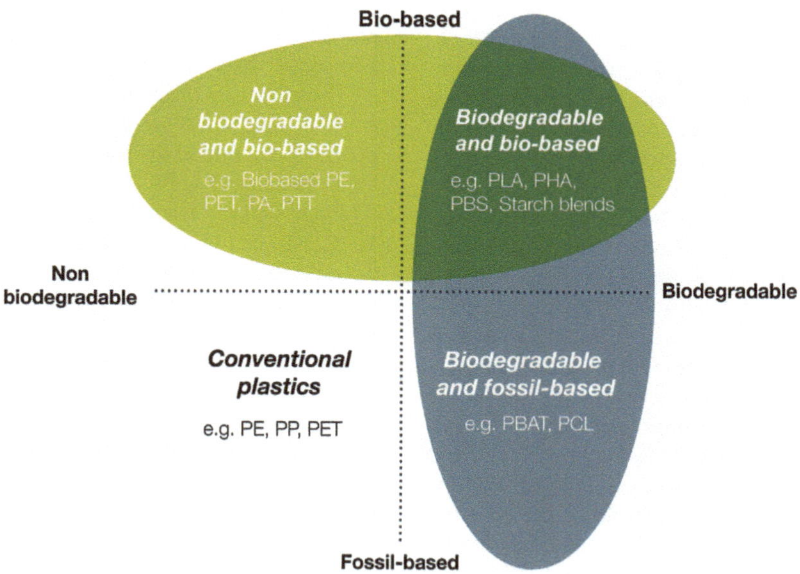

FIG. 8.1 Conventional petroleum-based plastics and bioplastics are that made up of biodegradable polymers (Ilyas and Sapuan, 2020).

Advantages

- Can be broken down by naturally occurring microorganism
- Reduces greenhouse gas emission levels
- Reduces carbon dioxide levels
- Does not release other harmful products after disposal
- Uses low energy during the production cycle
- Reduces the amount of waste we produce
- Current use of petroleum can be channeled to other needs
- Can mix and blend with conventional plastic
- Could create new export industries
- Could establish a new business opportunity and marketing platform
- Can degrade faster under certain conditions

Disadvantages

- Disposal of bioplastic waste must be through certain procedures
- Requires the some certain environmental conditions such as UV light, temperature and humidity) to associated with their disposal
- The costs of pesticides and herbicides are not taken into account during the manufacturing of biodegradable plastic
- The utilization of biodegradable plastics would reduce the number of plastic that can be recycling due to their limited properties
- High capital cost
- Requires croplands to produce items
- Some of the bioplastics only reduce their size to microsize plastic during decompostion process, but this does not solve the current ocean pollution problems, besides contaminated microplastics could expose marine organism to high concentrations of toxins
- May produce methane in landfills

FIG. 8.2 Advantages and disadvantages of biodegradable plastics.

polymer. PHA is a natural biopolymer material developed rapidly in recent 20 years. PHA are polyesters that were synthesized in nature by many microorganisms, including bacterial fermentation of lipids and sugars.

PHA has good mechanical, thermal processing, biodegradability, and biocompatibility properties. It can be used in agricultural materials, biomedical materials, and packaging applications. Recently, this polymer has attract huge attention in field of biomaterials.

Besides that, the blending polymer of starch and aliphatic polyester from sustainable and renewable resources to produce biodegradable plastics such as garbage bags and other products has been successfully studied and applied in American and European countries. One of the company named Novamont in Novara, Italy, had promoted a bioeconomy model based on the efficient use of resources and on territorial regeneration (Sahay and Ierapetritou, 2009). This company sets up biorefineries for the production of bioplastics and bioproducts of renewable resources conceived to protect environment, ecology, and wildlife. In 2018, Novamont had announced the opening of a revamped Mater-Biopolymer plant south of Rome to produce MATER-BI. This plant significantly boosting production capacity from 120,000 tons per year to 150,000. Remarkably, MATER-BI is widely used in Europe and the United States.

Polyhydroxyalkyl fatty acid esters (PHA) are biosynthetic polyesters consisting of a series of different repetitive unit structures. PHA has many interesting properties and has been utilized in many applications such as medical applications, fishing lines, and plastics bag. The biosynthesis of PHA monomer depends on many aspects, including the types of carbon sources on which biological growth depends, the types of metabolic pathways through which organisms convert these carbon sources into PHA monomers, and the substrate specificity of enzymes involved in PHA synthesis (Pérez-Arauz et al., 2019).

2.1 Properties of Polyhydroxyalkanoates

PHA is a kind of thermoplastic material with high degree of result. Its physical properties and chemical structure are basically similar to polypropylene (PP) and polyethylene (PE), and it is capable of drawing, pressing film, injection molding, etc. PHA is a kind of intracellular carbon source and energy storage, which can be naturally decomposed and utilized by many microorganisms and is biodegradable.

Soil burial experiments showed that PHA films with thickness of 0.07 mm could be degraded basically in about 6 weeks. The intracellular degradation of PHA was mainly through the formation of monomers or dimers by PHA depolymerase. There were two mechanisms of extracellular degradation, one was the automatic hydrolysis of melting bonds without supervision, and the other was decomposed by depolymerization of extracellular PHA. For example, as shown by Mergaert et al., 295 microorganisms have been found to degrade PHA in soil. In addition to the main characteristics of biodegradability, PHA has special properties such as bio-PHAe transport, optical activity, piezoelectricity, foam resistance, low permeability, etc. It can be widely used in industry, agriculture, medicine, scientific research, and other fields (Lu et al., 2009). In the PHA family, more than 90 different monomers have been discovered, except for a small number of 4-hydroxybutyric acid, 4-hydroxyvaleric acid, and 5-hydroxyvaleric acid, most of which are 3HA. At most, the fermentation mechanism and its properties are well known.

2.2 Advantages of Polyhydroxyalkanoates

PHA (and bioplastics in common) are exceptionally appealing materials for three essential reasons: they can be made from renewable sources, they can biodegrade, and they are biocompatible. To the primary point, it is exceptionally energizing that researchers are finding ways to gather and utilize fabric from sources like microbes amalgamation (PHA) and corn or sugarcane (other bioplastics like PLA). Already, crops had to be redirected for the production of bioplastics, but within the final decade or so, there has been a center on utilizing squander materials (such as banana peels, potato peelings, etc.) to deliver bioplastics instep. By utilizing squander items, utilization of rare assets can be maximized.

Biodegradability is the other key viewpoint that creates PHA an awfully promising material. Environmental contamination could be a hot subject with vital suggestions; ordinary petrochemical-based plastics have been at the exceptionally center of the controversy, fundamentally since they are so broad and do not degrade effectively. For occasion, pictures of marine creatures choked by or ingesting different plastics show the stark reality of contamination. Of course, fossil fuel-based plastics have done much to move forward the material lives of people, but when it comes to contamination, modern developments in biodegradability are exceptionally welcome.

Since PHA are biocompatible (which implies not destructive to living tissue), they can and have been used in an assortment of therapeutic and surgical applications. Looking forward, the potential moreover exists that PHA will be included in "wearable" inner innovation applications.

PHA can reduce the landfill needed to bury the plastic waste as the period time needed for PHA to biodegrade is fast and reliable, which reduces the impact to environment. Besides, biopolymers can decrease the natural affect derived from plastic waste transfer due to the truth that the biodegradation time for biopolymers within the land surface beneath standard conditions is roughly 2 months (Hassan et al., 2013).

PHA have the advantages to biodegrade in most of the situation, not only on land surface but also in water area due to the special process undergone during the biodegradation process. PHA can be biodegraded under both anaerobic and aerobic conditions by PHA degraders as shown in most situation, including the marine environment (Shah et al., 2008).

2.3 Application of Polyhydroxyalkanoates

Bacterial plastics have gained a very high attention worldwide and globally as the better choice in packaging materials. The applications include in medical devices, skin care, personal hygiene products, plastic packaging, and utilized in agricultural field as mulching films. It is an alternative option for conventional plastics that are being used in agriculture field, which are nonbiodegradable and nonenvironment friendly. This bacterial plastics are the better substitution for those examples mentioned previously because these conventional plastic are single-use application and the demand amounts are enormous, thus creating a lot of plastic waste. According to Chen (2010), the application of PHA is not only limited to bioplastics application, however, PHA have also been make full use as implant biomaterials, biofuels, fine chemicals, medicines, and for regulating bacterial metabolism as well as enhancing the quality of industrial microorganisms.

The scope of application was made possible by further chemically modifying the PHA's functional groups. The utilization of PHA in various industries has grown considerably wide. Currently, aluminum is used as cover for cupboard to prevent water from entering the product. Hence, to overcome this problem, PHA latex can be utilized to cover cardboard to make the surface of product water-resistant. Besides that, only small amount of PHA is required for this purpose, and thus it works out to be cost-effective alternative (Patricia et al., 2007).

Nonetheless, PHA is also extensively use in the tissue engineering. Tissue engineering involves the use of a tissue scaffold for the formation of new viable tissue for a medical purpose. Therefore, suitable material for this application must have properties such as support cell growth, allow tissue ingrowth, guide and organize the cells, degrade to nontoxic products, and biocompatibility. Thus, according to these criteria, PHA is seem completely fit into these criteria and make it the best candidate as the suitable material to be used in tissue engineering application.

Furthermore, PHA also can be utilized in controlled drug release systems. The capability of this material to work with a suitable host response and biodegradability properties make it useful for drug delivery in the medical field. Besides that, many researches have been conducted and stated that various variety of monomers can be added into PHA for modification. These modification would result in numerous changes of physical properties that range from strong elastomers to highly crystalline materials. The rate of decomposition can be indirectly controlled by precisely controlling the monomer composition of PHA. Catalyst reaction of the enzymatic degradation is normally done by bacterial PHA depolymerase.

Metabolix, a US-based company, blend P(3HB) and poly(3-hydroxyoctanoate) to produce a new compound of PHA and marketed their product in the market. This newly formed PHA compound is an elastomer that has been ratified by the Food and Drug Administration for usage as food additives. This example shows that the range of PHA application has been expanded widely as they could even found in the food industries.

Following by the application in food additive, the application of PHA can be found in the electronic industries as well. This function is possible due to the PHA piezoelectric nature. The electronic components or parts that can be produced by PHA are stretch and acceleration measuring instruments, shock wave sensors, gas lighters, lighters, material testing, pressure sensors for keyboards, oscillators: for atomization of liquids and ultrasonic therapy, loudspeakers, headphones, and acoustics: sound pressure measuring instruments, ultrasonic detectors, and microphone.

Last but not least, polystyrene (PS) waste was renowned as the waste materials that takes the longest degradation time or might be nondegradable. However, with the advanced scientific exploration, researchers were able to devise a novel way to alter the abundant PS waste into PHA biopolymer using combination of pyrolysis. This process is cost-effective and efficient and can be one of the ways to optimize the use of PS residues.

3 POLYLACTIC ACID

World issues such as environmental, economic, and health issues have been constantly pushing scientists, researchers, and manufacturers to slowly replace the extensive use of plastics in the current world. Most plastics produced today are made from nonrenewable resources such as petroleum. Unlike any other plastics, PLA is made up from renewable resources such as corn starch or sugar cane (Siakeng et al., 2019). PLA is known as bioplastic because it is obtained from biomass. Not only that PLA is biodegradable, it also has attributes like PS, PE, or PP. Due to its similar production process, it can be produced easily from the existing production plant for petrochemical industry plastics. This advantage makes it cost effective to be utilized. PLA is mainly produced through condensation and polymerization processes. Ring-opening polymerization process is a procedure that uses metal catalyst mix with lactide to form a bigger PLA molecules. Similarly, the condensation process goes through the same procedure with different temperature and produce different by-products (Singhvi et al., 2019). There are few types of PLA, e.g., racemic PLLA, regular PPLA, PDLA, and PDLLA. All of them are having different characteristics yet similar because they are derived from renewable sources. PLA is a thermoplastic polyester which means that it can be heated to their melting point at around 150–160°C, cooled, and heated again without any degradation. In contrast, the thermoset plastic can be heated only once and it is irreversible. Due to its characteristics, PLA has been the most studied and utilized biodegradable plastic in the human history. It is replacing conventional petrochemical-based polymers slowly and becomes the leading biomaterial for medical application as well as in other plastic industries (Farah et al., 2016).

3.1 Advantages of Polylactic Acid

PLA is tremendously used in research and daily life as biocompatible polymer due to properties without toxic or carcinogenic effects for human body (Rasal et al., 2010). For quite a long time, we have been cautioned of the hazardous synthetic substances that escaped when common plastics are burned. In biological perspective, PLA plastics do not develop these poisonous gas if ended up in the event of burning instead of finding their path to a composting facility operate in large scale.

Most of the polymers that exist nowadays are derived from nonrenewable resources especially petrochemical, which are just accessible in limited quantity all through the world. In the end, these fossil resources will be used up. Production of synthetic polymers, as well as to get rid of it by burning, will create a lot of CO_2, which contributes to the global warming (Pang et al., 2010).

PLA, which is produced from corn, is a resource that can be restored every year (Singhvi et al., 2019). PLA are getting more attention commercially since they are produced from corn-based starch, sequester critical amounts of CO_2 in respect to petrochemical-based materials, save energy, and degrade in a short time. Researchers have shown that PLA demonstrates lower fossil resources utilization, which reduces the risk of summer smog and global warming.

PLA will bring a lot of benefits such as its preparation from lactide monomer which can get from a renewable source in agriculture field (Gewin, 2003; Sawyer, 2003), low quantities of CO_2 used up (Dorgan et al., 2001), the contribution in reducing amount of landfill and developing economies of farm, and lastly better mechanical properties compared to PS and PET (Auras et al., 2005).

3.2 Disadvantages of Polylactic Acid

Table 8.1 give a summary on the degradation rate of some synthetic polyesters. It show that PLA is slower than PGA, PLGA, and PCLA copolymers in term of degradation rate. The reason behind this condition is the poor properties that make the water diffusion difficult to the semicrystallinity of PLA.

TABLE 8.1
Biodegradation Rate of Some Polyesters (Ikada and Tsuji, 2000; Zhu et al., 1991).

Polymers	Molecular Mass (kDa)	Degradation Rate
PLLA	100–300	(SC) 50% in 1–2 years
PGA	–	(C) 100% in 2–3 months
PLGA	40–100	(A) 100% in 50–100 years
PCL	40–80	(SC) 50% in 4 years
PCLA	100–500	(A) 100% in 3–12 months
PTMC	14	(SC) 9% in 30 weeks

To some certain extent, despite the fact that PLA is biodegradable, the duration taken will be so long. As indicated by Elizabeth (2006), PLA may well separate into its constituent parts, carbon dioxide and water in a controlled composting condition, that is, increase the surrounding temperature to 140°F. In any case, it will take much more time in a fertilizer container or in a landfill stuffed so firmly that no light and little oxygen are accessible to aid the procedure.

It is regretting to say that, most PLA plastic will not break down into natural components in any composting pile appearing in courtyard. Rather, these items should be transported to a commercial composting facility. In any case, as the business develops, we trust that the facility for commercial composting will tag along. Discarding PLA plastic items in a landfill would be considered as a suicide alternative.

3.3 Application of Polylactic Acid

There are three main applications for PLA plastics, which are domestic, medical, and packaging and 3D printing applications. Table 8.2 shows the applications of PLA and its usages.

4 STARCH

Starch-based biopolymer films have been extensively utilized in medicine as well as food packaging application, in which the biofilm should be edible in many cases, such as applications in medicine capsules, candy wrappers, etc. This type of biofilm has a potential to be used for controlling water permeability, as a barrier for volatile compounds and gases, and to maintain the food freshness. However, starch-based biopolymer has poor mechanical properties and high water vapor permeability. To overcome this drawback, starch-based biopolymers are (1) blend with other polymers, (2) reinforced with plasticizer, particle, or fiber fillers, or (3) modified to starch structure (Ilyas et al., 2020a,b, 2019a–d, 2018a–d; Sanyang et al., 2018). Nevertheless, the reinforcement of filler or plasticizer has some problems related to safety issues. Therefore, the selection of filler, plasticizer, polymer blend, and chemical used to modified starch must be nonallergic, nontoxic, fully biodegradable, and digestible and can be consumed by living things.

4.1 Properties of Starch

Physically, most native starches are semicrystalline, having a crystallinity of about 20%–45% (Abral et al., 2019b). The short-branched chains in the amylopectin

TABLE 8.2
The Application of PLA Plastics (Jamshidian et al., 2010).

Number	Applications	Usages
1	Domestics	Plates and saucers, cups, cutlery, fruit juices, fresh water, sports drinks, cold drink cups, transparent food containers, foodware, dairy containers, jelly and jam container, and edible oils container.
2	Medical	Medical devices such as plates, rods, pins, and screws.
3	Packaging	Vegetable bags, candy twist wrap, lidding film, salad, blister packaging, window envelope film, label film, shrink wrap material, and other packaging applications.
4	FDM machines (3D printing)	3D printable filament, lost PLA casting for molten metal, and other 3D printing medical device prototypes (both biodegradable and permanent).

are the main crystalline component in starch granular. Crystalline regions exist in the form of double helices with a length of ~5 nm. Besides that, the amylopectin segments in the crystalline regions are all parallel to the axis of the large helix. The molecular weight of amylose is about 100 times lower than that of amylopectin. Moreover, the ratio of amylose to amylopectin much depends on the age and source of the starch. In addition, the ratio of amylose and amylopectin can also be controlled by the extraction process method used. Starch granules also contain small amounts of lipids and proteins. Fig. 8.3 shows the chemical structures and physical diagram illustration of amylopectin starch and amylose starch (Generalic, 2019). Moreover, thermoplastic starch (TPS) is formed by disrupting the ordered structures within the starch molecular. Heating process is required along with shear force to disrupt the starch granules. This shear forced and heating process would cause swelling and nonirreversible transition of amorphous regions in the presence of plasticizer, under certain condition (Sanyang et al., 2016). Table 8.3 shows the chemical composition of commercial starches. Besides that, from Table 8.4, it can be observed that the highest amylose content is sugar palm starch (SPS). Table 8.5 shows the SPS properties in comparison with sago starch. The mechanical properties of SPS biofilm was observed higher compared with sago starch biofilm. Therefore, SPS biofilms have higher potential to be utilized as bio-based packaging.

4.2 Advantages and Disadvantages of Starch Biopolymer

The development of novel starch-based polymer materials using renewable resources became hot topic among the researchers to overcome the environmental problems caused by plastic wastes. Starch can be extracted from sugar palm tree, tapioca, wheat, tapioca, potato, cassava, bengkuang, and maize (Jumaidin et al., 2019a,b, 2020). Generally, such material is stored in plants tissues as one-way carbohydrates. It is made up of glucose and can be attained by melting starch. However, this polymer is not available in animal tissues. Starch possess several advantages such as abundance, renewability, easy availability, biodegradability, ease process, and cheap. Native starch-based biopolymer possess many disadvantages, such as high thermal degradation rates, low mechanical properties, high water-barrier properties and processability (Hazrol et al., 2020; Nazrin et al., 2020). The disadvantages of starch-based biopolymer are listed in Fig. 8.4. Current technique that is being used to process starch nowadays is solution technique. Solution casting technique is the easiest and most widely utilized at laboratory scale compared with other method such as hot press. However, this technique cannot be conducted at large-industrial scale as it encompasses too long drying time processed. Therefore, in order to overcome this problem, the fabrication of starch-based polymer by thermoplastic treatments (reactive extrusion, foaming extrusion, film/sheet extrusion, and injection molding) could be considered.

4.3 Application of Starch Biopolymer

Bio-based and biodegradable starch-based biopolymers have an extensive range of applications such as mulching film horticultural crops, drop ceiling tiles, pharmaceutical, biomedical, corrugated board adhesives, paper, horticulture, agriculture, consumer electronics, automotive, textiles, and packaging (Spaccini et al., 2016; Tan et al., 2016). Table 8.6 summarized the starch-based biopolymer, its manufacturing technique, and applications.

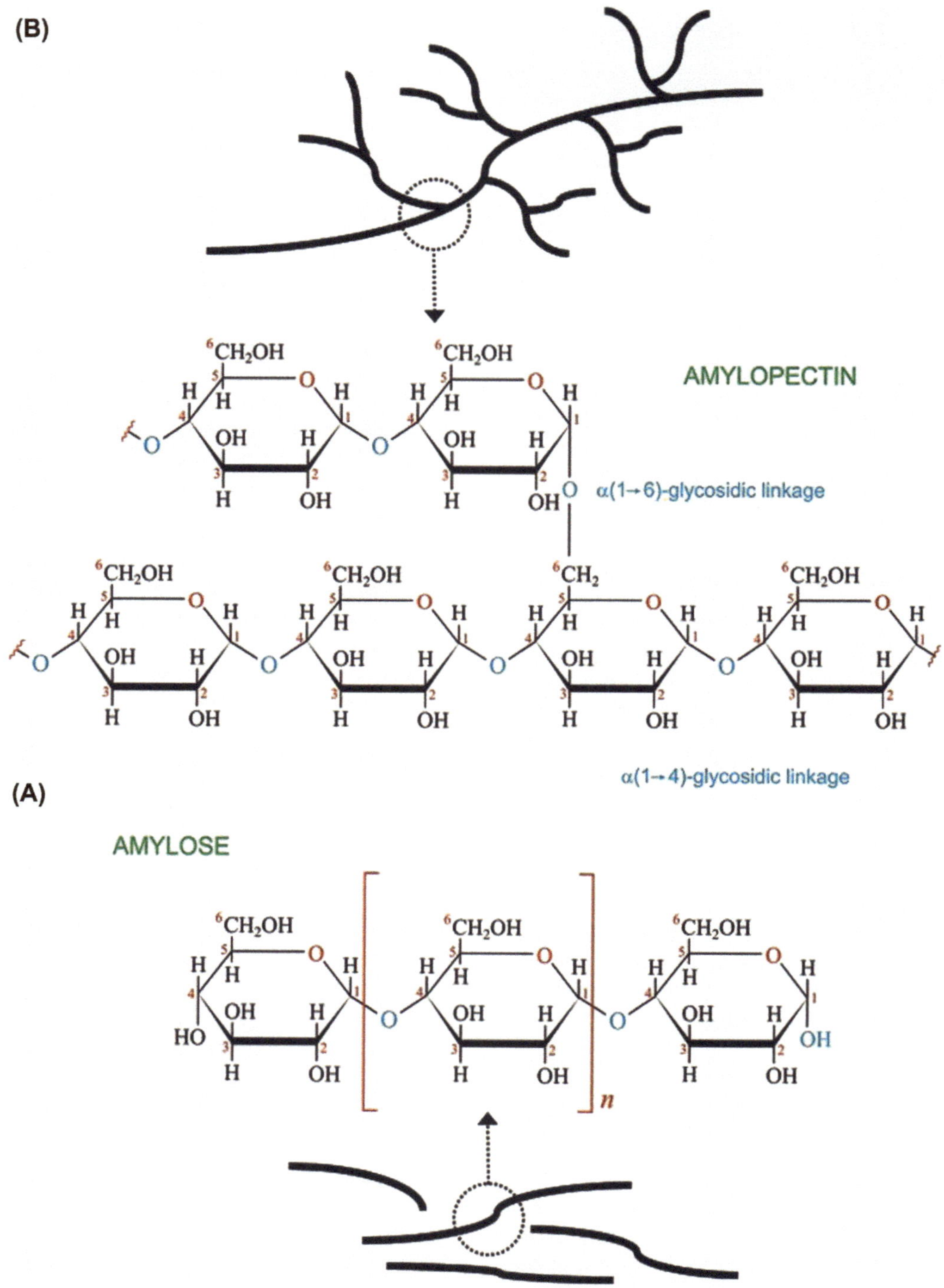

FIG. 8.3 Chemical structures and physical schematic representation of **(A)** amylose starch and **(B)** amylopectin starch.

TABLE 8.3
The Chemical Composition of Commercial Starches (Sanyang et al., 2016).

Starch	Density	Ash (%)	Amylose (%)	Water Content (%)
Wheat	1.44	0.2	26–27	13
Tapioca	1.446 −1.461	0.2	17	13
Maize	1.5	0.1	26–28	12–13
Potato	1.54 −1.55	0.4	20–25	18–19
Sago	–	0.2	24–27	10–20
Sugar palm starch	1.54	0.2	37.60	15

5 PROTEIN

5.1 Properties of Protein Biopolymer

Protein is considered as one of the most plentiful biological macromolecules in cells, occurring in an extensive variety of species and ranging in size from relatively small peptides to polymers with high molar mass. These molecules exhibit diverse biological functions (Durán et al., 2011), providing structure or biological activity in animals or plants. Besides that, proteins are well-known and compared with other macromolecules due to their structure that is based on approximately 20 amino acid monomers, rather than just a few or even one monomer, such as glucose in the case of cellulose and starch. Most proteins contain 100–500 amino acid residues (Fennema, 1985; Haghpanah et al., 2009).

The functional variety of proteins basically arises from their chemical structure. Depending on the sequential order of the amino acids, the protein will

TABLE 8.4
Properties of Sugar Palm Starch in Comparison With Sago Starch (Adawiyah et al., 2013).

Characterization	Parameters	Sugar Palm Starch	Sago Starch
Chemical composition	Amylose (%w/w) Fat (%w/w)	37.6 ± 1.46 0.27 ± 0.00	36.6 ± 1.55 0.24 ± 0.00
	Protein (%w/w)	0.10 ± 0.00	0.08 ± 0.00
	Moisture (%w/w)	9.03 ± 0.00	9.17 ± 0.00
	Ash (%w/w)	0.20 ± 0.00	0.16 ± 0.00
Gelatinization properties	Onset temperature (T_O) (°C) Peak temperature (T_P) (°C)	63.0 ± 0.12 67.7 ± 0.07	58.1 ± 0.28 67.3 ± 0.21
	Conclusion temperature (T_C) (°C)	74.6 ± 0.42	79.4 ± 0.88
	Range (T_C–T_O) (°C)	11.6 ± 0.49	21.3 ± 0.79
	ΔH (J/g)	15.4 ± 0.25	16.4 ± 0.24
Mechanical properties	Stress at 10% strain (kPa) Stress at shoulder point (kPa)	0.61 ± 0.10 23.0 ± 3.65	0.41 ± 0.04 15.5 ± 0.96
	Strain at shoulder point (%)	54.4 ± 3.54	59.1 ± 2.28
	Working until shoulder point (N mm)	20.2 ± 1.80	14.1 ± 0.82
	Breaking stress (kPa)	29.8 ± 2.64	–
	Breaking strain (%)	60.1 ± 2.61	–
	Work until breaking point (N mm)	29.6 ± 2.45	–
	Compressive force after breaking at 70% strain (N)	8.77 ± 0.59	9.97 ± 1.11
	Compressive force at 90% strain (N)	43.8 ± 2.34	44.0 ± 3.89
	Working until 90% strain (N mm)	108 ± 6.11	90.3 ± 8.37
	Adhesive force (N)	−3.64 ± 0.96	−8.99 ± 1.57

TABLE 8.5
Mechanical Properties of Starch-Based Biopolymer.

Film	Tensile Strength (TS, MPa)	Elasticity Modulus (EM, MPa)	Strain at Break (ε, %)	Year	References
Maize starch	0.24–20	51–315	–	2001	Anglès and Dufresne (2001)
Wheat starch	2.5–7.8	36–301	–	2005	Lu et al. (2005)
Potato starch	3	45	47	2006	Thunwall et al. (2006)
Rice starch	3.2	–	–	2006	Mehyar and Han (2006)
Pea starch	4.2	–	–	2006	Mehyar and Han (2006)
Amaranthus cruentus flour	0.8–3.0	–	74.2–620	2006	Colla et al. (2006)
Pea starch	1.4–5.8	8–98	38–51	2006	Zhang and Han (2006)
Maize starch	1–15	11–320	–	2006	Angellier et al. (2006)
Wheat starch	2.8–6.9	56–480	–	2006	Lu (2006)
Potato starch	13.7	460	–	2007	Kvien et al. (2007)
Corn starch	3	–	20	2008	Dai et al. (2008)
Pea starch	3.9–11.5	31.9–823.9	–	2008	Cao et al. (2008a,b)
Pea starch	3.9–11.9	31.9–498.2	–	2008	Cao et al. (2008a,b)
Maize starch	42	208–838	–	2008	Mathew et al. (2008)
Cassava	1.4–1.6	5–21	30–101	2009	Muller et al. (2009)
Cassava starch	4.8	84.3	–	2009	Teixeira et al. (2009)
Mango puree	8.76	322.05	–	2009	Azeredo et al. (2009)
Maize starch	6.75	220	–	2010	Kaushik et al. (2010)
Pea starch	2.5–12	20.4–210.3	–	2010	Liu et al. (2010)
Wheat starch	3.15–10.98	–	–	2010	Chang et al. (2010)
Corn starch	2.5–3.6	21–533	48–63	2011	Fu et al. (2011)
Rice starch	1.6–11	21–533	3–60	2011	Dias et al. (2011)
Maize starch	0.35	3.12	–	2011	Teixera et al. (2011)
Potato starch	–	460	–	2012	Chen et al. (2012)
Potato starch	17.5	1317.0	–	2013	Hietale et al. (2013)
Potato starch	5.01	160	–	2014	Nasri-Nasrabadi (2014)
Maize starch	2.35	53.6	–	2014	Karimi et al. (2014)
Maize starch	17.4	520	–	2014	Slavutsky and Bertuzzi (2014)
Cush-cush yam starch	1.88	13.9	19	2015	Gutiérrez et al. (2015)
Corn starch	38.0	141.0	–	2015	Babaee et al. (2015)
Potato starch	4.93	–	–	2016	Noshirvani et al. (2016)

Corn starch	11.2	12.4	–	2017	Llanos and Tadini (2018)
Sugar palm starch	4.8	59.97	38.10	2018	Ilyas et al. (2018a)
Yam bean starch	11.47	443	–	2018	Asrofi et al. (2018a)
Sugar palm Starch	11.5	178	–	2018	Ilyas et al. (2018a)
Tapioca starch	5.8	403	–	2018	Asrofi et al. (2018b)
Tapioca starch	12.48	479.8	–	2018	Syafri et al. (2018)
Sugar Palm Starch	10.68	121.26	–	2019	Ilyas et al. (2019c)
Bengkuang starch	–	–	–	2019	Syafri et al. (2019)

assume different structures along the polymer chain, based on disulfide cross-link interactions among the amino acid units, hydrophobic, electrostatic, hydrogen bonding, and van der Waals (Fennema, 1985). There are approximately billions of proteins with distinctive properties that can be produced by altering the chain length of polypeptides, the type and ratio of amino acids, and sequence of amino acid. Usually, oilseeds, milk, vegetables, cereals, eggs, and meats (including poultry and fish) have been the main sources of food proteins. According to Fennema (1985), the functional properties of proteins in foods are related to their structural and other physicochemical characteristics. Besides that, the structures of proteins can be altered by various chemical and physical processes such as metal ions, acids and alkalis, lipid interfaces, irradiation, pressure, mechanical treatment, and heat treatment (Fennema, 1985). Such agents have the capacity to change the proteins structures and affect their functional properties. In films development, these modifications are often used in the formation process to optimize proteins configuration and interactions, resulting in better film properties.

A wide variety of proteins from animal/vegetable sources can be used to produce films, as shown in Table 8.7. In addition to the use of proteins for films production, researchers are focusing on the study of some strategies to improve their performance and to provide bioactive properties. In the group of animal proteins, the most used are caseins, whey protein, collagen, gelatin, myofibrillar proteins, and egg proteins, and among the vegetable proteins, the most used are soy protein, gluten, and zein (Mhd Haniffa et al., 2016; Pérez-Gago and Rhim, 2014).

In addition, casein-based films and biomaterials obtained from caseinate can be found in many applications such as in mulching films, in coatings for vegetables and fruits, in edible films, and in packaging

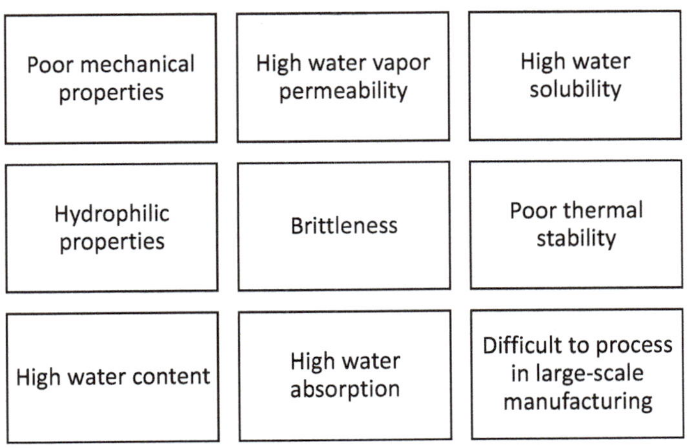

FIG. 8.4 Disadvantages of starch-based biopolymer.

TABLE 8.6
Starch-Based Biopolymer, Its Manufacturing Technique, and Applications.

Polymer Component	Manufacturing Techniques	Applications	References
Plasticized starch	Solution casting	Transparent materials	Nasseri and Mohammadi (2014)
Starch	Blending, solution casting	Air permeable, resistant, surface-sized paper, food packaging	Slavutsky and Bertuzzi (2014) and Yang et al. (2014)
Starch	Solution casting	Food packaging	Liu et al. (2010)
Cassava starch	Solution casting	Food packaging	Teixeira et al. (2009)
Sugar palm starch	Solution casting	Food packaging	Ilyas et al. (2018e)
Sugar palm starch	Solution casting	Food packaging	Ilyas et al. (2018b)
Sugar palm starch	Solution casting	Food packaging	Atikah et al. (2019) and Ilyas et al. (2018e)
Wheat starch	Solution casting	Food packaging	Lu et al. (2006)
Tuber native potato	Solution casting	Packaging	Montero et al. (2017)
Cereal corn Starch	Solution casting	Packaging	Montero et al. (2017)
Legume pea starch	Solution casting	Packaging	Montero et al. (2017)
Yam bean	Solution casting	Packaging	Asrofi et al. (2018b)
Yam bean	Solution casting	Packaging	Asrofi et al. (2018a)
Cassava bagasse Starch	Solution casting	Packaging	Teixeira et al. (2009)
Ramie starch	Solution casting	Packaging	Lu et al. (2006)
Potato	Solution casting	Packaging	Chen et al. (2012)
Cassava Starch	Solution casting	Packaging	Syafri et al. (2018)
Bengkuang Starch	Solution casting	Packaging	Syafri et al. (2019)

application. Table 8.8 shows the comparison of mechanical properties of some of the milk protein films formed with plasticizer.

5.2 Advantages and Disadvantages of Protein Biopolymer

There are several advantages of protein biopolymer, especially for food packaging application such as highly nutritional quality, good potential to adequately protect food product from their surrounding environment and excellent sensory properties (Gupta and Nayak, 2015). Proteins also act as a flavor and carrier of antioxidant, besides improving the quality of food and bacteriostats. The proteins such as milk proteins (whey proteins and casein) (Su et al., 2010), gluten (Zhong and Yuan, 2013), corn (Aydt et al., 1991), peanut (Aydt et al., 1991), sunflower seed (Martinez et al., 2005), whey protein (Jooyandeh, 2011), soy bean (Zhang et al., 2010), gelatin from collagen (Gómez-Guillén et al., 2011), soy protein (Tian et al., 2011), and wheat (Aydt et al., 1991) are suitable for the fabrication of protein biopolymer film due to its nutritional properties.

This type of biopolymer is also being used in nonedible packaging. Protein-based biopolymer also had impressed gas barrier properties compared with those prepared from polysaccharides and lipids (Cuq et al., 1998). Amazingly, the oxygen gas (O_2) permeability of the soy protein film was 670, 540, 500, and 260 times lower than that of pectin, starch, low density PE, and methyl cellulose, respectively, when they are not moist (Cuq et al., 1998). One of the disadvantage of soy protein film is low moisture barrier properties. This is because of their hydrophilic property and the considerable amount of hydrophilic plasticizer used in film preparation (Cuq et al., 1998).

TABLE 8.7
Some Research Using Different Protein Bases in the Preparation of Films Combining Strategies to Improve the Properties of Films.

Protein Type	Strategies	Effect Observed	References
VEGETABLE PROTEIN			
Amaranth	Native waxy and maize starch nanocrystals	Structure reinforcement	Condés et al. (2015)
Bitter vetch (*Vicia ervilia*) seed/corn zein	Bilayer	Improved structure	Arabestani et al. (2016)
Canola		–	Shi and Dumont (2014)
Pea	Lysozyme	Antimicrobial	Fabra et al. (2014)
Sesame meal	–	–	Sharma and Singh (2016)
Soy	Chestnut (*Castanea mollissima)* bur extracts	Antioxidant	Wang et al. (2016)
Soy protein isolate	Peanut protein nanoparticles	Structure reinforcement	Li et al. (2015)
Soy/agar	Blend/extrusion	Structure reinforcement	Garrido et al. (2016)
Sunflower protein	Clove essential oil	Structure reinforcement and improved food shelf life	Salgado et al. (2013)
ANIMAL PROTEIN			
Argentine anchovy (*Engraulis anchoita*)	Sorbic or benzoic acids	Antimicrobial	Rocha et al. (2014)
Chicken feet	–	–	Lee et al. (2015a)
Fish gelatin	Chitosan nanoparticles	Structure reinforcement	Hosseini et al. (2015)
Gelatin	Longan seed extract	Antioxidant	Sai-Ut et al. (2015)
Porcine meat and bone meal	Coriander oil	Antimicrobial	Lee et al. (2015b)
Shrimp (*Litopenaeus vannamei*) muscle	Cinnamaldehyde/thermal treatment	Cross-linking	Gómez-Estaca et al. (2014)
Whey protein	Gum tragacanth	Structure reinforcement	Tonyali et al. (2018)
Whey protein	Ultraviolet treatment	Structure reinforcement	Díaz et al. (2017)
Whey protein isolate	*Lactobacillus rhamnosus*	Antimicrobial	Beristain-Bauza et al. (2016)
Whitemouth croaker muscle	Pink pepper phenolic	Browning reduction	Romani et al. (2018)

5.3 Application of Protein Biopolymer

Several scientists had conducted studied on the protein-based edible films such as corn zein films on nut and fruit products, whey protein film, casein emulsion film, and soy protein film for food packaging applications (Calva-Estrada et al., 2019; Gennadios, 2004; Rhim et al., 2004; Schmid and Müller, 2019). The polymeric characteristics of the protein film have been used for edible food packaging application (Khwaldia et al., 2010; Oussalah et al., 2004; Su et al., 2010; Zhang et al., 2010), but for nonfood packaging application, the major problems are an advancement of mechanical properties (such as tensile modulus, shear strength, flexural, elasticity, strength, and toughness). Table 8.9 displays packaging and biomedical applications of protein biopolymer. Besides that, biocompatible materials from proteins biomaterials have been utilized to develop scaffolds for various biomedical applications such as drug delivery, tissue engineering, wound dressings, and membrane filters. Fig. 8.5 shows the images of CMC (carboxymethyl cellulose)/SPI (soy protein isolate) film. This film is made up by a continuous

TABLE 8.8
Mechanical Properties of the Milk Protein Film Formed in the Presence of Different Formulation of Plasticizers.

Film	Tensile Strength (MPa)	Elongation at Break (%)	References
Sodium caseinate/ glycerol (4: 1)	10.5	17.4–26.7	Siew et al. (1999)
Sodium caseinate/ glycerol (2: 1)	73.7 –84.2	10.9–11.7	
Sodium caseinate/ PEG (4.54: 1)	5.3	10.9–16.35	
Sodium caseinate/ PEG (1.9:1)	25.4	10.9–13.9	
Whey protein/ glycerol (5.7/1)	4.1	29.1	McHugh et al. (1994) and McHugh and Krochta (1994)
Whey protein/ glycerol (2.3/1)	30.8	13.9	
Whey protein/ sorbitol (2.3/1)	1.6	14.0	
Whey protein/ sorbitol (1/ 1)	8.7	14.7	

TABLE 8.9
Packaging and Biomedical Applications Made From Protein Biopolymer.

Protein Biopolymer	References
PACKAGING APPLICATION	
Gluten films	Hemsri et al. (2011), Zhong and Yuan (2013), Zuo et al. (2009)
Milk protein films	Banerjee et al. (1996), Chen (1995), Kinsella and Morr (1984), Su et al. (2010), Vachon et al. (2000), Yoo and Krochta (2011)
Soy protein films	Park et al. (2000), Rangavajhyala et al. (1997), Rhim et al. (2006), Stuchell and Krochta (1994), Su et al. (2010), Tian et al. (2011), Zhang et al. (2010)
Corn zein films	Khwaldia et al. (2010), Lim and Jane (1994), Pol et al. (2002)
Gelatin films	Avena-Bustillos et al. (2011), Carvalho and Grosso (2006), Sobral et al. (2001)
Silk protein films	Jiang et al. (2007), Zhang et al. (2004)
BIOMEDICAL APPLICATION	
Scaffold in tissue engineering	Haghpanah et al. (2009), Hu et al. (2010)
Drug delivery systems	Cho et al. (2008), Koutsopoulos et al. (2009), Qiu and Park (2001), Torchilin (2005)
Biosynthetic hybrid hydrogels scaffold	Almany and Seliktar (2005)

casting method. Remarkably, this film can be produced simply with absence of puncture and cracks. Besides that, the film produced has a good flexible and durability, as much as it is necessary to be rolled into forms for sensible applications (Su et al., 2010).

6 CHITIN AND CHITOSAN

6.1 Properties of Chitin and Chitosan

Chitin is a polysaccharide with linear chains comprising of 2-acetamide-2-deoxy-β-D-glucopyranose units, which are linked by glycosidic bonds β(1 → 4) (Ma et al., 2019). Chitin exists as the second most abundant organic substance in the biosphere, surpassed only by cellulose, but chitin surpasses the latter in terms of replacement rate, which is twice as high as cellulose (Deringer et al., 2016). Chitin is found in the skeletal structure of invertebrates, such as arthropods, annelids, mollusks, and coelenterates, and in the cell walls of

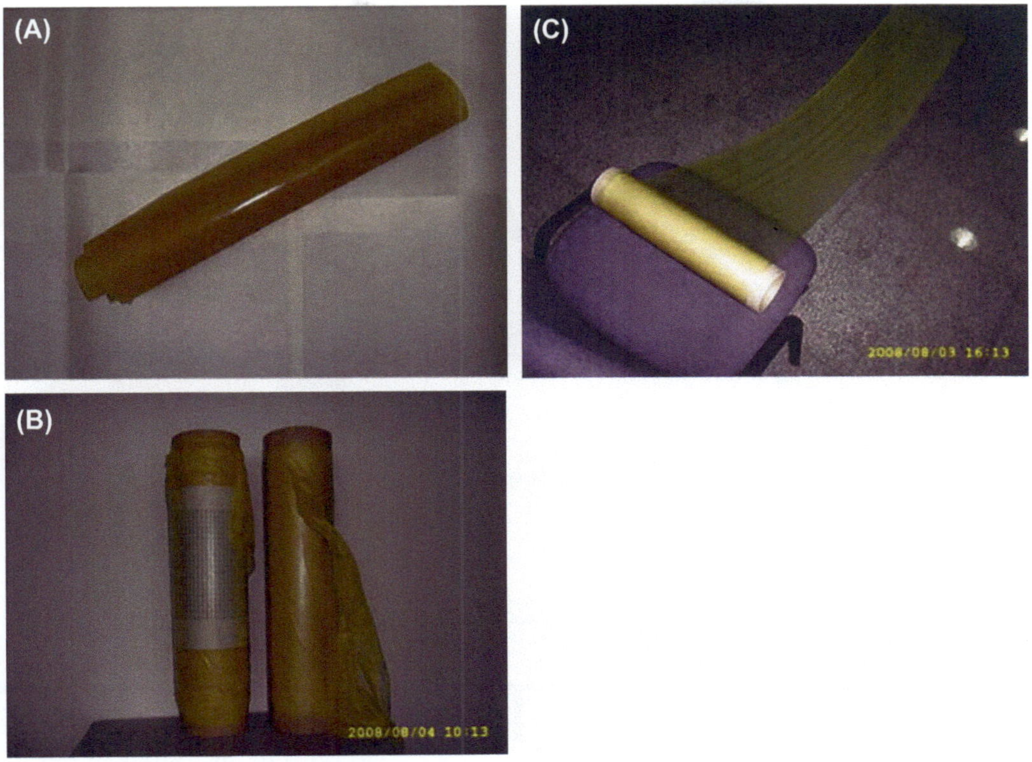

FIG. 8.5 Photographs of CMC/SPI film in expanded and rolled states fabricated by a continuous casting method (Su et al., 2010).

diatoms and some fungi (Nataraj et al., 2018). Depending on the organism, chitin adopts different polymorphic structures called α-, β- and γ-chitin. The different polymorphic structures of chitin correspond to different arrangements in solid state (Pighinelli, 2019). The α-chitin is found in rigid and resistant structures, such as the arthropod cuticle. This chitin is usually associated with proteins, inorganic materials, or both (Deringer et al., 2016). The β- and γ-chitin occur in flexible structures and are also resistant. The α-chitin is the most abundant form and is considered the most stable, considering that the conversion of β- and γ-chitin in the first one is irreversible (Deringer et al., 2016).

Chitosan is commonly made by alkaline deacetylation of α-chitin. Chitin and chitosan are important linear polysaccharides consisting, respectively, of 2-acetamide-2-deoxy-β-D–glucopyranose (GlcNAc) and 2-amino-2-deoxy-β-D-glucopyranose (GlcN) linked by β(1 → 4) (Nataraj et al., 2018). Fig. 8.6 shows the molecular structures of chitin and chitosan. Chitosan solubilization can be attributed to protonation of the $-NH_2$ in the C−2 of D-glucosamine units in acid medium, which results in the conversion into a polyelectrolyte (Roy et al., 2017). The polycationic character of chitosan in acid medium is due to its weak base characteristic, thus its amino groups are easily protonated (Nataraj et al., 2018). Moreover, the hydroxyl groups of carbons 3 and 6 can also be protonated, which increases chitosan reactivity (Nataraj et al., 2018).

6.2 Advantages and Disadvantages of Chitin and Chitosan

Chitin is also fascinating in cosmetology because it is tolerable by the skin. It acts as an effective hydrating agent and a film-forming tensor having two benefits that are often cited: it provides water and it minimizes dehydration. The molecular weight of most chitosan-based products is too high that they cannot penetrate the skin, which is a crucial characteristic of a skincare product. These materials consist of chitosan hydrochloride, chitosan acetate, chitosan lactate, carboxymethyl chitosan, quaternized derivatives, oligosaccharides, and chitin sulfate and carboxymethyl chitin (Pighinelli, 2019).

Chitosan films have revealed remarkable results, possessing good mechanical properties and having the

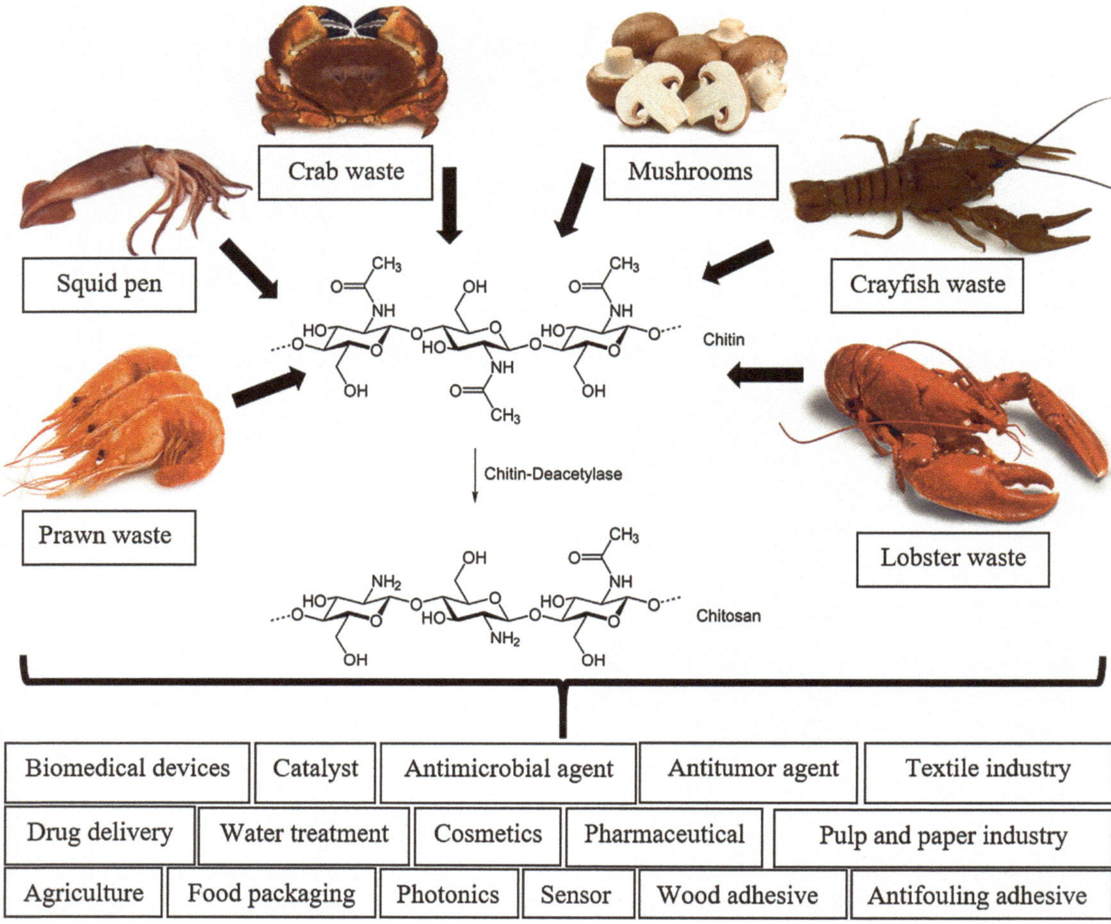

FIG. 8.6 Molecular structures of chitin and chitosan, their sources and applications.

advantage of the ability to integrate functional substances, for instance, as vitamins and carriers that release antimicrobial agents. Chitosan derivatives are advantageous due to the biocompatibility and safe, nontoxic to living tissues and their hydrophilic property, biodegradability, antibacterial activity, bioadhesivity, mucoadhesivity, and complexing property (Llanos and Tadini, 2018).

Chitosan has the ability to form gels in addition to having viscosity-related properties, complete biodegradability, and even antitumor influence, like alginate polysaccharide. Chitosan creates films permeable to air that aid cellular regeneration while protecting tissues from microbial attacks. Chitosan also stimulates the process of regeneration of tissues, making it a suitable material in artificial skin manufacturing for the applications in skin grafts on high degree burns and in surgical applications (suture threads) (Oryan and Sahvieh, 2017). In terms of health benefit, chitosan is able to trap lipids at its insolubilization pH along the digestive tract, which significantly reduces cholesterol level in the blood. Chitosan possesses bioadhesive properties that make it of interest in adhesive-sustained release formulation required. Mucoadhesivity permits to enhance the adsorption of drugs especially at neutral pH.

This natural polymer owns several inherent features making it an effective material for environmental purposes: (i) lower cost in comparison with activated carbon or organic resins available in market, (ii) outstanding pollutant-adhering capacities and outstanding selectivity, (iii) versatility, and (iv) biodegradability. Indeed, major applications of chitosan are based on its excellent capability to tightly bind a whole range of pollutants (Vidal and Moraes, 2019).

Researchers had summarized that the biosorbents are effective in pollutant elimination with the extra advantage of being cheap, nontoxic, and biocompatible.

Chitosan has gained substantial attention for developing microcapsules, which is advantageous as drug carriers due to their controlled release properties and biocompatibility (Rokhade et al., 2007). Presence of microcapsules also makes chitosan applicable for wall materials for textile finishing product encapsulation (Alonso et al., 2010). Numerous techniques have been used in the creation of chitosan microcapsules, such as spray drying and phase coacervation (Liu et al., 2011); the microcapsules yield is either single or multilayer, depending on microencapsulation method (Pothakamury and Barbosa-Cánovas, 1995).

Chitin and chitosan poses high organic solvent resistance, which is useful for separation membranes. This biomaterial is used with organic solvents, where chemical resistance is typical. Explicitly, chitin is highly acid resistant, meanwhile chitosan is highly alkaline resistant. These features contribute to the application of chitin and chitosan as separation membranes for a range of uses in response to specific requirements (Chaudhary et al., 2015).

However, chitosan films are unsuitable for packaging application due to the fact that they are highly permeable to water vapor. In addition, their hydrophilic character also make them to exhibit resistance to fat diffusion and selective gas permeability (Morin-Crini et al., 2019).

6.3 Application of Chitin and Chitosan

Chitin- or chitosan-based biomaterials are promising candidates to be used for wound healing, tissue engineering, and drug delivery. The fact that they are originated from the freely available natural sources has made them more economically stable compared with synthetic polymer materials. Moreover, their utilizations in biomedical field, such as scaffolds for tissue engineering, manages to minimize cost and manpower needed for second surgery to remove them, since chitin/chitosan-based materials are biodegradable. Besides, their biocompatibility also avoids the necessity for any treatments to be performed due to the rejection of implants fabricated from employing these materials. Currently existing chitin- and chitosan-based commercial products are as summarized in Table 8.10. The markets for chitin and chitosan are the United States, China, Norway, France, Poland, Japan, Germany, Korea, Canada, Australia, and the United Kingdom (Crini and Lichtfouse, 2016; Morin-Crini et al., 2019). In 2015, Japan has dominated (advanced in technology, commercialization, and utilization of these

TABLE 8.10
Commercial Products From Chitin and Chitosan and Their Applications.

Product	Manufacturer	Application	Patent/References
LipoSan Ultra	Primex	Weight loss	US 6130321, Johnson and Nichols (2000)
Slim MED	KitoZyme	Weight management and treatment	US 20040126444 A1, D'huart and Dallas (2004)
ChitoDot	Tricol biomedical, Inc.	Wound dressing and bleeding control	US 8269058 B2, McCarthy et al. (2012)
BST-Gel	Piramal Healthcare (Canada) Inc.	Chronic wound healing, bone filling, cartilage repair, invertebral disc regeneration	US 8747899, Chaput and Chenite (2014)
ChitoFlex PRO	Tricol biomedical, Inc.	Wound dressing and bleeding control	US 8668924 B2, McCarthy et al. (2014)
HemCon ChitoGauze PRO	Tricol Biomedical, Inc.	Wound dressing material	US 9205170 B2, Lucchesi and Xie (2015)
Talymed	Marine Polymer Technologies	Wound dressing material	US 9139664 B2, Finkielsztein and Vournakis (2015)
Protasan	NovaMatrix	Pharmaceutical application	WO 2015081304 A1, Francis et al. (2015)
Reaxon	Medovent	Nerve rejuvenation	

biopolymers) the industry market of chitin and chitosan accounting for 35%, which is equivalent to 700–800 tons per annual (Morin-Crini et al., 2019).

7 POLY(BUTYLENE SUCCINATE)

7.1 Properties of PBS

The world community is anticipated to grow to 9 billion by 2050, hence resulting in the increment of the plastic production and undoubtedly, plastic wastes (Emadian et al., 2017). These motivated researchers to explore and introduce bioplastics, such as PCL, polylactide (PLA), PBS, and TPS. PBS is identified as the most notable biopolymer that is produced via polycondensation of butanediol and succinic acid having unique features, such as good melt processability, great toughness, high chemical resistance, high heat distortion temperature, biodegradable, good mechanical properties, high chemical, and thermal resistance (Boonprasith et al., 2013). This biopolymer is derived from natural sources (Jamaluddin et al., 2016). PBS is a thermoplastic polymer resin in the polyester family. PBS comprises of polymerized units of butylene succinate with repeating $C_8H_{12}O_4$ units. There are two ways to synthesize PBS:

1. Transesterification process (from succinate diesters)
2. Direct esterification process (from the diacid).

The typical route in PBS production is direct esterification of succinic acid with 1,4-butanediol, which involves two steps. First, excess of diol is esterified with diacid to from PBS oligomers with water as by-product of process. Next, these PBS oligomers are transesterified under vacuum condition with the presence of catalyst (zirconium, germanium, titanium, or tin) in order to produce a high molecular weight (Mw) polymer that is PBS as shown in Fig. 8.7.

The physical properties of poly(butylene succinate-co-butylene adipate) copolyesters can vary with comonomer content, as tabulated in Table 8.11 and Fig. 8.8.

7.2 Advantages and Disadvantages of PBS

PBS are synthesized using succinic acid. Succinic acid is also obtainable from the fermentation of microorganisms on renewable feedstocks, for instance, glucose, starch, xylose, etc (Song and Lee, 2006). *Actinobacillus succinogenes*, *Anaerobiospirillum succiniciproducens*, *Mannheimia succiniciproducens*, and recombinant *Escherichia coli* are well-known and well-established bacterial production strains that can produce succinic acid (Gigli et al., 2016; Song and Lee, 2006; Xu and Guo, 2010).

Biodegradability is another useful characteristic of PBS biopolymer, since the need of surgery to remove the carrier/implant can be prevented as it self-degrades when its desired function has ended. It is worth stating that not only the biomaterial but also the degradation products are nonhazardous for the host.

Conversely, PBS has attention-grabbing physiomechanical properties and can be simply synthesized by melt polycondensation at reasonable costs. Added value is given by the ability to achieve both succinic acid and 1,4-butanediol from renewable resources, which marks PBS a completely bio-based and biodegradable polymer.

7.3 Application of PBS

As PBS disintegrates naturally into water and CO_2, it offers as a biodegradable substitute to some commonly used plastics. The range of PBS application areas is still expanding and several areas can be recognized, but it remains hard to know explicitly in which specific object PBS is really used. In packaging field, PBS is converted into films, bags, or boxes, for both food and cosmetic packaging. PBS also could be found in disposable merchandises, such as tableware or medical articles. Next, in agriculture, PBS is useful in the manufacturing of mulching films or delayed release materials for pesticide and fertilizer. PBS is also promising to find market shares in fishery (for fishing nets), forestry, civil

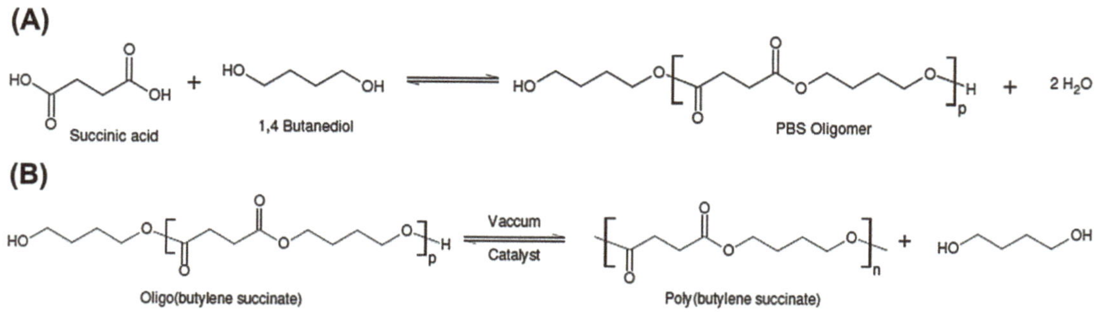

FIG. 8.7 PBS synthetization process using direct esterification process: **(A)** first step and **(B)** second step (Jamaluddin et al., 2016).

TABLE 8.11
Thermal Properties and Degree of Crystallinity of PBSA Random Polyesters (Xu and Guo, 2010).

Polymer	ΔH_m (J/g)	ΔH_m{{\tf="PSMPi4"\char56}} (J/g)	T_m (°C)	T_g (°C)[a]	Crystallinity[b] (%)	Crystallinity[c] (%)
PBS	67.4	110.3	112	−18	61.1	39.66
PBSA-5[d]	96.0	110.3	108	−21	87.0	54.47
PBSA-10	72.5	110.3	103	−23	65.7	45.83
PBSA-15	79.8	110.3	99	−27	72.3	45.27
PBSA-20	59.5	110.3	92	−34	53.9	46.83

[a] The glass transition temperature (T_g) was adopted from the tanδ peak measured by dynamic thermal analysis.
[b] The degree of crystallinity was the ratio of melting enthalpy determined by DSC to the melting enthalpy of completely crystalline PBS (100 J/g).
[c] The degree of crystalline was calculated from WAXD results.
[d] The number indicates the molar percentage of adipic acid in the total feed acids for synthesis of PBSA copolyester.

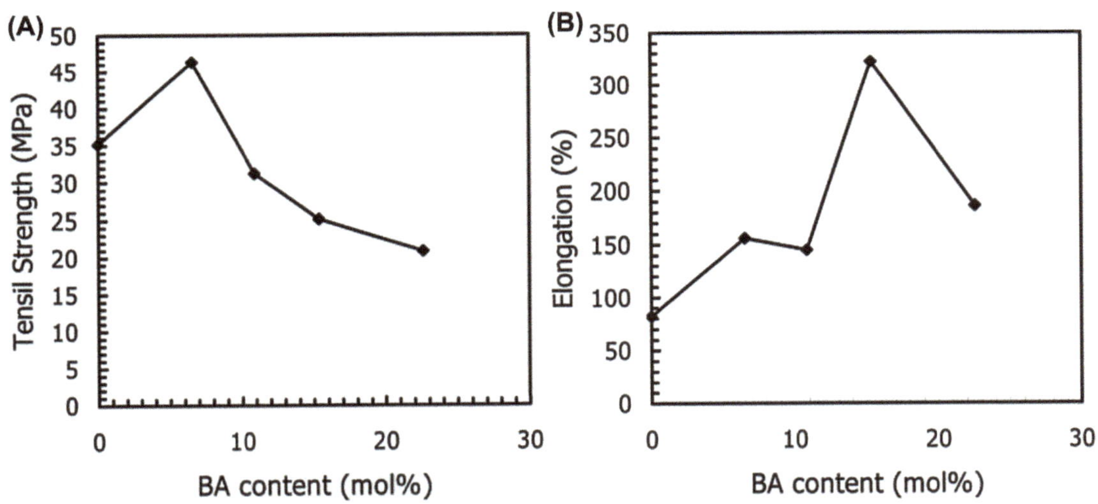

FIG. 8.8 The mechanical properties of PBSA at different contents of butylene adipate (BA) content: **(A)** Tensile strength and **(B)** Elongation at break (Xu and Guo, 2010).

engineering, or other fields in which recovery and recycling of materials after utilization is challenging. In medical field, PBS is adopted as biodegradable drug encapsulation system and is also being studied for implants.

8 SUMMARY AND FUTURE PERSPECTIVES

The progress of bio-based polymers as substitution for petroleum-based synthetics has been an area of attention due to the nondegradable and nonrenewable nature of synthetic plastics, as well as a primary research challenge for scientists. With the future fossil fuel crisis, the exploration and expansion of alternative chemical/material alternatives is crucial in minimizing mankind's reliance on fossil fuel resources. Some of the probable substitute candidates are PHA, PLA, starch, protein, chitin, chitosan, and PBS. Bio-based polymers are generally defined as polymers manufactured from renewable resources, comprising of three different categories: (1) natural polymers originated from biomass such as the agropolymers from agroresources, e.g., starch, cellulose, protein, chitin, and chitosan; (2) polymers formed by microorganisms, e.g., PHA; and (3) synthetic biopolymers, which are chemically synthesized using monomers obtained from agroresources, e.g., poly(lactic acid) and PBS, which is biodegradable as well. Overall, bio-based polymers are still new, though, they are in continuous development. Research attempts are obviously intent on addressing challenges constraining the use of bio-based polymers, comprising reduction the production and processing costs, improving barrier and mechanical properties, or introducing extra functions as active and smart packaging. Being a carbon neutral and valuable polymer manufactured from many renewable carbon sources by microorganisms, PHA is said to be a sustainable and environmental-friendly material. However, nowadays, PHA is not cost competitive compared with fossil-derived products. PLA is also produced from renewable sources having high tensile strength and modulus and can be processed via conventional processing methods. Starch complies all the principle aspects; hence, it is suitable for edible coatings/films. Chitin is the most abundant natural amino polysaccharide and is predicted to be produced yearly almost as much as cellulose. It has grabbed great attention not only as an underutilized resource but also as a new useful material of great potential in many fields, and current progress in chitin chemistry is quite remarkable. PBS is a renowned aliphatic polyester, provided its fascinating thermomechanical properties and the proven biodegradability, combined with acceptable raw material and production costs. As a conclusion, biodegradable plastics are often reflected as savior products. Shifting to these materials would lessen carbon dioxide, greenhouse gas emission levels, energy of manufacturing, and amount of waste produced and create opportunity for new plastic industries.

REFERENCES

Abral, H., Ariksa, J., Mahardika, M., Handayani, D., Aminah, I., Sandrawati, N., Pratama, A.B., Fajri, N., Sapuan, S.M., Ilyas, R.A., 2020a. Transparent and antimicrobial cellulose film from ginger nanofiber. Food Hydrocolloids 98, 105266. https://doi.org/10.1016/j.foodhyd.2019.105266.

Abral, H., Ariksa, J., Mahardika, M., Handayani, D., Aminah, I., Sandrawati, N., Sapuan, S.M., Ilyas, R.A., 2019a. Highly transparent and antimicrobial PVA based bio-nanocomposites reinforced by ginger nanofiber. Polymer Testing 106186. https://doi.org/10.1016/j.polymertesting.2019.106186.

Abral, H., Atmajaya, A., Mahardika, M., Hafizulhaq, F., Kadriadi, Handayani, D., Sapuan, S.M., Ilyas, R.A., 2020b. Effect of ultrasonication duration of polyvinyl alcohol (PVA) gel on characterizations of PVA film. Journal of Materials Research and Technology 1–10. https://doi.org/10.1016/j.jmrt.2019.12.078.

Abral, H., Basri, A., Muhammad, F., Fernando, Y., Hafizulhaq, F., Mahardika, M., Sugiarti, E., Sapuan, S.M., Ilyas, R.A., Stephane, I., 2019b. A simple method for improving the properties of the sago starch films prepared by using ultrasonication treatment. Food Hydrocolloids 93, 276–283. https://doi.org/10.1016/j.foodhyd.2019.02.012.

Adawiyah, D.R., Sasaki, T., Kohyama, K., 2013. Characterization of arenga starch in comparison with sago starch. Carbohydrate Polymers 92, 2306–2313. https://doi.org/10.1016/j.carbpol.2012.12.014.

Aisyah, H.A., Paridah, M.T., Sapuan, S.M., Khalina, A., Berkalp, O.B., Lee, S.H., Lee, C.H., Nurazzi, N.M., Ramli, N., Wahab, M.S., Ilyas, R.A., 2019. Thermal properties of woven kenaf/carbon fibre-reinforced epoxy hybrid composite panels. International Journal of Polymer Science 1–8. https://doi.org/10.1155/2019/5258621.

Almany, L., Seliktar, D., 2005. Biosynthetic hydrogel scaffolds made from fibrinogen and polyethylene glycol for 3D cell cultures. Biomaterials 26, 2467–2477. https://doi.org/10.1016/j.biomaterials.2004.06.047.

Alonso, D., Gimeno, M., Sepúlveda-Sánchez, J.D., Shirai, K., 2010. Chitosan-based microcapsules containing grapefruit seed extract grafted onto cellulose fibers by a non-toxic procedure. Carbohydrate Research 345, 854–859. https://doi.org/10.1016/j.carres.2010.01.018.

Angellier, H., Molina-Boisseau, S., Dole, P., Dufresne, A., 2006. Thermoplastic starch–waxy maize starch nanocrystals nanocomposites. Biomacromolecules 7, 531–539. https://doi.org/10.1021/bm050797s.

Anglès, M.N., Dufresne, A., 2001. Plasticized starch/tunicin whiskers nanocomposite materials. 2. Mechanical behavior. Macromolecules 34, 2921–2931. https://doi.org/10.1021/ma001555h.

Arabestani, A., Kadivar, M., Amoresano, A., Illiano, A., Di Pierro, P., Porta, R., 2016. Bitter vetch (*Vicia ervilia*) seed protein concentrate as possible source for production of bilayered films and biodegradable containers. Food Hydrocolloids 60, 232–242. https://doi.org/10.1016/j.foodhyd.2016.03.029.

Asrofi, M., Abral, H., Kasim, A., Pratoto, A., Mahardika, M., Hafizulhaq, F., 2018a. Mechanical properties of a water hyacinth nanofiber cellulose reinforced thermoplastic starch bionanocomposite: effect of ultrasonic vibration during processing. Fibers 6, 40. https://doi.org/10.3390/fib6020040.

Asrofi, M., Abral, H., Kasim, A., Pratoto, A., Mahardika, M., Hafizulhaq, F., 2018b. Characterization of the sonicated yam bean starch bionanocomposites reinforced by nanocellulose water hyacinth fiber (WHF): the effect of various fiber loading. Journal of Engineering Science and Technology 13, 2700–2715.

Asyraf, M.R.M., Ishak, M.R., Sapuan, S.M., Yidris, N., 2020. Woods and composites cantilever beam: a comprehensive review of experimental and numerical creep methodologies. Journal of Materials Research and Technology. https://doi.org/10.1016/j.jmrt.2020.01.0.

Atikah, M.S.N., Ilyas, R.A., Sapuan, S.M., Ishak, M.R., Zainudin, E.S., Ibrahim, R., Atiqah, A., Ansari, M.N.M., Jumaidin, R., 2019. Degradation and physical properties of sugar palm starch/sugar palm nanofibrillated cellulose bionanocomposite. Polimery 64, 27–36. https://doi.org/10.14314/polimery.2019.10.5.

Atiqah, A., Jawaid, M., Sapuan, S.M., Ishak, M.R., Ansari, M.N.M., Ilyas, R.A., 2019. Physical and thermal properties of treated sugar palm/glass fibre reinforced thermoplastic polyurethane hybrid composites. Journal of Materials Research and Technology 8, 3726–3732. https://doi.org/10.1016/j.jmrt.2019.06.032.

Auras, R.A., Singh, S.P., Singh, J.J., 2005. Evaluation of oriented poly(lactide) polymers vs. existing PET and oriented PS for fresh food service containers. Packaging Technology and Science 18, 207–216. https://doi.org/10.1002/pts.692.

Avena-Bustillos, R.J., Chiou, B., Olsen, C.W., Bechtel, P.J., Olson, D.A., McHugh, T.H., 2011. Gelation, oxygen permeability, and mechanical properties of mammalian and fish gelatin films. Journal of Food Science 76, E519–E524. https://doi.org/10.1111/j.1750-3841.2011.02312.x.

Aydt, T.P., Weller, C.L., Testin, R.F., 1991. Mechanical and barrier properties of edible corn and wheat protein films. Transactions of the American Society of Agricultural Engineers 34, 207–211. https://doi.org/10.13031/2013.31646.

Azammi, A.M.N., Ilyas, R.A., Sapuan, S.M., Ibrahim, R., Atikah, M.S.N., Asrofi, M., Atiqah, A., 2020. Characterization studies of biopolymeric matrix and cellulose fibres based composites related to functionalized fibre-matrix interface. In: Interfaces in Particle and Fibre Reinforced Composites. Elsevier, London, pp. 29–93. https://doi.org/10.1016/B978-0-08-102665-6.00003-0.

Azeredo, H.M.C., Mattoso, L.H.C., Wood, D., Williams, T.G., Avena-Bustillos, R.J., McHugh, T.H., 2009. Nanocomposite edible films from mango puree reinforced with cellulose nanofibers. Journal of Food Science 74. https://doi.org/10.1111/j.1750-3841.2009.01186.x.

Babaee, M., Jonoobi, M., Hamzeh, Y., Ashori, A., 2015. Biodegradability and mechanical properties of reinforced starch nanocomposites using cellulose nanofibers. Carbohydrate Polymers 132, 1–8. https://doi.org/10.1016/j.carbpol.2015.06.043.

Banerjee, R., Chen, H., Wu, J., 1996. Milk protein-based edible film mechanical strength changes due to ultrasound process. Journal of Food Science 61, 824–828. https://doi.org/10.1111/j.1365-2621.1996.tb12211.x.

Beristain-Bauza, S.C., Mani-López, E., Palou, E., López-Malo, A., 2016. Antimicrobial activity and physical properties of protein films added with cell-free supernatant of Lactobacillus rhamnosus. Food Control 62, 44–51. https://doi.org/10.1016/j.foodcont.2015.10.007.

Boonprasith, P., Wootthikanokkhan, J., Nimitsiriwat, N., 2013. Mechanical, thermal, and barrier properties of nanocomposites based on poly(butylene succinate)/thermoplastic starch blends containing different types of clay. Journal of Applied Polymer Science 130, 1114–1123. https://doi.org/10.1002/app.39281.

Calva-Estrada, S.J., Jiménez-Fernández, M., Lugo-Cervantes, E., 2019. Protein-based films: advances in the development of biomaterials applicable to food packaging. Food Engineering Reviews 11, 78–92. https://doi.org/10.1007/s12393-019-09189-w.

Cao, X., Chen, Y., Chang, P.R., Muir, A.D., Falk, G., 2008a. Starch-based nanocomposites reinforced with flax cellulose nanocrystals. Express Polymer Letters 2, 502–510. https://doi.org/10.3144/expresspolymlett.2008.60.

Cao, X., Chen, Y., Chang, P.R., Stumborg, M., Huneault, M.A., 2008b. Green composites reinforced with hemp nanocrystals in plasticized starch. Journal of Applied Polymer Science 109, 3804–3810. https://doi.org/10.1002/app.28418.

Carvalho, R.A. de, Grosso, C.R.F., 2006. Properties of chemically modified gelatin films. Brazilian Journal of Chemical Engineering 23, 45–53. https://doi.org/10.1590/S0104-66322006000100006.

Chang, P.R., Jian, R., Zheng, P., Yu, J., Ma, X., 2010. Preparation and properties of glycerol plasticized-starch (GPS)/cellulose nanoparticle (CN) composites. Carbohydrate Polymers 79, 301–305. https://doi.org/10.1016/j.carbpol.2009.08.007.

Chaput, C., Chenite, A., June 10, 2014. Injectable in situ self-forming mineral-polymer hybrid composition and uses thereof. US Patent 8747899.

Chaudhary, J.P., Vadodariya, N., Nataraj, S.K., Meena, R., 2015. Chitosan-based aerogel membrane for robust oil-in-water emulsion separation. ACS Applied Materials & Interfaces 7, 24957–24962. https://doi.org/10.1021/acsami.5b08705.

Chen, D., Lawton, D., Thompson, M.R., Liu, Q., 2012. Biocomposites reinforced with cellulose nanocrystals derived from potato peel waste. Carbohydrate Polymers 90, 709–716. https://doi.org/10.1016/j.carbpol.2012.06.002.

Chen, G.-Q., 2010. Nodax^{TM} class PHA copolymers: their properties and applications - industrial production of PHA. Plastics From Bacteria: Natural Functions and Applications 14, 121–132. https://doi.org/10.1007/978-3-642-03287.

Chen, H., 1995. Functional properties and applications of edible films made of milk proteins. Journal of Dairy Science 78, 2563–2583. https://doi.org/10.3168/jds.S0022-0302(95)76885-0.

Cho, K., Wang, X., Nie, S., Chen, Z., Shin, D.M., 2008. Therapeutic nanoparticles for drug delivery in cancer. Clinical Cancer Research 14, 1310–1316. https://doi.org/10.1158/1078-0432.CCR-07-1441.

Colla, E., do Amaral Sobral, P.J., Menegalli, F.C., 2006. Amaranthus cruentus flour edible films: influence of stearic acid addition, plasticizer concentration, and emulsion stirring speed on water vapor permeability and mechanical properties. Journal of Agricultural and Food Chemistry 54, 6645–6653. https://doi.org/10.1021/jf0611217.

Condés, M.C., Añón, M.C., Mauri, A.N., Dufresne, A., 2015. Amaranth protein films reinforced with maize starch nanocrystals. Food Hydrocolloids 47, 146–157. https://doi.org/10.1016/j.foodhyd.2015.01.026.

Crini, G., Lichtfouse, E., 2016. Sustainable Agriculture Reviews, Sustainable Agriculture Reviews, Sustainable Agriculture Reviews. Springer International Publishing, Cham. https://doi.org/10.1007/978-3-319-26777-7.

Cuq, B., Gontard, N., Guilbert, S., 1998. Proteins as agricultural polymers for packaging production. Cereal Chemistry Journal 75, 1–9. https://doi.org/10.1094/CCHEM.1998.75.1.1.

D'huart, J., Dallas, C., July 1, 2004. Cactaceae-based formulation having the property of fixing fats, and method for obtaining same. US Patent 2004/0126444 A1.

Dai, H., Chang, P.R., Yu, J., Ma, X., 2008. N,N-Bis(2-hydroxyethyl)formamide as a new plasticizer for thermoplastic starch. Starch - Stärke 60, 676–684. https://doi.org/10.1002/star.200800017.

Deringer, V.L., Englert, U., Dronskowski, R., 2016. Nature, strength, and cooperativity of the hydrogen-bonding network in α-chitin. Biomacromolecules 17, 996–1003. https://doi.org/10.1021/acs.biomac.5b01653.

Dias, A.B., Müller, C.M.O., Larotonda, F.D.S., Laurindo, J.B., 2011. Mechanical and barrier properties of composite films based on rice flour and cellulose fibers. Lebensmittel-Wissenschaft und -Technologie- Food Science and Technology 44, 535–542. https://doi.org/10.1016/j.lwt.2010.07.006.

Díaz, O., Candia, D., Cobos, Á., 2017. Whey protein film properties as affected by ultraviolet treatment under alkaline conditions. International Dairy Journal 73, 84–91. https://doi.org/10.1016/j.idairyj.2017.05.009.

Dorgan, J.R., Lehermeier, H.J., Palade, L.-I., Cicero, J., 2001. Polylactides: properties and prospects of an environmentally benign plastic from renewable resources. Macromolecular Symposia 175, 55–66. https://doi.org/10.1002/1521-3900(200110)175:1<55::AID-MASY55>3.0.CO;2-K.

Durán, N., Lemes, A.P., Durán, M., Freer, J., Baeza, J., 2011. A minireview of cellulose nanocrystals and its potential integration as Co-product in bioethanol production. Journal of the Chilean Chemical Society 56, 672–677. https://doi.org/10.4067/S0717-97072011000200011.

Elizabeth, R., 2006. Corn plastic to the rescue. Smithsonian Magazine 84–88.

Emadian, S.M., Onay, T.T., Demirel, B., 2017. Biodegradation of bioplastics in natural environments. Waste Management 59, 526–536. https://doi.org/10.1016/j.wasman.2016.10.006.

Fabra, M.J., Sánchez-González, L., Chiralt, A., 2014. Lysozyme release from isolate pea protein and starch based films and their antimicrobial properties. Lebensmittel-Wissenschaft und -Technologie- Food Science and Technology 55, 22–26. https://doi.org/10.1016/j.lwt.2013.08.001.

Farah, S., Anderson, D.G., Langer, R., 2016. Physical and mechanical properties of PLA, and their functions in widespread applications — a comprehensive review. Advanced Drug Delivery Reviews 107, 367–392. https://doi.org/10.1016/j.addr.2016.06.012.

Fennema, O., 1985. Chemical changes in food during processing—an overview. In: Chemical Changes in Food During Processing. Springer Netherlands, Dordrecht, pp. 1–16. https://doi.org/10.1007/978-94-017-1016-9_1.

Finkielsztein, S., Vournakis, J., September 22, 2015. Hemostatic compositions and therapeutic regimens. US 9139664 B2.

Francis, R., Prestwich, G., Hunt, G., June 4, 2015. System and method of delivering a hyaluronic acid composition and a copper composition for treatment of dermatologic conditions. WO 2015081304 A1.

Fu, Z., Wang, L., Li, D., Wei, Q., Adhikari, B., 2011. Effects of high-pressure homogenization on the properties of starch-plasticizer dispersions and their films. Carbohydrate Polymers 86, 202–207. https://doi.org/10.1016/j.carbpol.2011.04.032.

Garrido, T., Etxabide, A., Guerrero, P., de la Caba, K., 2016. Characterization of agar/soy protein biocomposite films: effect of agar on the extruded pellets and compression moulded films. Carbohydrate Polymers 151, 408–416. https://doi.org/10.1016/j.carbpol.2016.05.089.

Garside, M., 2019. Global Plastic Production 1950–2018. Statista. https://www.statista.com/statistics/282732/global-production-of-plastics-since-1950/.

Generalic, E., 2019. "Starch" Croatian-English Chemistry Dictionary & Glossary. KTF-Split. https://glossary.periodni.com/glossary.php?en=starch.

Gennadios, A., 2004. Edible films and coatings from proteins. In: Proteins in Food Processing. https://doi.org/10.1533/9781855738379.3.442.

Gewin, V., 2003. Genetically modified corn— environmental benefits and risks. PLoS Biology 1, e8. https://doi.org/10.1371/journal.pbio.0000008.

Gigli, M., Fabbri, M., Lotti, N., Gamberini, R., Rimini, B., Munari, A., 2016. Poly(butylene succinate)-based polyesters for biomedical applications: a review. European Polymer Journal 75, 431–460. https://doi.org/10.1016/j.eurpolymj.2016.01.016.

Gómez-Estaca, J., Montero, P., Gómez-Guillén, M.C., 2014. Shrimp (*Litopenaeus vannamei*) muscle proteins as source to develop edible films. Food Hydrocolloids 41, 86–94. https://doi.org/10.1016/j.foodhyd.2014.03.032.

Gómez-Guillén, M.C., Giménez, B., López-Caballero, M.E., Montero, M.P., 2011. Functional and bioactive properties of collagen and gelatin from alternative sources: a review. Food Hydrocolloids 25, 1813–1827. https://doi.org/10.1016/j.foodhyd.2011.02.007.

Gupta, P., Nayak, K.K., 2015. Characteristics of protein-based biopolymer and its application. Polymer Engineering & Science 55, 485–498. https://doi.org/10.1002/pen.23928.

Gutiérrez, T.J., Tapia, M.S., Pérez, E., Famá, L., 2015. Structural and mechanical properties of edible films made from native and modified cush-cush yam and cassava starch. Food Hydrocolloids 45, 211–217. https://doi.org/10.1016/j.foodhyd.2014.11.017.

Haghpanah, J.S., Yuvienco, C., Civay, D.E., Barra, H., Baker, P.J., Khapli, S., Voloshchuk, N., Gunasekar, S.K., Muthukumar, M., Montclare, J.K., 2009. Artificial protein block copolymers blocks comprising two distinct self-assembling domains. ChemBioChem 10, 2733–2735. https://doi.org/10.1002/cbic.200900539.

Halimatul, M.J., Sapuan, S.M., Jawaid, M., Ishak, M.R., Ilyas, R.A., 2019a. Effect of sago starch and plasticizer content on the properties of thermoplastic films: mechanical testing and cyclic soaking-drying. Polimery 64, 32–41. https://doi.org/10.14314/polimery.2019.6.5.

Halimatul, M.J., Sapuan, S.M., Jawaid, M., Ishak, M.R., Ilyas, R.A., 2019b. Water absorption and water solubility properties of sago starch biopolymer composite films filled with sugar palm particles. Polimery 64, 27–35. https://doi.org/10.14314/polimery.2019.9.4.

Hassan, M.A., Yee, L.N., Yee, P.L., Ariffin, H., Raha, A.R., Shirai, Y., Sudesh, K., 2013. Sustainable production of polyhydroxyalkanoates from renewable oil-palm biomass. Biomass and Bioenergy 50, 1–9. https://doi.org/10.1016/j.biombioe.2012.10.014.

Hazrol, M.D., Sapuan, S.M., Ilyas, R.A., Othman, M.L., Sherwani, S.F.K., 2020. Electrical properties of sugar palm nanocrystalline cellulose, reinforced sugar palm starch nanocomposites. Polimery 55, 33–40. https://doi.org/10.14314/polimery.2020.5.4.

Hemsri, S., Asandei, A.D., Grieco, K., Parnas, R.S., 2011. Biopolymer composites of wheat gluten with silica and alumina. Composites Part A: Applied Science and Manufacturing 42, 1764–1773. https://doi.org/10.1016/j.compositesa.2011.07.032.

Hietala, M., Mathew, A.P., Oksman, K., 2013. Bionanocomposites of thermoplastic starch and cellulose nanofibers manufactured using twin-screw extrusion. European Polymer Journal 49, 950–956. https://doi.org/10.1016/j.eurpolymj.2012.10.016.

Hosseini, S.F., Rezaei, M., Zandi, M., Farahmandghavi, F., 2015. Fabrication of bio-nanocomposite films based on fish gelatin reinforced with chitosan nanoparticles. Food Hydrocolloids 44, 172–182. https://doi.org/10.1016/j.foodhyd.2014.09.004.

Hu, X., Wang, X., Rnjak, J., Weiss, A.S., Kaplan, D.L., 2010. Biomaterials derived from silk–tropoelastin protein systems. Biomaterials 31, 8121–8131. https://doi.org/10.1016/j.biomaterials.2010.07.044.

Ikada, Y., Tsuji, H., 2000. Biodegradable polyesters for medical and ecological applications. Macromolecular Rapid Communications 21, 117–132. https://doi.org/10.1002/(sici)1521-3927(20000201)21:3<117::aid-marc117>3.3.co;2-o.

Ilyas, R.A., Sapuan, S.M., Ishak, M.R., Zainudin, E.S., 2017. Effect of delignification on the physical, thermal, chemical, and structural properties of sugar palm fibre. BioResources 12, 8734–8754. https://doi.org/10.15376/biores.12.4.8734-8754.

Ilyas, R.A., Sapuan, S.M., 2020. The preparation methods and processing of natural fibre bio-polymer composites. Current Organic Synthesis 16, 1068–1070. https://doi.org/10.2174/157017941608200120105616.

Ilyas, R.A., Sapuan, S.M., Atiqah, A., Ibrahim, R., Abral, H., Ishak, M.R., Zainudin, E.S., Nurazzi, N.M., Atikah, M.S.N., Ansari, M.N.M., Asyraf, M.R.M., Supian, A.B.M., Ya, H., 2020a. Sugar palm (*Arenga pinnata* [*Wurmb.*] Merr) starch films containing sugar palm nanofibrillated cellulose as reinforcement: water barrier properties. Polymer Composites 41, 459–467. https://doi.org/10.1002/pc.25379.

Ilyas, R.A., Sapuan, S.M., Ibrahim, R., Abral, H., Ishak, M.R., Zainudin, E.S., Asrofi, M., Atikah, M.S.N., Huzaifah, M.R.M., Radzi, A.M., Azammi, A.M.N., Shaharuzaman, M.A., Nurazzi, N.M., Syafri, E., Sari, N.H., Norrrahim, M.N.F., Jumaidin, R., 2019d. Sugar palm (*Arenga pinnata* (*Wurmb.*) Merr) cellulosic fibre hierarchy: a comprehensive approach from macro to nano scale. Journal of Materials Research and Technology 8, 2753–2766. https://doi.org/10.1016/j.jmrt.2019.04.011.

Ilyas, R.A., Sapuan, S.M., Ibrahim, R., Abral, H., Ishak, M.R., Zainudin, E.S., Atikah, M.S.N., Mohd Nurazzi, N., Atiqah, A., Ansari, M.N.M., Syafri, E., Asrofi, M., Sari, N.H., Jumaidin, R., 2019a. Effect of sugar palm nanofibrillated cellulose concentrations on morphological, mechanical and physical properties of biodegradable films based on agro-waste sugar palm (*Arenga pinnata* (*Wurmb.*) Merr) starch. Journal of Materials Research and Technology 8, 4819–4830. https://doi.org/10.1016/j.jmrt.2019.08.028.

Ilyas, R.A., Sapuan, S.M., Ibrahim, R., Abral, H., Ishak, M.R., Zainudin, E.S., Atiqah, A., Atikah, N., Syafri, E., Asrofi, M., Jumaidin, R., 2020b. Thermal, biodegradability and water barrier properties of bio-nanocomposites based on plasticised sugar palm starch and nanofibrillated celluloses from sugar palm fibres. Journal of Biobased Materials and Bioenergy 14, 1–13. https://doi.org/10.1166/jbmb.2020.1951.

Ilyas, R.A., Sapuan, S.M., Ibrahim, R., Atikah, M.S.N., Atiqah, A., Ansari, M.N.M., Norrrahim, M.N.F., 2019b. Production, processes and modification of nanocrystalline cellulose from agro-waste: a review. In: Nanocrystalline Materials. IntechOpen, pp. 3–32. https://doi.org/10.5772/intechopen.87001.

Ilyas, R.A., Sapuan, S.M., Ishak, M.R., Zainudin, E.S., 2019c. Sugar palm nanofibrillated cellulose (*Arenga pinnata* (*Wurmb.*) Merr): effect of cycles on their yield, physic-chemical, morphological and thermal behavior. International Journal of Biological Macromolecules 123, 379–388. https://doi.org/10.1016/j.ijbiomac.2018.11.124.

Ilyas, R.A., Sapuan, S.M., Ishak, M.R., 2018. Isolation and characterization of nanocrystalline cellulose from sugar palm fibres (Arenga Pinnata). Carbohydrate Polymers 181, 1038–1051. https://doi.org/10.1016/j.carbpol.2017.11.045.

Ilyas, R.A., Sapuan, S.M., Ishak, M.R., Zainudin, E.S., 2018a. Development and characterization of sugar palm nanocrystalline cellulose reinforced sugar palm starch bionanocomposites. Carbohydrate Polymers 202, 186–202. https://doi.org/10.1016/j.carbpol.2018.09.002.

Ilyas, R.A., Sapuan, S.M., Ishak, M.R., Zainudin, E.S., 2018b. Sugar palm nanocrystalline cellulose reinforced sugar palm starch composite: degradation and water-barrier properties. In: IOP Conference Series: Materials Science and Engineering. https://doi.org/10.1088/1757-899X/368/1/012006.

Ilyas, R.A., Sapuan, S.M., Ishak, M.R., Zainudin, E.S., 2018e. Water transport properties of bio-nanocomposites reinforced by sugar palm (arenga pinnata) nanofibrillated cellulose. Journal of Advanced Research in Fluid Mechanics and Thermal Sciences Journal 51, 234–246.

Ilyas, R.A., Sapuan, S.M., Ishak, M.R., Zainudin, E.S., Atikah, M.S.N., 2018c. Characterization of sugar palm nanocellulose and its potential for reinforcement with a starch-based composite. In: Sugar Palm Biofibers, Biopolymers, and Biocomposites, first ed. CRC Press/Taylor & Francis Group, Boca Raton, FL, pp. 189–220. https://doi.org/10.1201/9780429443923-10.

Ilyas, R.A., Sapuan, S.M., Sanyang, M.L., Ishak, M.R., Zainudin, E.S., 2018d. Nanocrystalline cellulose as reinforcement for polymeric matrix nanocomposites and its potential applications: a review. Current Analytical Chemistry 14, 203–225. https://doi.org/10.2174/1573411013666171003155624.

Jamaluddin, N., Razaina, M.T., Ishak, Z.M., 2016. Mechanical and morphology behaviours of polybutylene (succinate)/thermoplastic polyurethaneblend. Procedia Chemistry 19, 426–432. https://doi.org/10.1016/j.proche.2016.03.034.

Jamshidian, M., Tehrany, E.A., Imran, M., Jacquot, M., Desobry, S., 2010. Poly-Lactic Acid: Production, Applications, Nanocomposites, and Release Studies. Comprehensive Reviews in Food Science and Food Safety 9, 552–571. https://doi.org/10.1111/j.1541-4337.2010.00126.x.

Jiang, C., Wang, X., Gunawidjaja, R., Lin, Y.-H., Gupta, M.K., Kaplan, D.L., Naik, R.R., Tsukruk, V.V., 2007. Mechanical properties of robust ultrathin silk fibroin films. Advanced Functional Materials 17, 2229–2237. https://doi.org/10.1002/adfm.200601136.

Johnson, E., Nichols, E., October 2000. High tap density chitosan, and methods of production. US 6130321 A.

Jooyandeh, H., 2011. Whey protein films and coatings: a review. Pakistan Journal of Nutrition 10, 296–301. https://doi.org/10.3923/pjn.2011.296.301.

Jumaidin, R., Ilyas, R.A., Saiful, M., Hussin, F., Mastura, M.T., 2019a. Water transport and physical properties of sugarcane bagasse fibre reinforced thermoplastic potato starch biocomposite. Journal of Advanced Research in Fluid Mechanics and Thermal Sciences 61, 273–281.

Jumaidin, R., Khiruddin, M.A.A., Asyul Sutan Saidi, Z., Salit, M.S., Ilyas, R.A., 2020. Effect of cogon grass fibre on the thermal, mechanical and biodegradation properties of thermoplastic cassava starch biocomposite. International Journal of Biological Macromolecules 146, 746–755. https://doi.org/10.1016/j.ijbiomac.2019.11.011.

Jumaidin, R., Saidi, Z.A.S., Ilyas, R.A., Ahmad, M.N., Wahid, M.K., Yaakob, M.Y., Maidin, N.A., Rahman, M.H.A., Osman, M.H., 2019b. Characteristics of cogon grass fibre reinforced thermoplastic cassava starch biocomposite: water absorption and physical properties. Journal of Advanced Research in Fluid Mechanics and Thermal Sciences 62 (62), 43–52.

Karimi, S., Tahir, P., Dufresne, A., Karimi, A., Abdulkhani, A., 2014. A comparative study on characteristics of nanocellulose reinforced thermoplastic starch biofilms prepared with different techniques. Nordic Pulp and Paper Research Journal 29, 41–45.

Kaushik, A., Singh, M., Verma, G., 2010. Green nanocomposites based on thermoplastic starch and steam exploded cellulose nanofibrils from wheat straw. Carbohydrate Polymers 82, 337–345. https://doi.org/10.1016/j.carbpol.2010.04.063.

Khwaldia, K., Arab-Tehrany, E., Desobry, S., 2010. Biopolymer coatings on paper packaging materials. Comprehensive Reviews in Food Science and Food Safety 9, 82–91. https://doi.org/10.1111/j.1541-4337.2009.00095.x.

Kinsella, J.E., Morr, C.V., 1984. Milk proteins: physicochemical and functional properties. CRC Critical Reviews in Food Science & Nutrition 21, 197–262. https://doi.org/10.1080/10408398409527401.

Koutsopoulos, S., Unsworth, L.D., Nagai, Y., Zhang, S., 2009. Controlled release of functional proteins through designer self-assembling peptide nanofiber hydrogel scaffold. Proceedings of the National Academy of Sciences 106, 4623–4628. https://doi.org/10.1073/pnas.0807506106.

Kvien, I., Sugiyama, J., Votrubec, M., Oksman, K., 2007. Characterization of starch based nanocomposites. Journal of Materials Science 42, 8163–8171. https://doi.org/10.1007/s10853-007-1699-2.

Lee, J.-H., Lee, J., Song, K.B., 2015a. Development of a chicken feet protein film containing essential oils. Food Hydrocolloids 46, 208–215. https://doi.org/10.1016/j.foodhyd.2014.12.020.

Lee, J.-H., Won, M., Song, K.B., 2015b. Physical properties and antimicrobial activities of porcine meat and bone meal protein films containing coriander oil. Lebensmittel-Wissenschaft und -Technologie- Food Science and Technology 63, 700–705. https://doi.org/10.1016/j.lwt.2015.03.043.

Li, X., Ji, N., Qiu, C., Xia, M., Xiong, L., Sun, Q., 2015. The effect of peanut protein nanoparticles on characteristics of protein- and starch-based nanocomposite films: a comparative study. Industrial Crops and Products 77, 565–574. https://doi.org/10.1016/j.indcrop.2015.09.026.

Lim, S., Jane, J., 1994. Storage stability of injection-molded starch-zein plastics under dry and humid conditions. Journal of Environmental Polymer Degradation 2, 111–120. https://doi.org/10.1007/BF02074779.

Liu, D., Zhong, T., Chang, P.R., Li, K., Wu, Q., 2010. Starch composites reinforced by bamboo cellulosic crystals. Bioresource Technology 101, 2529–2536. https://doi.org/10.1016/j.biortech.2009.11.058.

Liu, L., Yang, J.-P., Ju, X.-J., Xie, R., Liu, Y.-M., Wang, W., Zhang, J.-J., Niu, C.H., Chu, L.-Y., 2011. Monodisperse core-shell chitosan microcapsules for pH-responsive burst release of hydrophobic drugs. Soft Matter 7, 4821. https://doi.org/10.1039/c0sm01393e.

Llanos, J.H.R., Tadini, C.C., 2018. Preparation and characterization of bio-nanocomposite films based on cassava starch or chitosan, reinforced with montmorillonite or bamboo nanofibers. International Journal of Biological Macromolecules 107, 371–382. https://doi.org/10.1016/j.ijbiomac.2017.09.001.

Lu, J., Tappel, R.C., Nomura, C.T., 2009. Mini-Review: biosynthesis of poly(hydroxyalkanoates). Polymer Reviews 49, 226–248. https://doi.org/10.1080/15583720903048243.

Lu, Y., Weng, L., Cao, X., 2006. Morphological, thermal and mechanical properties of ramie crystallites—reinforced plasticized starch biocomposites. Carbohydrate Polymers 63, 198–204. https://doi.org/10.1016/j.carbpol.2005.08.027.

Lu, Y., Weng, L., Cao, X., 2005. Biocomposites of plasticized starch reinforced with cellulose crystallites from cottonseed linter. Macromolecular Bioscience 5, 1101–1107. https://doi.org/10.1002/mabi.200500094.

Lucchesi, L., Xie, H., December 8, 2015. Wound dressing devices and methods. US 9205170 B2.

Ma, X., Qiao, C., Wang, X., Yao, J., Xu, J., 2019. Structural characterization and properties of polyols plasticized chitosan films. International Journal of Biological Macromolecules 135, 240–245. https://doi.org/10.1016/j.ijbiomac.2019.05.158.

Martinez, K., Baeza, R., Millan, F., Pilosof, A., 2005. Effect of limited hydrolysis of sunflower protein on the interactions with polysaccharides in foams. Food Hydrocolloids 19, 361–369. https://doi.org/10.1016/j.foodhyd.2004.10.002.

Mathew, A.P., Thielemans, W., Dufresne, A., 2008. Mechanical properties of nanocomposites from sorbitol plasticized starch and tunicin whiskers. Journal of Applied Polymer Science 109, 4065–4074. https://doi.org/10.1002/app.28623.

McCarthy, S., Gregory, K., Wiesmann, W., Campbell, T., March 11, 2014. Wound dressing and method for controlling severe, life threatening bleeding. US 8668924 B2.

McCarthy, S., McGrath, B., Winata, E., September 18, 2012. Absorbable tissue dressing assemblies, systems, and methods formed from hydrophilic polymer sponge structures such as chitosan. US 8269058 B2.

Mchugh, T.H., Aujard, J.-F., Krochta, J.M., 1994. Plasticized whey protein edible films: water vapor permeability properties. Journal of Food Science 59, 416–419. https://doi.org/10.1111/j.1365-2621.1994.tb06980.x.

McHugh, T.H., Krochta, J.M., 1994. Sorbitol- vs glycerol-plasticized whey protein edible films: integrated oxygen permeability and tensile property evaluation. Journal of Agricultural and Food Chemistry 42, 841–845. https://doi.org/10.1021/jf00040a001.

Mehyar, G.F., Han, J.H., 2006. Physical and mechanical properties of high-amylose rice and pea starch films as affected by relative humidity and plasticizer. Journal of Food Science 69, E449–E454. https://doi.org/10.1111/j.1365-2621.2004.tb09929.x.

Mhd Haniffa, M., Ching, Y., Abdullah, L., Poh, S., Chuah, C., 2016. Review of bionanocomposite coating films and their applications. Polymers 8, 246. https://doi.org/10.3390/polym8070246.

Montero, B., Rico, M., Rodríguez-Llamazares, S., Barral, L., Bouza, R., 2017. Effect of nanocellulose as a filler on biodegradable thermoplastic starch films from tuber, cereal and legume. Carbohydrate Polymers 157, 1094–1104. https://doi.org/10.1016/j.carbpol.2016.10.073.

Morin-Crini, N., Lichtfouse, E., Torri, G., Crini, G., 2019. Fundamentals and applications of chitosan. In: Crini, G., Lichtfouse, E. (Eds.), Sustainable Agriculture Reviews, Sustainable Agriculture Reviews. Springer International Publishing, Cham, pp. 49–123. https://doi.org/10.1007/978-3-030-16538-3_2.

Müller, C.M.O., Laurindo, J.B., Yamashita, F., 2009. Effect of cellulose fibers addition on the mechanical properties and water vapor barrier of starch-based films. Food Hydrocolloids 23, 1328–1333. https://doi.org/10.1016/j.foodhyd.2008.09.002.

Nasri-Nasrabadi, B., Behzad, T., Bagheri, R., 2014. Preparation and characterization of cellulose nanofiber reinforced thermoplastic starch composites. Fibers and Polymers 15, 347–354. https://doi.org/10.1007/s12221-014-0347-0.

Nasseri, R., Mohammadi, N., 2014. Starch-based nanocomposites: a comparative performance study of cellulose whiskers and starch nanoparticles. Carbohydrate Polymers 106, 432–439. https://doi.org/10.1016/j.carbpol.2014.01.029.

Nataraj, D., Sakkara, S., Meghwal, M., Reddy, N., 2018. Crosslinked chitosan films with controllable properties for commercial applications. International Journal of Biological Macromolecules 120, 1256–1264. https://doi.org/10.1016/j.ijbiomac.2018.08.187.

Nazrin, A., Sapuan, S.M., Zuhri, M.Y.M., Ilyas, R.A., Syafiq, R., Sherwani, S.F.K., 2020. Nanocellulose Reinforced Thermoplastic Starch (TPS), Polylactic Acid (PLA), and Polybutylene Succinate (PBS) for Food Packaging Applications. Frontiers in Chemistry 8, 1–12. https://doi.org/10.3389/fchem.2020.00213.

Norizan, M.N., Abdan, K., Ilyas, R.A., Biofibers, S.P., 2020. Effect of fiber orientation and fiber loading on the mechanical and thermal properties of sugar palm yarn fiber reinforced unsaturated polyester resin composites. Polimery 65, 34–43. https://doi.org/10.14314/polimery.2020.2.5.

Noshirvani, N., Ghanbarzadeh, B., Fasihi, H., Almasi, H., 2016. Starch-PVA nanocomposite film incorporated with cellulose nanocrystals and MMT: a comparative study. International Journal of Food Engineering 12, 37–48. https://doi.org/10.1515/ijfe-2015-0145.

Nurazzi, N.M., Khalina, A., Sapuan, S.M., Ilyas, R.A., Rafiqah, S.A., Hanafee, Z.M., 2019a. Thermal properties of treated sugar palm yarn/glass fiber reinforced unsaturated polyester hybrid composites. Journal of Materials Research and Technology 9 (2), 1606–1618. https://doi.org/10.1016/j.jmrt.2019.11.086.

Nurazzi, N.M., Khalina, A., Sapuan, S.M., Ilyas, R.A., 2019b. Mechanical properties of sugar palm yarn / woven glass fiber reinforced unsaturated polyester composites : effect of fiber loadings and alkaline treatment. Polimery 64, 12–22. https://doi.org/10.14314/polimery.2019.10.3.

Oryan, A., Sahvieh, S., 2017. Effectiveness of chitosan scaffold in skin, bone and cartilage healing. International Journal of Biological Macromolecules 104, 1003–1011. https://doi.org/10.1016/j.ijbiomac.2017.06.124.

Oussalah, M., Caillet, S., Salmiéri, S., Saucier, L., Lacroix, M., 2004. Antimicrobial and antioxidant effects of milk protein-based film containing essential oils for the preservation of whole beef muscle. Journal of Agricultural and Food Chemistry 52, 5598–5605. https://doi.org/10.1021/jf049389q.

Pang, X., Zhuang, X., Tang, Z., Chen, X., 2010. Polylactic acid (PLA): research, development and industrialization. Biotechnology Journal 5, 1125–1136. https://doi.org/10.1002/biot.201000135.

Park, H.J., Kim, S.H., Lim, S.T., Shin, D.H., Choi, S.Y., Hwang, K.T., 2000. Grease resistance and mechanical properties of isolated soy protein-coated paper. Journal of the American Oil Chemists' Society 77, 269–273. https://doi.org/10.1007/s11746-000-0044-2.

Patricia, P., Norma, A.S., Guzm, D.L., Fripiat, J.J., 2007. In: Mesoporous Silica From Rice Hull Ash, vol. 619, pp. 614–619. https://doi.org/10.1002/jctb.

Pérez-Arauz, A.O., Aguilar-Rabiela, A.E., Vargas-Torres, A., Rodríguez-Hernández, A.-I., Chavarría-Hernández, N., Vergara-Porras, B., López-Cuellar, M.R., 2019. Production and characterization of biodegradable films of a novel polyhydroxyalkanoate (PHA) synthesized from peanut oil. Food Packaging and Shelf Life 20, 100297. https://doi.org/10.1016/j.fpsl.2019.01.001.

Pérez-Gago, M.B., Rhim, J.-W., 2014. Edible coating and film materials. In: Innovations in Food Packaging. Elsevier, pp. 325–350. https://doi.org/10.1016/B978-0-12-394601-0.00013-8.

Pighinelli, L., 2019. Methods of chitin production a short review. American Journal of Biomedical Science & Research 3, 307–314. https://doi.org/10.34297/AJBSR.2019.03.000682.

Pol, H., Dawson, P., Acton, J., Ogale, A., 2002. Soy protein isolate/corn-zein laminated films: transport and mechanical properties. Journal of Food Science 67, 212–217. https://doi.org/10.1111/j.1365-2621.2002.tb11386.x.

Pothakamury, U.R., Barbosa-Cánovas, G.V., 1995. Fundamental aspects of controlled release in foods. Trends in Food Science & Technology 6, 397–406. https://doi.org/10.1016/S0924-2244(00)89218-3.

Qiu, Y., Park, K., 2001. Environment-sensitive hydrogels for drug delivery. Advanced Drug Delivery Reviews 53, 321–339. https://doi.org/10.1016/S0169-409X(01)00203-4.

Rangavajhyala, N., Ghorpade, V., Hanna, M., 1997. Solubility and molecular properties of heat-cured soy protein films. Journal of Agricultural and Food Chemistry 45, 4204–4208. https://doi.org/10.1021/jf9702048.

Rasal, R.M., Janorkar, A.V., Hirt, D.E., 2010. Poly(lactic acid) modifications. Progress in Polymer Science 35, 338–356. https://doi.org/10.1016/j.progpolymsci.2009.12.003.

Rhim, J.-W., Lee, J.-H., Hong, S.-I., 2006. Water resistance and mechanical properties of biopolymer (alginate and soy protein) coated paperboards. Lebensmittel-Wissenschaft und -Technologie- Food Science and Technology 39, 806–813. https://doi.org/10.1016/j.lwt.2005.05.008.

Rhim, J.W., Weller, C.L., Gennadios, A., 2004. Effects of soy protein coating on shell strength and quality of shell eggs. Food Science and Biotechnology 13 (4), 455–459.

Rocha, M. da, Loiko, M.R., Tondo, E.C., Prentice, C., 2014. Physical, mechanical and antimicrobial properties of Argentine anchovy (*Engraulis anchoita*) protein films incorporated with organic acids. Food Hydrocolloids 37, 213–220. https://doi.org/10.1016/j.foodhyd.2013.10.017.

Rochman, C.M., Browne, M.A., Halpern, B.S., Hentschel, B.T., Hoh, E., Karapanagioti, H.K., Rios-Mendoza, L.M., Takada, H., Teh, S., Thompson, R.C., 2013. Classify plastic waste as hazardous. Nature 494, 169–171. https://doi.org/10.1038/494169a.

Rokhade, A.P., Shelke, N.B., Patil, S.A., Aminabhavi, T.M., 2007. Novel interpenetrating polymer network microspheres of chitosan and methylcellulose for controlled release of theophylline. Carbohydrate Polymers 69, 678–687. https://doi.org/10.1016/j.carbpol.2007.02.008.

Romani, V.P., Hernández, C.P., Martins, V.G., 2018. Pink pepper phenolic compounds incorporation in starch/protein blends and its potential to inhibit apple browning. Food Packaging and Shelf Life 15, 151–158. https://doi.org/10.1016/j.fpsl.2018.01.003.

Roy, J.C., Salaün, F., Giraud, S., Ferri, A., Chen, G., Guan, J., 2017. Solubility of chitin: solvents, solution behaviors and their related mechanisms. In: Solubility of Polysaccharides. InTech. https://doi.org/10.5772/intechopen.71385.

Sahay, N., Ierapetritou, M., 2009. Nihar SCM. IFAC Proceedings Volumes (IFAC-PapersOnline) 7, 405–410. https://doi.org/10.1002/aic.

Sai-Ut, S., Benjakul, S., Rawdkuen, S., 2015. Retardation of lipid oxidation using gelatin film incorporated with longan seed extract compared with BHT. Journal of Food Science & Technology 52, 5842–5849. https://doi.org/10.1007/s13197-014-1631-0.

Salgado, P.R., López-Caballero, M.E., Gómez-Guillén, M.C., Mauri, A.N., Montero, M.P., 2013. Sunflower protein films incorporated with clove essential oil have potential application for the preservation of fish patties. Food Hydrocolloids 33, 74–84. https://doi.org/10.1016/j.foodhyd.2013.02.008.

Sanyang, M.L., Ilyas, R.A., Sapuan, S.M., Jumaidin, R., 2018. Sugar palm starch-based composites for packaging applications. In: Bionanocomposites for Packaging Applications. Springer International Publishing, Cham, pp. 125–147. https://doi.org/10.1007/978-3-319-67319-6_7.

Sanyang, M.L., Sapuan, S.M., Jawaid, M., Ishak, M.R., Sahari, J., 2016. Recent developments in sugar palm (*Arenga pinnata*) based biocomposites and their potential industrial applications: a review. Renewable and Sustainable Energy Reviews 54, 533–549. https://doi.org/10.1016/j.rser.2015.10.037.

Sawyer, D.J., 2003. Bioprocessing—no longer a field of dreams. Macromolecular Symposia 201, 271–282. https://doi.org/10.1002/masy.200351130.

Schmid, M., Müller, K., 2019. Whey protein-based packaging films and coatings. Whey Proteins. https://doi.org/10.1016/b978-0-12-812124-5.00012-6.

Shah, A.A., Hasan, F., Hameed, A., Ahmed, S., 2008. Biological degradation of plastics: a comprehensive review. Biotechnology Advances 26, 246–265. https://doi.org/10.1016/j.biotechadv.2007.12.005.

Sharma, L., Singh, C., 2016. Sesame protein based edible films: development and characterization. Food Hydrocolloids 61, 139–147. https://doi.org/10.1016/j.foodhyd.2016.05.007.

Shi, W., Dumont, M.-J., 2014. Processing and physical properties of canola protein isolate-based films. Industrial Crops and Products 52, 269–277. https://doi.org/10.1016/j.indcrop.2013.10.037.

Siakeng, R., Jawaid, M., Ariffin, H., Sapuan, S.M., Asim, M., Saba, N., 2019. Natural fiber reinforced polylactic acid composites: a review. Polymer Composites 40, 446–463. https://doi.org/10.1002/pc.24747.

Siew, D.C.W., Heilmann, C., Easteal, A.J., Cooney, R.P., 1999. Solution and film properties of sodium caseinate/glycerol and sodium caseinate/polyethylene glycol edible coating systems. Journal of Agricultural and Food Chemistry 47, 3432–3440. https://doi.org/10.1021/jf9806311.

Singhvi, M.S., Zinjarde, S.S., Gokhale, D.V., 2019. Polylactic acid: synthesis and biomedical applications. Journal of Applied Microbiology 127, 1612–1626. https://doi.org/10.1111/jam.14290.

Slavutsky, A.M., Bertuzzi, M.A., 2014. Water barrier properties of starch films reinforced with cellulose nanocrystals obtained from sugarcane bagasse. Carbohydrate Polymers 110, 53–61. https://doi.org/10.1016/j.carbpol.2014.03.049.

Sobral, P.J.A., Menegalli, F.C., Hubinger, M.D., Roques, M.A., 2001. Mechanical, water vapor barrier and thermal properties of gelatin based edible films. Food Hydrocolloids 15, 423–432. https://doi.org/10.1016/S0268-005X(01)00061-3.

Song, H., Lee, S.Y., 2006. Production of succinic acid by bacterial fermentation. Enzyme and Microbial Technology 39, 352–361. https://doi.org/10.1016/j.enzmictec.2005.11.043.

Spaccini, R., Todisco, D., Drosos, M., Nebbioso, A., Piccolo, A., 2016. Decomposition of bio-degradable plastic polymer in a real on-farm composting process. Chemical and Biological Technologies in Agriculture 3 (1), 4. https://doi.org/10.1186/s40538-016-0053-9.

Stuchell, Y.M., Krochta, J.M., 1994. Enzymatic treatments and thermal effects on edible soy protein films. Journal of Food Science 59, 1332–1337. https://doi.org/10.1111/j.1365-2621.1994.tb14709.x.

Su, J.-F., Huang, Z., Yuan, X.-Y., Wang, X.-Y., Li, M., 2010. Structure and properties of carboxymethyl cellulose/soy protein isolate blend edible films crosslinked by Maillard reactions. Carbohydrate Polymers 79, 145–153. https://doi.org/10.1016/j.carbpol.2009.07.035.

Syafri, E., Kasim, A., Abral, H., Sudirman, Sulungbudi, G.T., Sanjay, M.R., Sari, N.H., 2018. Synthesis and characterization of cellulose nanofibers (CNF) ramie reinforced cassava starch hybrid composites. International Journal of Biological Macromolecules 120, 578–586. https://doi.org/10.1016/j.ijbiomac.2018.08.134.

Syafri, E., Sudirman, M., Yulianti, E., Deswita, Asrofi, M., Abral, H., Sapuan, S.M., Ilyas, R.A., Fudholi, A., 2019. Effect of sonication time on the thermal stability, moisture absorption, and biodegradation of water hyacinth (*Eichhornia crassipes*) nanocellulose-filled bengkuang (*Pachyrhizus erosus*) starch biocomposites. Journal of Materials Research and Technology 8, 6223–6231. https://doi.org/10.1016/j.jmrt.2019.10.016.

Tan, Z., Yi, Y., Wang, H., Zhou, W., Yang, Y., Wang, C., 2016. Physical and degradable properties of mulching films prepared from natural fibers and biodegradable polymers. Applied Sciences 6, 147. https://doi.org/10.3390/app6050147.

Teixeira, E. de M., Pasquini, D., Curvelo, A.A.S.S., Corradini, E., Belgacem, M.N., Dufresne, A., 2009. Cassava bagasse cellulose nanofibrils reinforced thermoplastic cassava starch. Carbohydrate Polymers 78, 422–431. https://doi.org/10.1016/j.carbpol.2009.04.034.

Teixeira, E.D.M., Lotti, C., Corrêa, A.C., Teodoro, K.B.R., Marconcini, J.M., Mattoso, L.H.C., 2011. Thermoplastic corn starch reinforced with cotton cellulose nanofibers. Journal of Applied Polymer Science 120, 2428–2433. https://doi.org/10.1002/app.33447.

Thiagamani, S.M.K., Rajini, N., Siengchin, S., Varada Rajulu, A., Hariram, N., Ayrilmis, N., 2019. Influence of silver nanoparticles on the mechanical, thermal and antimicrobial properties of cellulose-based hybrid nanocomposites. Composites Part B: Engineering 165, 516–525. https://doi.org/10.1016/j.compositesb.2019.02.006.

Thunwall, M., Boldizar, A., Rigdahl, M., 2006. Compression molding and tensile properties of thermoplastic potato starch materials. Biomacromolecules 7, 981–986. https://doi.org/10.1021/bm050804c.

Tian, H., Xu, G., Yang, B., Guo, G., 2011. Microstructure and mechanical properties of soy protein/agar blend films: effect of composition and processing methods. Journal of Food Engineering 107, 21–26. https://doi.org/10.1016/j.jfoodeng.2011.06.008.

Tonyali, B., Cikrikci, S., Oztop, M.H., 2018. Physicochemical and microstructural characterization of gum tragacanth added whey protein based films. Food Research International 105, 1–9. https://doi.org/10.1016/j.foodres.2017.10.071.

Torchilin, V.P., 2005. Recent advances with liposomes as pharmaceutical carriers. Nature Reviews Drug Discovery 4, 145–160. https://doi.org/10.1038/nrd1632.

UNEP, 2018. Single-use Plastics: A Roadmap for Sustainability. United Nation Environment Programme.

Vachon, C., Yu, H.-L., Yefsah, R., Alain, R., St-Gelais, D., Lacroix, M., 2000. Mechanical and structural properties of milk protein edible films cross-linked by heating and γ-irradiation. Journal of Agricultural and Food Chemistry 48, 3202–3209. https://doi.org/10.1021/jf991055r.

Vidal, R.R.L., Moraes, J.S., 2019. Removal of organic pollutants from wastewater using chitosan: a literature review. International Journal of Environmental Science and Technology 16, 1741–1754. https://doi.org/10.1007/s13762-018-2061-8.

Wang, H., Hu, D., Ma, Q., Wang, L., 2016. Physical and antioxidant properties of flexible soy protein isolate films by incorporating chestnut (*Castanea mollissima*) bur extracts. Lebensmittel-Wissenschaft und -Technologie- Food Science and Technology 71, 33–39. https://doi.org/10.1016/j.lwt.2016.03.025.

Xu, J., Guo, B.H., 2010. Poly(butylene succinate) and its copolymers: research, development and industrialization. Biotechnology Journal 5, 1149–1163. https://doi.org/10.1002/biot.201000136.

Yang, S., Tang, Y., Wang, J., Kong, F., Zhang, J., 2014. Surface treatment of cellulosic paper with starch-based composites reinforced with nanocrystalline cellulose. Industrial & Engineering Chemistry Research 53, 13980–13988. https://doi.org/10.1021/ie502125s.

Yoo, S., Krochta, J.M., 2011. Whey protein-polysaccharide blended edible film formation and barrier, tensile, thermal and transparency properties. Journal of the Science of Food and Agriculture 91, 2628–2636. https://doi.org/10.1002/jsfa.4502.

Zhang, C., Guo, K., Ma, Y., Ma, D., Li, X., Zhao, X., 2010. Original article: incorporations of blueberry extracts into soybean-protein-isolate film preserve qualities of packaged lard. International Journal of Food Science and Technology 45, 1801–1806. https://doi.org/10.1111/j.1365-2621.2010.02331.x.

Zhang, Y.-Q., Tao, M.-L., Shen, W.-D., Zhou, Y.-Z., Ding, Y., Ma, Y., Zhou, W.-L., 2004. Immobilization of L-asparaginase on the microparticles of the natural silk sericin protein and its characters. Biomaterials 25, 3751–3759. https://doi.org/10.1016/j.biomaterials.2003.10.019.

Zhang, Y., Han, J.H., 2006. Mechanical and thermal characteristics of pea starch films plasticized with monosaccharides and polyols. Journal of Food Science 71, E109–E118. https://doi.org/10.1111/j.1365-2621.2006.tb08891.x.

Zhong, N., Yuan, Q., 2013. Preparation and properties of molded blends of wheat gluten and cationic water-borne polyurethanes. Journal of Applied Polymer Science 128, 460–469. https://doi.org/10.1002/app.38198.

Zhu, K.J., Hendren, R.W., Jensen, K., Pitt, C.G., 1991. Synthesis, properties, and biodegradation of poly(1,3-trimethylene carbonate). Macromolecules 24, 1736–1740. https://doi.org/10.1021/ma00008a008.

Zuo, M., Song, Y., Zheng, Q., 2009. Preparation and properties of wheat gluten/methylcellulose binary blend film casting from aqueous ammonia: a comparison with compression molded composites. Journal of Food Engineering 91, 415–422. https://doi.org/10.1016/j.jfoodeng.2008.09.019.

CHAPTER 9

Electrospun Cellulose Acetate Nanofiber: Characterization and Applications

SUSHMITA MAJUMDER • AHMED SHARIF • MD ENAMUL HOQUE

1 INTRODUCTION

Electrospinning has been widely recognized as a productive and adept technique for the fabrication of polymer nanofibers (Hoque et al., 2013; Huang et al., 2003; Li and Xia, 2004; Nuge et al., 2017; Teo and Ramakrishna, 2006; Wang et al., 2005). The polymer fiber materials exhibit several extraordinary characteristics, which comprise of very large surface-area-to-volume ratio that can be as large as 10^3 times of the nanofiber, enhanced mechanical properties (e.g., tensile strength and stiffness), and ease in functionalizing the surface features compared with known form of the material when the diameters are scaled down from micrometers to submicrons or nanometers. The increase in surface area induces nanoeffect as the number of atoms on the surface of the fiber is augmented, which leads to better functionality. Also, this technique allows greater control over the process and thus aids in the manipulation of fiber properties according to the need. These exciting features render the polymer nanofibers to be ideal for different important and sophisticated applications (Huang et al., 2003). Even though other methods of fabricating nanofiber assemblies exist, such as phase separation, template synthesis, and drawing, these techniques have their own drawbacks, which make electrospinning to stand alone in fiber production by virtue of its ease in use, greater control, and flexibility of the process.

If flexibility is concerned, electrospinning is able to produce nonwoven and continuous nanofibers from a vast span of materials. Of the important classes of materials, electrospinning is able to fabricate nanofibers of polymers, metals, semiconductors, composites, and ceramics. Recent researches focus on finding different electrospinnable materials and exploring the electrospinning conditions to yield fibers from them. However, in order to realize the potential of electrospun nanofibers, it is essential to fabricate fibers of various alignments, as the fiber arrangement will have a considerable effect on the device performance. For example, ordered assemblies and nanogrooves have been shown to induce cell proliferation and morphology. Electrospinning eliminates the difficulty to form various assemblies through physical manipulation due to the size of nanofibers as this technique is able to fabricate various nanofiber assemblies in situ. Hence, electrospinning supersedes other large-scale nanofiber production techniques because this method allows manipulation of nanofiber assemblies such as morphology and orientation so that the requirement for a specific application is met and also its performance is improved (Teo and Ramakrishna, 2006; Wang et al., 2005).

2 OVERVIEW OF ELECTROSPINNING

A typical electrospinning setup consists of basically three vital constituents: a high-voltage power source, a spinneret or syringe pump, and a collector (an earthed conductor). A schematic diagram of the fundamental setup to interpret electrospinning is shown in Fig. 9.1. Generally, a polymer solution is fed into the syringe, and an electric field is applied by placing one electrode at the needle head and the other at the collector (Greiner and Wendorff, 2007; Si et al., 2012). The needle tip to collector distance and the feed rate are fixed beforehand. When voltage is applied, the polymer solution at the needle tip gets charged. This caused the solution to experience two different forces acting upon it: electrostatic repulsion force due to similar charges and the surface tension. The surface tension acts opposite

Advanced Processing, Properties, and Applications of Starch and Other Bio-Based Polymers. https://doi.org/10.1016/B978-0-12-819661-8.00009-3

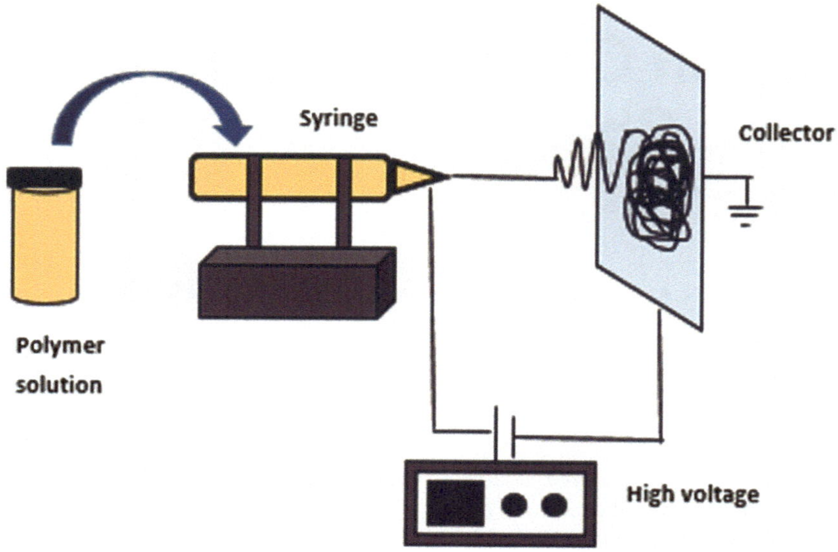

FIG. 9.1 Electrospinning setup with its basic components.

in direction to the repulsive force, and when the voltage attains a certain critical value, the surface tension is overcome by the electrostatic repulsion of similar charges in the polymer solution (Chew et al., 2006; Sill and von Recum, 2008). Taylor cone, a conical fluidic structure, is developed at the tip of the needle under the influence of the electrostatic interactions (Li and Xia, 2004).

3 OPTIMIZING PARAMETERS OF ELECTROSPINNING

Electrospinning is governed by the regulation of certain parameters that can be grouped as process parameters, solution parameters, and ambient parameters. The solution parameters consist of molecular weight of the polymer, concentration of the solution, viscosity, feed rate, surface tension, and electrical conductivity of the solution (Deitzel et al., 2001; Matabola and Moutloali, 2013; Nezarati et al., 2013; Si et al., 2012). There has to be an optimum concentration of the solution to initiate electrospinning. Below this concentration, beads will be produced instead of fibers (Haider et al., 2013). However, an excess in concentration can result in jet instability and also hinder fiber formation. Likewise, the molecular weight of the polymer is another factor to be considered as it refers to the number of chain entanglement in the solution, which also corresponds to the viscosity of the solution. Greater the chain entanglement, the higher the viscosity. If viscosity is low, no fiber formation takes place as it causes spraying of polymer jet, which deposits in the collector as droplets. Electrical conductivity is associated with the type of polymer and solvents used and also on the availability of ionizable salts in the solution. Higher electrical conductivity is achieved when there are sufficient charges in the solution, and it leads to the fiber of smaller diameter (Bi, 2013). Among the process parameters, voltage adjustment is very crucial as jet initiation from the needle tip only takes place under a critical voltage, and this voltage can differ from one type of solution to another (Thie, 2012; Mckee et al., 2004, 2006). Similarly, the feed rate and needle to collector distance are all to be adjusted (Matabola and Moutloali, 2013). The distance should be such that the polymer gets enough time to dry up before accumulating at the collector. The ambient parameters such as temperature and humidity can also be controlled to modify fiber properties (Deitzel et al., 2002; Nezarati et al., 2013).

4 POLYMERS IN ELECTROSPINNING

Of the classes of materials to electrospin, polymers have made the most remarkable contribution regarding their manufacturing and processing ease and applications. Fibers have been reported to be electrospun from a number of polymers of which many are either natural or synthetic polymers. It can be also a blend of these two. Natural polymers include collagen, chitosan, cellulose acetate, fibrinogen, casein, silk fibroin, and many more. Natural polymers, which are also termed as biopolymers, are preferable for biomedical applications as they offer better biocompatibility and low immunogenicity. Furthermore, they have an inherent tendency to attach cells due to their presence of specific protein sequences (Zhong et al., 2006; Xu et al., 2004; Kadler et al., 1996).

The ability for natural polymers to fulfill the requirements of a specific market creates a mounting slot for them. Therefore, enhanced products can be manufactured when the inherent and fundamental properties of biopolymers are pooled with the stirring nanoeffects that nanofibrous mats have to present. Biopolymer nanofibers have myriad purposes as nanocomposite strengthening fibers for nanotechnology, particle filters, filters for metal recovery, sutures, as templates, and in biologically and chemically protective clothing. Another feature of the nanofibers, porosity, can be changed and effects such as the wetting properties, number of attaching points for cells, and rate of degradation can all be altered. Thus, textiles for medical purposes, filtration for chemicals, fuel cell membranes, electrochemical cells, catalysis, and nanoreinforcements would advantage from using a nanofibrous mat with augmented porosity (Ki et al., 2007; Phys, 2011).

5 BACKGROUND OF CELLULOSE ACETATE IN ELECTROSPINNING

Cellulose is a widely available polymer with the advantages of being renewable and biodegradable. However, electrospinning cellulose possesses great challenges as it is insoluble in most of the common solvents due to the presence of inter- and intramolecular hydrogen bonds. Moreover, the solvents in which cellulose is soluble such as dimethylsulfoxide/paraformaldehyde are not electrospinnable (Kulpinski, 2005). Hence, derivatives of cellulose are explored. Of the cellulose derivatives, cellulose acetate, an esterification product of cellulose, offers superior properties and favorable conditions for electrospinning.

Cellulose acetate was first electrospun in 1998 using acetone as a solvent, which resulted in beaded morphology. This was attributed to the low viscosity of the solution and the low boiling point of acetone (Jaeger et al., 1998). Based on this earlier study, in 2002, another study was conducted using acetic acid, acetone, and dimethylacetamide (DMAc) as solvents. The result showed that a binary system of acetone/DMAc fabricated the most homogenous fibers with diameters varying between 100 nm and 1 μm (Liu and Hsieh, 2002). Later, a new solvent system of acetone/water was attempted for cellulose acetate, which resulted in uniform fibers (Son et al., 2004a). Similar studies on cellulose acetate were also conducted in 2005, where the focus was made on improving mechanical properties of the fiber by heat treatment. In 2007, cellulose acetate was electrospun containing vitamin A and E with diameters 247–265 nm for cosmetic applications. Moreover, the mats were functionalized with drugs for topical drug delivery, and due to good swelling of the fiber mat, it released most of it (Ma et al., 2005; Taepaiboon et al., 2007).

New dimensions were added when cellulose acetate membranes were prepared by coelectrospinning or blend electrospinning. Polyhexamethylene biguanide, an antibacterial agent, was integrated into the electrospun fibers. The existence of cellulose acetate in the nanofiber membrane enhanced its hydrophilicity and permeability to moisture and air (Liu et al., 2012). Sophistication in research on electrospinning of cellulose acetate went on with its surface modification by developing poly[(vinylbenzyl) trimethylammonium chloride)] brushes via a chemical sequence consisting of multisteps for the purpose DNA adsorption (Demirci et al., 2014). In 2016, cellulose acetate nanofiber was further studied to modify its property by perfluoro imparting superhydrophobicity to it. Thus, it was established as an effective way for oil/water separation (Arslan et al., 2016). Recently, an extensive study was made on establishing a systematic way to study solvents for cellulose acetate electrospinning by adopting and analyzing the Teas chart (Majumder et al., 2019). Hence, the potential of cellulose acetate can be very well comprehended by analyzing the abovementioned applications of cellulose acetate nanofiber mat.

6 CHARACTERIZATIONS OF CELLULOSE ACETATE NANOFIBER

After electrospinning, the cellulose acetate nanofibers are characterized to observe and interpret their physical, chemical, and thermal properties with various characterization equipments such as differential scanning calorimetry (DSC), viscosity, Fourier transform infrared spectroscopy (FTIR), thermogravimetric analysis (TGA), X-ray photoelectron spectroscopy (XPS), field emission scanning electron microscopy (FESEM), swelling analysis, contact angle measurement, X-ray diffractometry (XRD), tensile test, thin-layer chromatography (TLC), nuclear magnetic resonance (NMR), dynamic light scattering (DLS), ultraviolet-visible spectroscopy (UV-Vis), atomic force microscopy (AFM), and Raman spectroscopy.

6.1 FESEM Study for Cellulose Acetate

Presently, scanning electron microscopy (SEM) is profoundly exploited in the pursuit of the interpretation of cellulose acetate nanofibers. The combination of larger depth of field, higher magnification, compositional information, and greater resolution makes the SEM one of the most comprehensively exercised tools in high-tech industries and research areas. Recently, SEM study was carried out on electrospun cellulose acetate in the combination of different solvents [15]. Cellulose acetate was electrospun combining different solvents (Reneker and Yarin, 2008; Reneker et al., 2007). When DMAc was used as the solvent, it resulted in only beads. Reports showed that when acetic acid was used with DMAc and acetone separately, continuous fabrication of fiber was possible. It was observed that 15% of cellulose acetate when mixed with 3:1 acetic acid/DMAC and electrospun yielded fibers with beads as showed in Fig. 9.2A. Again, this fiber formation is not supported if the concentration of cellulose acetate is lowered below 15% in acetic acid/DMAc. This is not always the scenario if a different solvent system is used. For example, 12.5%–20% cellulose acetate with 2:1 acetone/DMAc supported continuous fiber formation without any beads as represented in Fig. 9.2B.

The change in concentration has a remarkable effect on fiber diameter. 15% cellulose acetate with 2:1 acetone/DMAc resulted in fiber diameter of around 400 nm in Fig. 9.3A. When the concentration was reduced to 12.5%, the diameter was reduced by about half, Fig. 9.3B, and lowering the concentration further to 10%, it yielded 100 nm fibers with beads of diameter 1 μm showed in Fig. 9.3C (Liu and Hsieh, 2002).

From the same research, they also found that binary solvent systems appeared to be better than the constituent single solvents for electrospinning cellulose acetate. If acetone was to be used as a solvent, it resulted in different morphology. Ribbon or flat-shaped fibers were obtained as depicted in Fig. 9.4A whereas using DMAc resulted in no fibers but only beads as in Fig. 9.4B (Ghorani et al., 2013; Choktaweesap et al., 2007).

6.2 Rheological Analysis of Cellulose Acetate Solution

The rheological behavior or viscosity of cellulose acetate solution is directly linked with the electrospinning phenomenon and morphology. Viscosity plays a key role in any electrospinning technique as a certain degree of a viscous solution is needed for polymer jet to initiate; otherwise, droplets are emerged due to Rayleigh instability caused by surface tension. A high viscoelastic system favors smoother fibers as it is responsible for polymer chain entanglement in the solution. Without chain entanglement, fiber jet cannot form due to surface tension. Viscoelasticity also impedes shape changes of fiber which usually takes place during the process due to the effect of surface tension. Earlier works reported the viscosity as a function of shear rate from a rheometer and solution concentration cellulose acetate (10%–27.5%) in 2:1 acetone/DMAc. Fig. 9.5A indicated that at low concentration, the solution flow behavior showed shear thickening effect, which meant that the viscosity increased with increasing shear rate. However, with higher concentration it showed shear-thinning effect, i.e., the viscosity dropped with increasing shear rate. Fig. 9.5B showed the apparent viscosity at zero shear rate with respect to solution concentration to determine the overlapping polymer concentration that favored bead-free fiber formation. From the figure, it was analyzed that below the overlapping concentration fiber formation was suppressed while continuous fibers formed without beads above the overlapping potential (Vallejos et al., 2012).

6.3 Swelling Behavior Study

The swelling test of electrospun fiber mat was also carried out in literature to determine their effectiveness as topical or transdermal drug delivery. The electrospun fiber mat was cut into 2.5 × 2.5 cm^2 and immersed into 50 mL distilled water for 24 h. The difference between

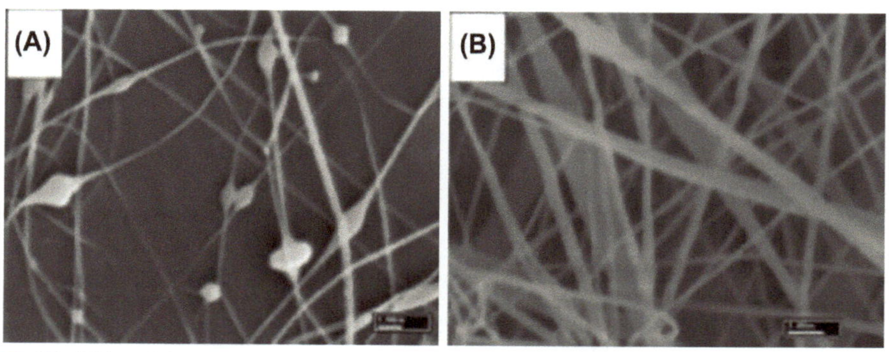

FIG. 9.2 **(A)** 15% cellulose acetate with 3:1 acetic acid/DMAc and **(B)** 12.5% cellulose acetate with 2:1 acetone/DMAc (Liu and Hsieh, 2002).

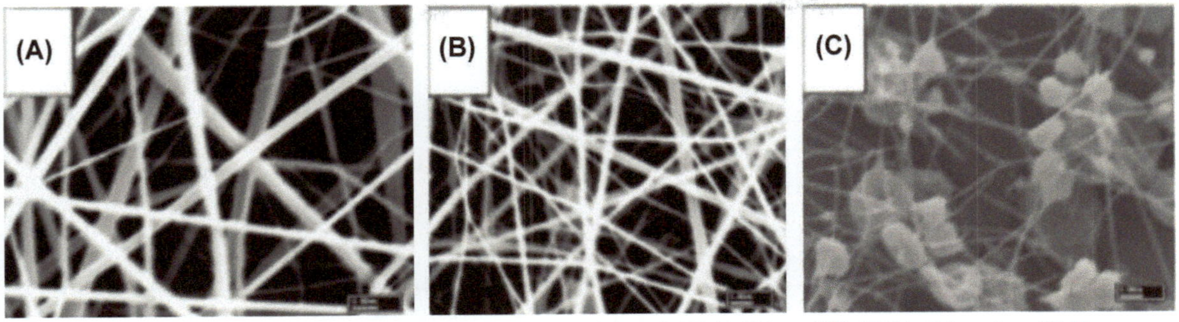

FIG. 9.3 FESEM of cellulose acetate of concentration **(A)** 15%, **(B)** 12.5%, and **(C)** 10% in 2:1 acetone/DMAc (Liu and Hsieh, 2002).

the dry and wet weight of the fiber gives the swelling capacity of the fibers. Solution concentrations of 12%, 14%, 16%, and 20% cellulose acetate in 2:1 acetone/DMAc showed a regular increase in water absorption or swelling. This was attributed to the fact that with increasing solution concentration, the fiber diameter increased. This increase in diameter corresponded to larger volume per unit length of the fiber where a greater number of water molecules can be accommodated. Thus, the potential of cellulose acetate fiber to serve for the drug delivery purpose is commendable (Tungprapa et al., 2007b).

6.4 FTIR Study of Cellulose Acetate Fiber

Structural analysis of cellulose acetate fibers is carried out by FTIR where information on different bonds associated can be interpreted based on the absorbed peak positions. Fig. 9.6 showed FTIR spectra of pure cellulose, cellulose acetate, and deacetylated cellulose acetate fiber.

Cellulose acetate has two distinctive FTIR peaks at 1730 and 1220 cm^{-1}, conforming to C=O and C–O–C groups, respectively. The disappearance of these two peaks in the spectra of deacetylated CA indicated that the acetate group was excluded. During the comparison of the spectra of pure cellulose with that of deacetylated CA, little variances can be noticed (Song and Hinestroza, 2012; Vetrivel et al., 2018).

Another study showed the bonding information of cellulose acetate and regenerated cellulose acetate in Fig. 9.7. The broad absorption peak at 3478 cm^{-1} indicated the presence of the hydroxyl group. The characteristic peak at 1738 cm^{-1} represented the overlapped ester carbonyls in cellulose acetate. After deacetylation, the broad peak at 3478 cm^{-1} appeared more prominent due to the presence of more hydroxyl groups an also the characteristic peak at 1738 cm^{-1} disappeared in regenerated cellulose acetate (Ghorani et al., 2013).

6.5 XPS Analysis of Cellulose Acetate

Cellulose acetate fiber obtained by electrospinning can be modified with $AgNO_3$ in order to impart antibacterial properties in it. The chemical interactions between cellulose acetate and silver nanoparticles (NPs) were

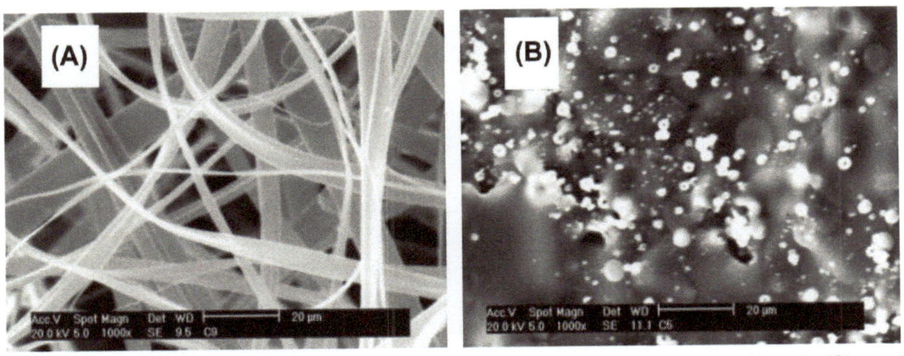

FIG. 9.4 Cellulose acetate in **(A)** acetone showed flat fibers and **(B)** DMAc showed beads (Ghorani et al., 2013).

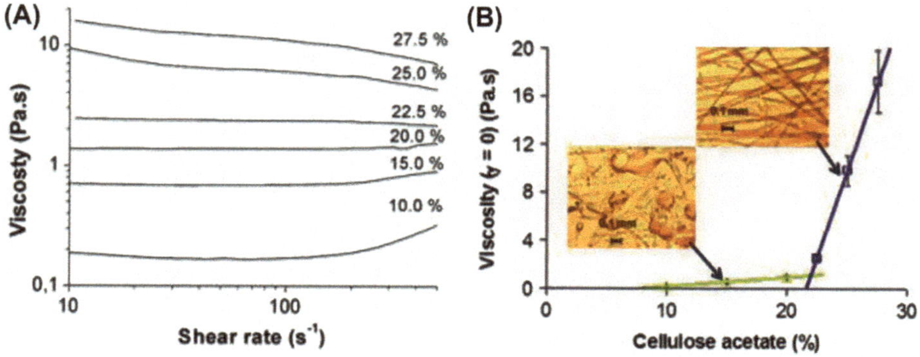

FIG. 9.5 **(A)** Apparent viscosity vs. shear rate of CA solutions and **(B)** effect of CA concentration on the apparent viscosity at 0 1/s shear rates (Vallejos et al., 2012).

analyzed in previous reports where it was found that, as in Fig. 9.8, the photoemission of C 1s of cellulose acetate was not affected by the introduction of $AgNO_3$ and thus implying that there was no interaction between the carbon atom and silver NPs. However, the photoemission spectra of O 1s were moved to higher energy resolving into two peaks. The lower peak matched with that of cellulose acetate and the higher one at 531.7 eV was ascribed to the interactions between silver NPs and carbonyl oxygen atoms (Son et al., 2004b).

6.6 Thermal Analysis of Cellulose Acetate Fiber by DSC and TGA

DSC and TGA were employed to characterize the thermal properties of CA fibers. For TGA, the employed heating rate was 10°C/min, while the thermal degradation temperature varied from 30 to 700°C. Similarly, for DSC, the heating rate was selected to be 5°C/min for the testing samples in a nitrogen atmosphere at a gas flow rate of 20 mL/min.

Fig. 9.9 presents that all three samples showed an endothermic peak between 30 and 110°C. This event

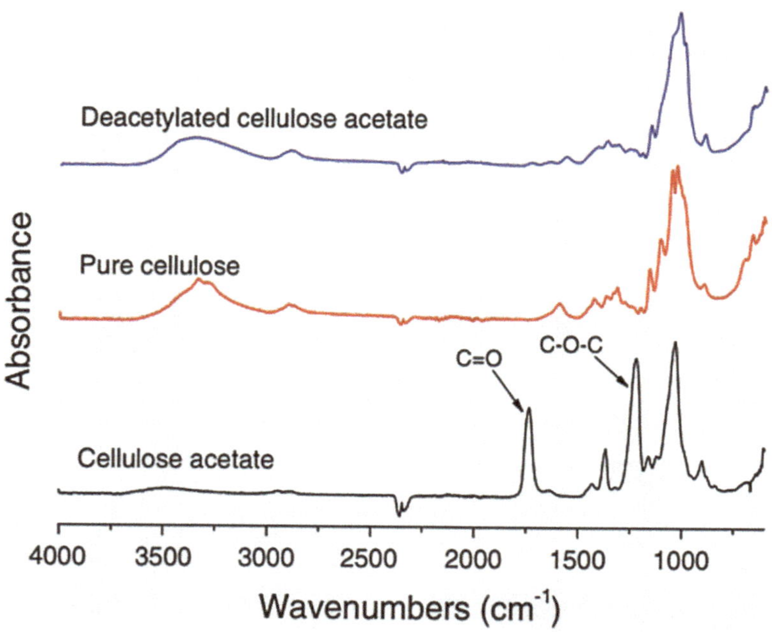

FIG. 9.6 FTIR spectra of cellulose acetate fibers, deacetylated cellulose acetate fibers, and pure cellulose (Song and Hinestroza, 2012).

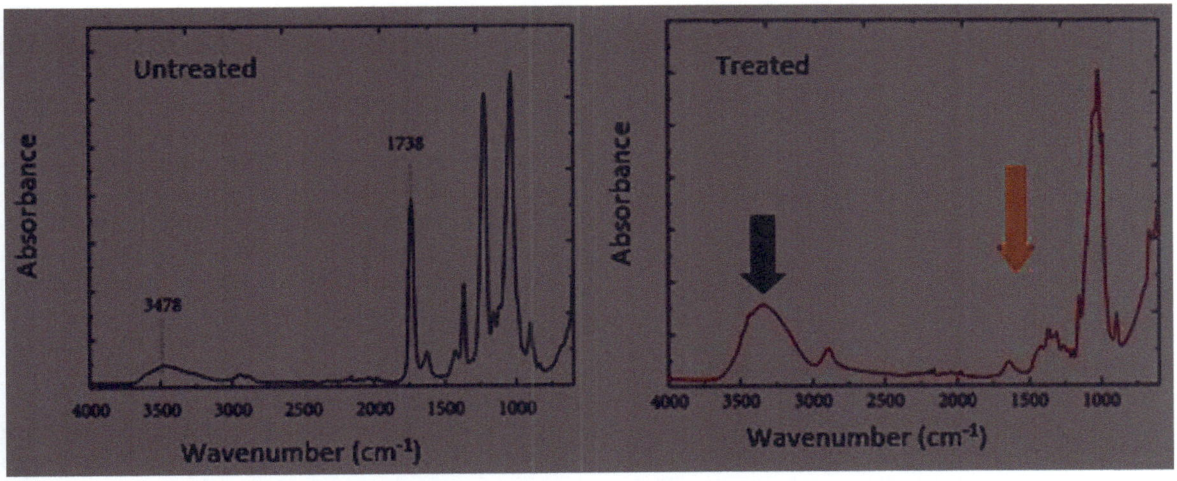

FIG. 9.7 FTIR spectra of cellulose acetate and regenerated cellulose acetate (Ghorani et al., 2013).

was attributed to water desorption phenomena. However, a strong glass transition peak was not observed in all of the samples, and it was attributed to the intrusion of crystalline regions. The melting peaks were observed in all three samples and the fusion enthalpy was calculated. It was observed that the three samples had the same melting temperature, around 237°C. The electrospun cellulose acetate membrane had the fusion enthalpy of about 15.2 J/g, which meant that cellulose acetate possessed a more crystalline geometry (Wu et al., 2014).

Fig. 9.10 presented the TGA curve for the previously mentioned three samples, which along the whole temperature range showed similar changes. The decomposition was observed between 350 and 410°C, chiefly due to a cellulose degradation process, for example, dehydration, depolymerization, and the decomposition of glucosyl. The electrospun structure showed the highest temperature difference of 43.7°C and thus it was concluded that higher temperature would be required to disintegrate this stable structure (Wu et al., 2014).

6.7 Hydrophilicity Study of Electrospun Cellulose Acetate

The surface hydrophilicity or hydrophobicity of the electrospun cellulose acetate and the casted cellulose acetate was compared by contact angle measurement study. The

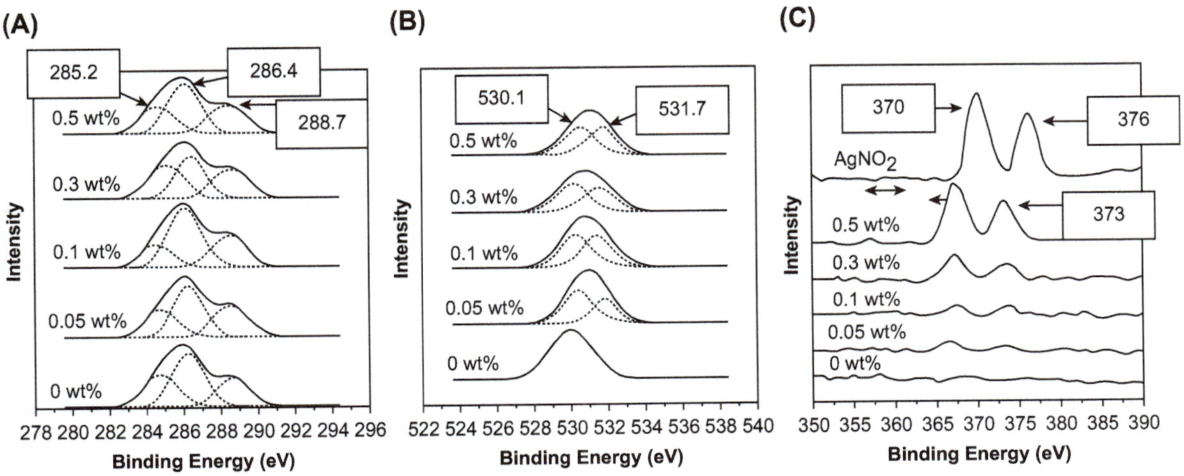

FIG. 9.8 **(A)** C1s **(B)** O1s, and **(C)** Ag3d XPS spectra for ultrafine cellulose acetate fibers with different amounts of $AgNO_3$ (Son et al., 2004b).

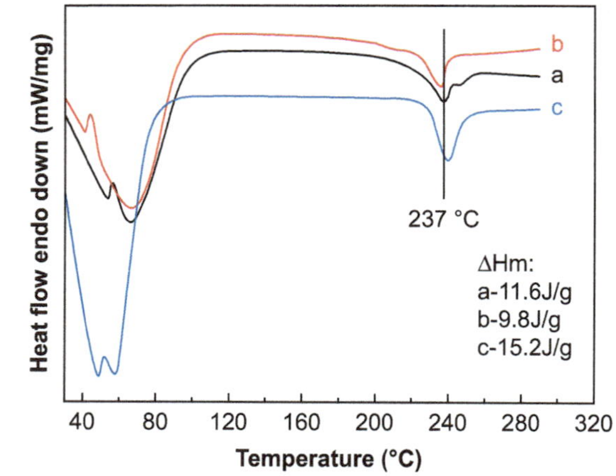

FIG. 9.9 DSC thermograms of cellulose acetate materials: **(A)** the as-obtained material without electrospinning; **(B)** the casted film, and **(C)** the electrospun fibrous membrane (Wu et al., 2014).

casted film showed a contact angle of 74.4 degree, while the electrospun fiber mesh showed 127.2 degree as in Fig. 9.11. Thus, the cast structure was hydrophilic in nature, unlike the electrospun structure which was hydrophobic. This phenomenon was approximately due to the rearrangement of CA molecules during the electrospinning process (Wu et al., 2014).

6.8 X-Ray Diffractometry of Cellulose Acetate Fiber

The XRD patterns for cellulose acetate were recorded to analyze the structural features of the fiber. Generally, the process was carried out at the tube voltage and tube current values of 40 kV and 40 mA, respectively.

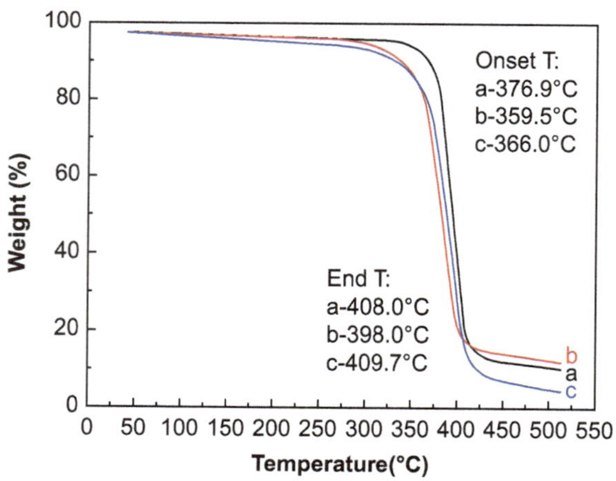

FIG. 9.10 TGA curves of cellulose acetate materials: **(A)** the as-obtained material without electrospinning; **(B)** the casted film; and **(C)** the electrospun fibrous membrane (Wu et al., 2014).

Fig. 9.12 showed the XRD patterns of pure cellulose acetate electrospun fibers and the fiber modified with silk fibroin. The two broad peaks at $2\Theta = 10.2$ degree and $2\Theta = 21.7$ degree indicated the amorphous structure of cellulose acetate. However, on adding silk fibroin to 5 wt.% cellulose acetate, a broad peak $2\Theta = 9.1$ degree appeared, whose intensity enhanced on increasing the concentration of cellulose acetate. This was attributed to the β-sheet conformation that occurred when hydrogen bonding took place between hydroxyl groups of cellulose acetate and amido and carboxyl groups of silk fibroin (Weitao et al., 2011; Taha et al., 2012).

6.9 UV-Vis Spectroscopy

Electrospun cellulose acetate fiber offers itself as a viable substrate for embedding metal NPs into it for various applications such as imparting antibacterial properties, desalination, and purification purposes. The presence of the NPs was ascertained with UV-Vis spectroscopy, which generated peak according to the elemental composition of the fiber. Also, the atom and mass percentages were evaluated by this characterization technique. A study conducted the UV-Vis of cellulose acetate fiber with 0.5 wt% $AgNO_3$ after irradiating with UV light of wavelengths 254 and 365 nm. Fig. 9.13 showed the UV-Vis spectrum for different irradiation times. A narrow and symmetrical absorption peak corresponded to the narrow size and uniform distribution of silver NPs. The intensity of plasmon peak increased and shifted to longer wavelengths for the period of 240 min of irradiation time, which bore testimony to the fact that the size and number of silver NPs increased up to this time. No peak broadening was observed for higher time periods due to the depletion of silver ions (Son et al., 2006; Martínez-Castañon et al., 2008).

6.10 NMR Spectroscopy Analysis

NMR spectroscopy is another vital characterization technique to investigate the integrity of loaded materials into the cellulose acetate fiber mat. In a study, curcumin loaded cellulose acetate fiber was fabricated by electrospinning, which has antiinflammatory and antitoxic properties. This analysis of integrity is crucial because due to the high voltage applied during electrospinning, the chemical integrity of the loaded curcumin might get affected. Thus, NMR was performed on the pure cellulose acetate and pure curcumin, and the peaks were compared with the electrospun fiber mat of cellulose acetate containing 5 wt.% curcumin. The presence of peaks of both the chemicals confirmed the uncontaminated integrity of curcumin (Suwantong et al., 2007).

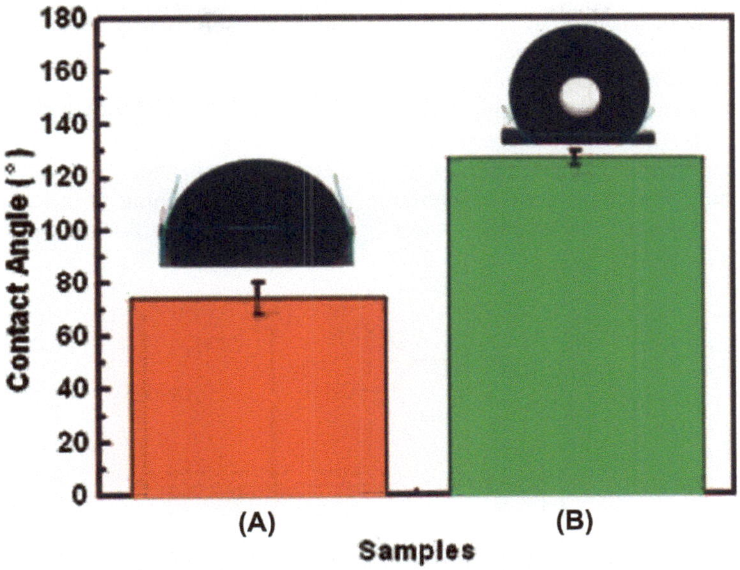

FIG. 9.11 Contact angle of cellulose acetate materials: **(A)** the casted film and **(B)** the electrospun fibrous membrane (Wu et al., 2014).

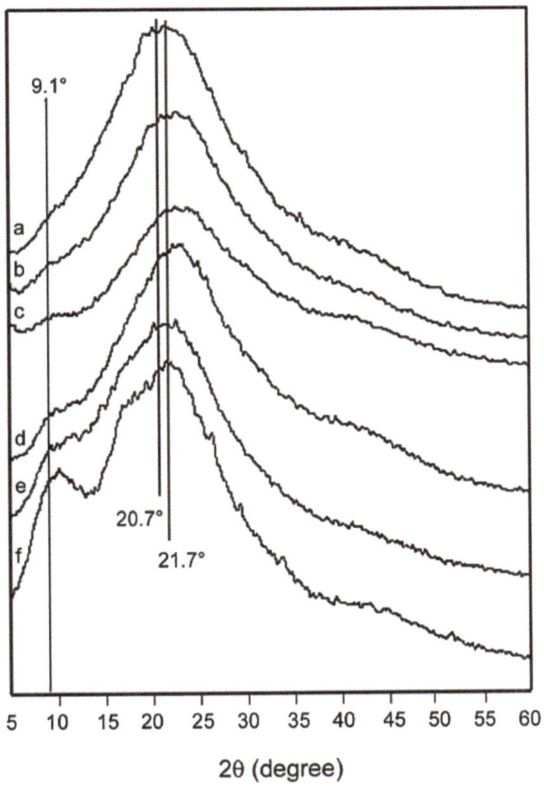

FIG. 9.12 X-ray diffraction spectra of the silk fibroin/cellulose acetate nanofibers with different contents of cellulose acetate: **(A)** 0%, **(B)** 5%, **(C)** 10%, **(D)** 30%, **(E)** 40%, and **(F)** 100% (all by weights) (Weitao et al., 2011).

6.11 Tensile Testing of Cellulose Acetate Fiber

The properties (e.g., mechanical) of the electrospun fiber were analyzed with the tensile test. The strength of the fiber mat depends upon a number of factors such as electrospinning solution concentration, fiber

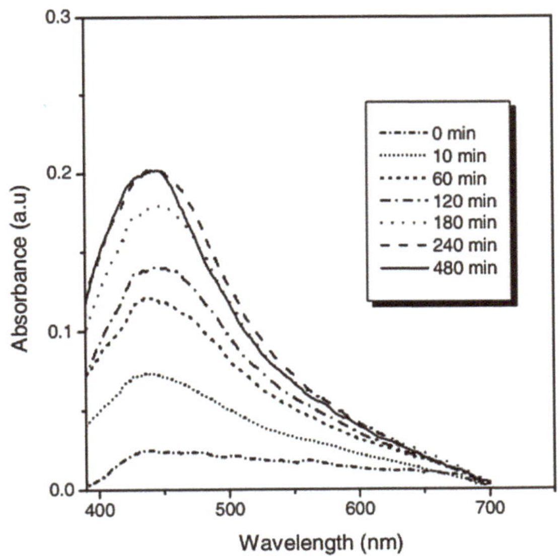

FIG. 9.13 UV-Vis spectra of cellulose acetate fiber with 0.5 wt.% $AgNO_3$ (Son et al., 2006).

alignment in the direction of tensile load, fiber's components, and so on. Cellulose acetate fiber modified with polyethylene glycol (PEG) for producing phase change fibers was tested, and the stress-strain curve is depicted in Fig. 9.14. Also Fig. 9.15 estimated the ultimate tensile strength and tensile strain for cellulose acetate to be 8.5 MPa and 10%, respectively. However, with increasing PEG content, the strength value decreased due to the breakage of continuous phase structure (Chen et al., 2011).

6.12 TLC of Electrospun Cellulose Acetate Fiber

Significant utilization of electrospun cellulose acetate fiber for the screening of steroidal wastes is acknowledged for TLC. For TLC procedures, standard solutions

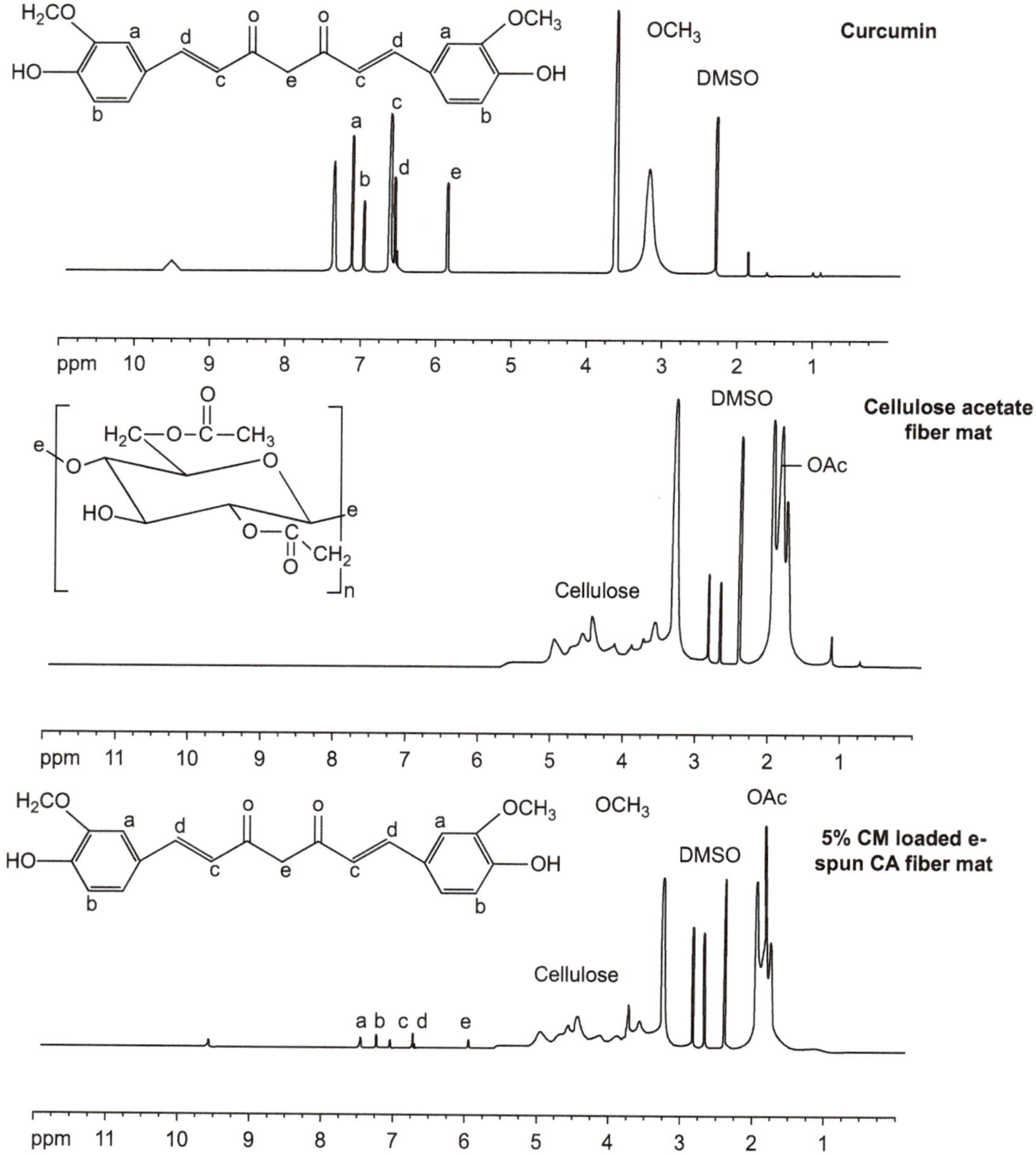

FIG. 9.14 NMR spectra of curcumin and electrospun cellulose acetate fiber mats from the base cellulose acetate solution and the solution containing 5% curcumin (Suwantong et al., 2007).

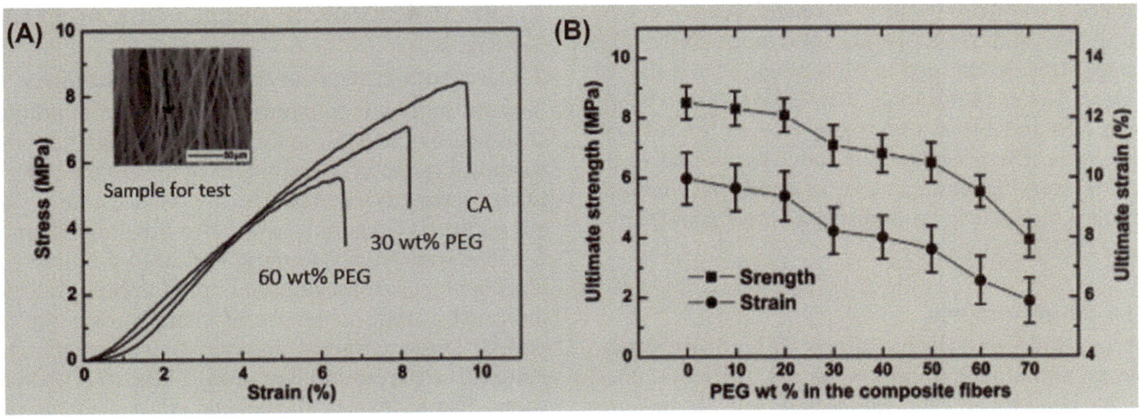

FIG. 9.15 **(A)** The stress-strain curves of electrospun fibers; **(B)** ultimate strength and ultimate strain of the fibers versus PEG content in the fibers (Chen et al., 2011).

having reference standards of dexamethasone at 1 mg/mL were prepared in ethanol. The cellulose acetate nanofiber plates of dimension 8.5 cm × 4 cm were used to separate dexamethasone and prednisolone by spotting 1 μL of sample or reference standard solutions from 1 cm of the bottom edge. Binary mixtures such as comprised of ethanol/water or methanol/water at various ratio combinations were used as mobile phases. When the front of the mobile phase detached away from the origin by 7 cm, the separation process was run until that time followed by its completion. After that, the spot was detected by spraying the plate with a mixture of 2% w/v sodium hydroxide and 0.2% w/v tetrazolium blue in methanol. Next, it was heated for 2 min when the violet-blue spots of steroids were observed under visible light. The movement of the steroids on the TLC plate was characterized by h_{Rf} value, defined as 100 R_f value, the ratio of the distance shifted by the solute to the distance shifted by the mobile phase front. The quality of chromatographic separation between the two steroids is represented by the resolution (Rs) parameter, which is calculated as the distance between two zone centers (d) divided by the average widths (W) of the zones (Rojanarata et al., 2013):

$$Rs = \frac{2d}{(W_1+W_2)}$$

6.13 DLS Analysis for Stability Analysis

The grafted NPs in cellulose acetate fibers need to be stabilized to prevent their agglomeration. If these particles agglomerate, their efficacy for the intended application decreases. Thus, the stability of the NPs in the fiber was ascertained via DLS. In a study, oleic acid (OA) and dimercaptosuccinic acid (DMSA) were used separately to stabilize iron oxide (Fe_3O_4) NPs in cellulose acetate fiber and their physiological stability was determined by DLS. Fig. 9.16A and B represented the two groups

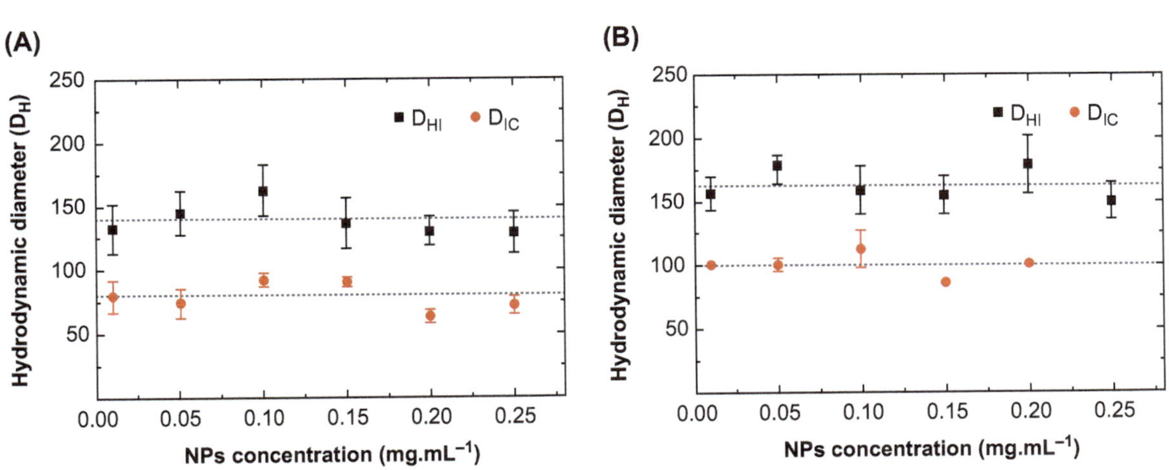

FIG. 9.16 The hydrodynamic diameter of **(A)** OA-Fe_3O_4 NPs and **(B)** DMSA-Fe_3O_4 NPs (Matos et al., 2018).

of calculated hydrodynamic diameter (D_H) for OA-Fe_3O_4 NPs and DMSA-Fe_3O_4 NPs, respectively. The average sizes determined at physiological pH value of 7 were 164 ± 13 nm and 101 ± 9 nm for DMSA-Fe_3O_4 NPs and 140 ± 13 nm and 79 ± 11 nm for OA-Fe_3O_4 NPs. The polydispersity index was found to be 0.2 for every experimental curve produced, demonstrating a moderately monodisperse suspension (Matos et al., 2018).

6.14 AFM Analysis

Porous cellulose acetate nanofibers are particularly suitable for filtration purposes. With a view, to produce porous nanofibers, cellulose acetate was electrospun with dichloromethane (DCM)/acetone mixture and characterized with AFM to determine the surface roughness due to porosity. AFM was conducted with a scan rate of 0.1 Hz in intermittent contact mode in the air using TAP 300 (Budget Sensors) type of cantilever having the resonant frequency of 200–400 kHz and a force constant of 20–75 N/m. The AFM images in Fig. 9.17 showed that the surface roughness and porosity increased with increasing concentration of DC. When the ratio of DCM/acetone was1:1, the porosity had a depth of 10–15 nm, whereas on increasing the ratio to 2:1, the porosity depth increased to 50 nm. This porosity was achieved because of the high volatility of DCM, which rapidly evaporated during electrospinning leading to a porous structured fiber (Celebioglu and Uyar, 2011).

6.15 Raman Spectroscopy

Hybrid nanocomposite fiber of cellulose acetate/graphene was characterized by Raman spectroscopy to study the interfacial interactions between cellulose acetate molecular chains and graphene. In Fig. 9.18, Raman spectra of the pure cellulose acetate nanofibers showed two characteristic bands at 1376 and 1748 cm^{-1} due to $-CH3$ (antisymmetric deformation vibrations of the methyl groups) and $-C{=}O$ vibrations, respectively. Similarly, Raman spectra of the cellulose acetate/graphene hybrid resultant nanofibers also showed the same characteristic bands at 1376 and 1748 cm^{-1}. Furthermore, a new band at 1572 cm^{-1} (corresponding to G band of graphene) was observed except for pure cellulose acetate nanofiber, suggesting that the graphene was well incorporated into the cellulose acetate fibers during electrospinning (Gopiraman et al., 2013).

7 APPLICATIONS OF CELLULOSE ACETATE FIBER

7.1 Immobilization of Bioactive Substance

Immobilization or entrapment of a number of polysaccharides, biocatalysts, and anticancer drugs has been recently arrested into the interior of the electrospun cellulose acetate fiber mat, which played a role in cell adhesion, proliferation, and differentiation. Vitamins A and E were surface-immobilized onto cellulose acetate fiber mat as round and smooth morphology, which indicated a uniform release of vitamins over the test period. Drug release is another dimension that has rendered electrospun fiber mat a desirable field of study. The controlled therapeutic effect of drugs at a particular location has been possible by virtue of this fiber mat. Indomethacin, naproxen, ibuprofen, and sulindac have been loaded onto electrospun ultrafine fiber of diameter 263–297 nm mats of cellulose acetate. These smooth fibers bulged in acetate buffer system postincubated at 37 °C for 24 h (Tungprapa et al., 2007a).

7.2 Cell Culture and Tissue Engineering

Tissue engineering requires the fabrication of polymeric scaffolds that provide a platform for cell attachment and growth. Electrospun cellulose acetate has been widely investigated for this purpose as the interaction of this fiber with extracellular matrix component at the nanoscale has provided with benefits of cell growth. Recently, saponification of electrospun cellulose acetate mat with diameters ranging from 200 nm to 1.5 μm has been fruitfully implemented for the production of innovative scaffolds with a view of nucleating bioactive calcium phosphate (CaP) crystals as a function of surface chemistry applicable to future bone healing purpose (Rodríguez et al., 2011).

7.3 Biosensor Application

Nanofibers with outstanding photosensitivity are the pillar of constructing biosensors or optical devices. If these are irradiated with appropriate light energy, there occurs a change in the molecular orientation, which causes a change UV-visible absorption spectra. Here, the establishment of the photochromic property of electrospun complex of cellulose acetate 1′,3′,3′-trimethyl-6-nitrospiro(2H-1-benzopyran-2,2′-indoline) (NO2SP) in acetone has heralded in a novel application field of the electrospun materials (Shuiping et al., 2010).

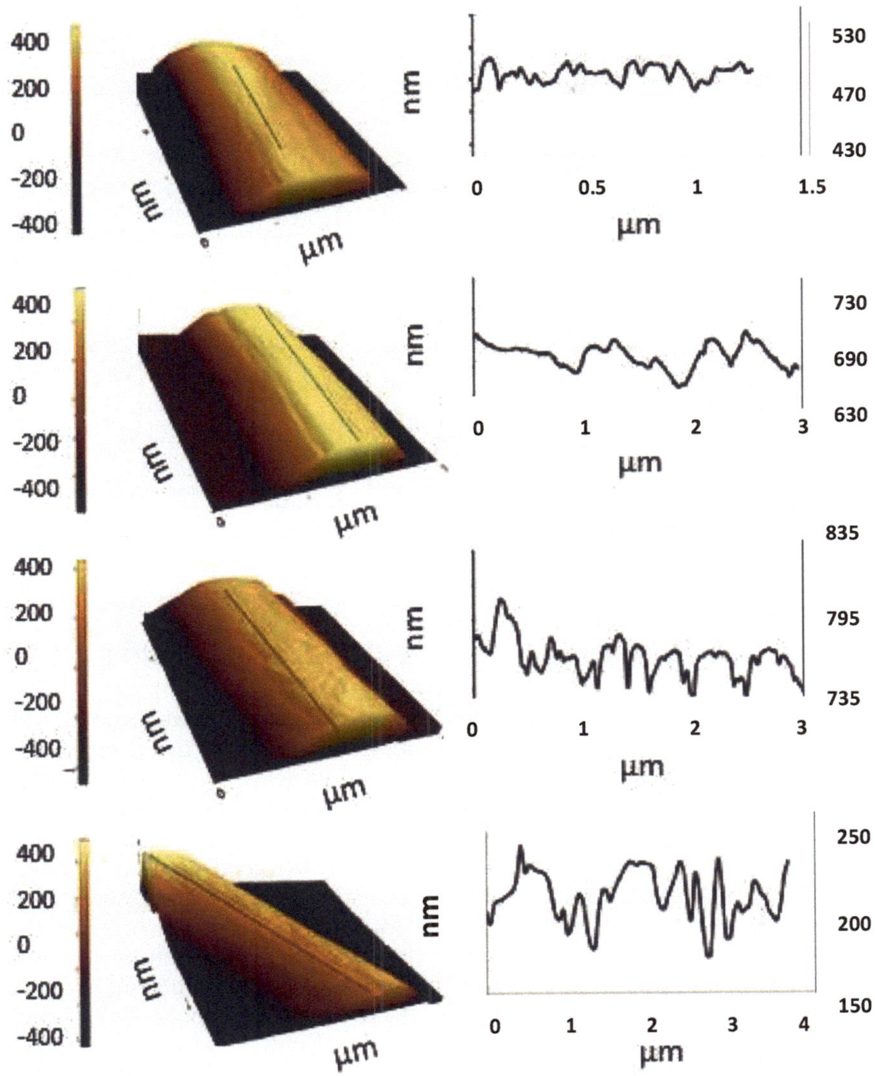

FIG. 9.17 AFM image and longitudinal axis cross-sectional sketch of the cellulose acetate fibers obtained from different DCM/Acetone ratio: **(A and B)** 10% (w/v) cellulose acetate in 1/1 DCM/Acetone; **(C and D)** 10% cellulose acetate in 2/1 DCM/Acetone; **(E and F)** 10% cellulose acetate in 3/1 DCM/Acetone; and **(G and H)** 7.5 cellulose acetate in 9/1 DCM/Acetone (Celebioglu and Uyar, 2011).

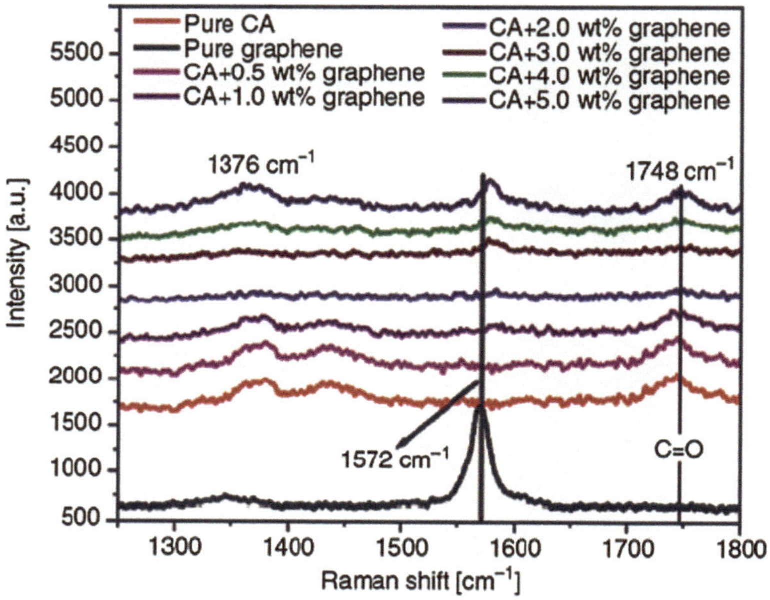

FIG. 9.18 Raman spectra of graphene-incorporated CA hybrid nanofibers (Gopiraman et al., 2013).

7.4 Nanomaterials Loaded Antimicrobial Mat

Food, textiles, and healthcare products are contaminated by microbial agents on a regular basis. As a result, applications of electrospun fiber mat with dispersed antimicrobial agents on its surface brings out new possibilities of combating this problem (Hoque et al., 2018). By various methods, antimicrobial agents such as silver and zinc NPs are embedded onto the fiber mat, which has high exposure due to the large surface area. The fibers with Ag NPs with an average size of 21 nm manifested superior antibacterial action against Gram-positive *Staphylococcus aureus* and Gram-negative *Escherichia coli, Klebsiella pneumoniae*, and *Pseudomonas aeruginosa* examined by the process called nonwoven fabric attachment method (Son et al., 2004b).

7.5 Temperature Adaptable Fabrics

Electrospun cellulose acetate/PEG fiber mat for thermal energy storage and release has been investigated. PEG acted as the material of model phase change, while cellulose acetate served the role of supporting framework. The phase change material imparted a desirable thermal storage and release profile (Chen et al., 2007). Most importantly, the repetitive test of heating-cooling thermal cycles to kindle dramatic temperature change conditions illustrated the remarkable capability of the electrospun fibers to adjust their interior temperature with the altered ambient temperature (Konwarh et al., 2013).

8 CONCLUSIONS AND FUTURE DIRECTIONS

This chapter explained the beneficial role of electrospinning cellulose acetate in different applications spanning from drug delivery vehicles to flexible medical devices like biosensors. Cellulose acetate appeared to be the most convenient of all cellulose derivatives due to its solubility in a good amount of solvents and offers suitable conditions for electrospinning. Characterizations have been carried out to ascertain the properties and features of cellulose acetate fiber, membrane, and films. FESEM revealed the variation in cellulose acetate fiber diameter and appearance due to change in electrospinning solvents, whereas rheological measurement explained the role of solution viscosity in yielding fiber. That cellulose acetate fiber mat can be utilized effectively as a wound dressing has been verified with the swelling test where the mat retained a greater percentage of fluid and XPS helped to evaluate the interactions between cellulose acetate and silver NPs for antibacterial activity. Other characterizations such as TGA, TLC, NMR, and AFM proved to be very useful in determining the usefulness of cellulose acetate fiber, membrane, or film in particular applications.

Electrospun cellulose acetate fibers have proved to be applicable in a wide variety of areas and in the future can play a promising role in addressing pressing scientific challenges of the society. High-performance scaffolds can be fabricated from electrospinning cellulose acetate and their biocompatibility can be assessed based on different conditions. Cellulose acetate nanofibers with hydrogels have the potential to be an effective wound dressing. Hence, the wound dressing efficiency of the electrospun cellulose acetate fiber mat can be compared after treating with hydrogels, silver sulfadiazine, antibiotics, NPs, plant extracts, proteins, and antimicrobial peptides and mark their biocompatibility. However, with rising alarm for antibiotic-resistant infection diseases, more powerful wound dressings are in a call. Research can be directed toward the incorporation of antimicrobial peptides into electrospun cellulose acetate mat and study their response to the host body.

REFERENCES

Arslan, O., Aytac, Z., Uyar, T., 2016. Superhydrophobic, hybrid, electrospun cellulose acetate nano fibrous mats for oil/water separation by tailored surface modification. ACS Applied Materials and Interfaces 8, 19747–19754. https://doi.org/10.1021/acsami.6b05429.

Bi, M., 2013. RSC Advances 4332–4340. https://doi.org/10.1039/c3ra22626c.

Celebioglu, A., Uyar, T., 2011. Electrospun porous cellulose acetate fibers from volatile solvent mixture. Materials Letters 65, 2291–2294. https://doi.org/10.1016/j.matlet.2011.04.039.

Chen, C., Wang, L., Huang, Y., 2011. Electrospun phase change fibers based on polyethylene glycol/cellulose acetate blends. Applied Energy 88, 3133–3139. https://doi.org/10.1016/j.apenergy.2011.02.026.

Chen, C., Wang, L., Huang, Y., 2007. Electrospinning of thermo-regulating ultrafine fibers based on polyethylene glycol/cellulose acetate composite. Polymer 48, 5202–5207. https://doi.org/10.1016/j.polymer.2007.06.069.

Chew, S.Y., Hufnagel, T.C., Lim, C.T., Leong, K.W., 2006. Mechanical properties of single electrospun drug-encapsulated nanofibres. Nanotechnology 17, 3880–3891. https://doi.org/10.1088/0957-4484/17/15/045.

Choktaweesap, N., Arayanarakul, K., Aht-Ong, D., Meechaisue, C., Supaphol, P., 2007. Electrospun gelatin fibers: effect of solvent system on morphology and fiber diameters. Polymer Journal 39, 622–631. https://doi.org/10.1295/polymj.PJ2006190.

Deitzel, J.M., Kleinmeyer, J., Harris, D.E.A., Tan, N.B., 2001. The effect of processing variables on the morphology of electrospun nanofibers and textiles. Polymer 42, 261–272. https://doi.org/10.1016/S0032-3861(00)00250-0.

Deitzel, J.M., Kosik, W., Mcknight, S.H., Tan, N.B., Desimone, J.M., Crette, S., 2002. Electrospinning of polymer nanofibers with specific surface chemistry. Polymer 43, 1025–1029. https://doi.org/10.1016/S0032-3861(01)00594-8.

Demirci, S., Celebioglu, A., Uyar, T., 2014. Surface modification of electrospun cellulose acetate nanofibers via RAFT polymerization for DNA adsorption. Carbohydrate Polymers 113, 200–207. https://doi.org/10.1016/j.carbpol.2014.06.086.

Ghorani, B., Russell, S.J., Goswami, P., 2013. Controlled morphology and mechanical characterisation of electrospun cellulose acetate fibre webs. International Journal of Polymer Science 2013. https://doi.org/10.1155/2013/256161.

Gopiraman, M., Fujimori, K., Zeeshan, K., Kim, B.S., Kim, I.S., 2013. Structural and mechanical properties of cellulose acetate/graphene hybrid nanofibers: spectroscopic investigations. Express Polymer Letters 7, 554–563. https://doi.org/10.3144/expresspolymlett.2013.52.

Greiner, A., Wendorff, J.H., 2007. Electrospinning: a fascinating method for the preparation of ultrathin fibers. Angewandte Chemie International Edition 46, 5670–5703. https://doi.org/10.1002/anie.200604646.

Haider, S., Al-Zeghayer, Y., Ali, F.A.A., Haider, A., Mahmood, A., Al-Masry, W.A., Imran, M., Aijaz, M.O., 2013. Highly aligned narrow diameter chitosan electrospun nanofibers. Journal of Polymer Research 20, 105. https://doi.org/10.1007/s10965-013-0105-9.

Hoque, M.E., Nuge, T., Yeow, T.K., Nordin, N., 2013. Electrospinning of Gelatin Nanofibre: Current Trends in Tissue Engineering Applications. Journal of Applied Mechanical Engineering 2 (4), e122. https://doi.org/10.4172/2168-9873.1000e122.

Hoque, M.E., Peiris, A.M., Atiqure Rahman, S.M., Abdul Wahab, M., 2018. New generation antibacterial nanofibrous membrane for potential water filtration [WWW document]. Current Analytical Chemistry 14 (3), 278–284. http://www.eurekaselect.com/156296/article.

Huang, Z.-M., Zhang, Y.-Z., Kotaki, M., Ramakrishna, S., 2003. A review on polymer nanofibers by electrospinning and their applications in nanocomposites. Composites Science and Technology 63, 2223–2253. https://doi.org/10.1016/S0266-3538(03)00178-7.

Jaeger, R., Bergshoeft, M.M., Martin, C., Vancso, S.G.J., 1998. Macromolecular Symposia 127, 141–150 (1998) 150, 141–150.

Kadler, K.E., Holmes, D.F., Trotter, J.A., Chapman, J.A., 1996. Collagen fibril formation. Biochemical Journal 316, 1–11. https://doi.org/10.1042/bj3160001.

Ki, C.S., Gang, E.H., Um, I.C., Park, Y.H., 2007. Nanofibrous membrane of wool keratose/silk fibroin blend for heavy metal ion adsorption. Journal of Membrane Science 302, 20–26. https://doi.org/10.1016/j.memsci.2007.06.003.

Konwarh, R., Karak, N., Misra, M., 2013. Electrospun cellulose acetate nanofibers: the present status and gamut of biotechnological applications. Biotechnology Advances 31, 421–437. https://doi.org/10.1016/j.biotechadv.2013.01.002.

Kulpinski, P., 2005. Cellulose nanofibers prepared by the N -methylmorpholine- N -oxide method. Journal of Applied Polymer Science 98, 1855–1859. https://doi.org/10.1002/app.22123.

Li, D., Xia, Y., 2004. Electrospinning of nanofibers: reinventing the wheel? Advanced Materials 16, 1151–1170. https://doi.org/10.1002/adma.200400719.

Liu, H., Hsieh, Y.L., 2002. Ultrafine fibrous cellulose membranes from electrospinning of cellulose acetate. Journal of Polymer Science Part B: Polymer Physics 40, 2119–2129. https://doi.org/10.1002/polb.10261.

Liu, X., Lin, T., Gao, Y., Xu, Z., Huang, C., Yao, G., Jiang, L., Tang, Y., Wang, X., 2012. Antimicrobial electrospun nanofibers of cellulose acetate and polyester urethane composite for wound dressing. Journal of Biomedical Materials Research Part B: Applied Biomaterials 100, 1556–1565. https://doi.org/10.1002/jbm.b.32724.

Ma, Z., Kotaki, M., Ramakrishna, S., 2005. Electrospun cellulose nanofiber as affinity membrane. Journal of Membrane Science 265, 115–123. https://doi.org/10.1016/j.memsci.2005.04.044.

Majumder, S., Matin, A., Sharif, A., Arafat, M.T., 2019. Understanding solubility, spinnability and electrospinning behaviour of cellulose acetate using different solvent systems. Bulletin of Materials Science 42, 171. https://doi.org/10.1007/s12034-019-1857-6.

Matabola, K.P., Moutloali, R.M., 2013. The influence of electrospinning parameters on the morphology and diameter of poly (vinyledene fluoride) nanofibers- effect of sodium chloride. Journal of Materials Science 48, 5475–5482. https://doi.org/10.1007/s10853-013-7341-6.

Matos, R.J.R., Chaparro, C.I.P., Silva, J.C., Almeida, M., Paulo, J., Soares, P.I.P., 2018. Electrospun composite cellulose acetate/iron oxide nanoparticles non- woven membranes for magnetic hyperthermia applications. Carbohydrate Polymers 198, 9–16. https://doi.org/10.1016/j.carbpol.2018.06.048.

Martínez-Castañon, G.A., Nino-Martinez, N., Martinez-Gutierrez, F., Martinez-Mendoza, J.R., Ruiz, F., 2008. Synthesis and antibacterial activity of silver nanoparticles with different sizes. Journal of Nanoparticle Research 10, 1343–1348. https://doi.org/10.1007/s11051-008-9428-6.

Mckee, M.G., Layman, J.M., Cashion, M.P., Long, T.E., 2006. Phospholipid Nonwoven Electrospun Membranes, Science, 311, pp. 353–355. https://doi.org/10.1126/science.1119790.

Mckee, M.G., Wilkes, G.L., Colby, R.H., Long, T.E., 2004. Correlations of solution rheology with electrospun fiber formation of linear and branched polyesters. Macromolecules 1760–1767.

Nezarati, R.M., Eifert, M.B., Cosgriff-Hernandez, E., 2013. Effects of humidity and solution viscosity on electrospun fiber morphology. Tissue Engineering Part C: Methods 19, 810–819. https://doi.org/10.1089/ten.tec.2012.0671.

Nuge, T., Tshai, K.Y., Lim, S.S., Nordin, N., Hoque, M.E., 2017. Preparation and characterization of CU-, FE-, AG-, ZN- and NI- doped gelatin nanofibers for possible applications in antibacterial nanomedicine. Journal of Engineering Science and Technology 12, 68–81.

Phys, J.A., 2011. Porous fiber formation in polymer-solvent system undergoing solvent evaporation. Journal of Applied Physics 043512. https://doi.org/10.1063/1.2259812.

Reneker, D.H., Yarin, A.L., 2008. Electrospinning jets and polymer nanofibers. Polymer 49, 2387–2425. https://doi.org/10.1016/j.polymer.2008.02.002.

Reneker, D.H., Yarin, A.L., Zussman, E., Xu, H., 2007. Electrospinning of nanofibers from polymer solutions and melts. Advances in Applied Mechanics 41, 43–346. https://doi.org/10.1016/S0065-2156(06)41002-4.

Rodríguez, K., Renneckar, S., Gatenholm, P., 2011. Biomimetic calcium phosphate crystal mineralization on electrospun cellulose-based scaffolds. ACS Applied Materials and Interface 3, 681–689. https://doi.org/10.1021/am100972r.

Rojanarata, T., Plianwong, S., Su-Uta, K., Opanasopit, P., Ngawhirunpat, T., 2013. Electrospun cellulose acetate nanofibers as thin layer chromatographic media for eco-friendly screening of steroids adulterated in traditional medicine and nutraceutical products. Talanta 115, 208–213. https://doi.org/10.1016/j.talanta.2013.04.078.

Shuiping, L., Lianjiang, T., Weili, H., Xiaoqiang, L., Yanmo, C., 2010. Cellulose acetate nanofibers with photochromic property: fabrication and characterization. Materials Letters 64, 2427–2430. https://doi.org/10.1016/j.matlet.2010.08.018.

Si, Y., Ren, T., Li, Y., Ding, B., Yu, J., 2012. Fabrication of magnetic polybenzoxazine-based carbon nanofibers with Fe_3O_4 inclusions with a hierarchical porous structure for water treatment. Carbon 50, 5176–5185. https://doi.org/10.1016/j.carbon.2012.06.059.

Sill, T.J., von Recum, H.A., 2008. Electrospinning: applications in drug delivery and tissue engineering. Biomaterials 29, 1989–2006. https://doi.org/10.1016/j.biomaterials.2008.01.011.

Son, W.K., Youk, J.H., Lee, T.S., Park, W.H., 2004a. Electrospinning of ultrafine cellulose acetate fibers: studies of a new solvent system and deacetylation of ultrafine cellulose acetate fibers. Journal of Polymer Science Part B: Polymer Physics 42, 5–11. https://doi.org/10.1002/polb.10668.

Son, W.K., Youk, J.H., Lee, T.S., Park, W.H., 2004. Preparation of antimicrobial ultrafine cellulose acetate fibers with silver nanoparticles. Macromolecular Rapid Communications 25, 1632–1637. https://doi.org/10.1002/marc.200400323.

Son, W.K., Youk, J.H., Park, W.H., 2006. Antimicrobial cellulose acetate nanofibers containing silver nanoparticles. Carbohydrate Polymers 65, 430–434. https://doi.org/10.1016/j.carbpol.2006.01.037.

Song, J., Birbach, N.L., Hinestroza, J.P., 2012. Deposition of silver nanoparticles on cellulosic fibers via stabilization of carboxymethyl groups. Cellulose 19, 411–424. https://doi.org/10.1007/s10570-011-9647-3.

Suwantong, O., Opanasopit, P., Ruktanonchai, U., Supaphol, P., 2007. Electrospun cellulose acetate fiber mats containing curcumin and release characteristic of the herbal substance. Polymer 48, 7546–7557. https://doi.org/10.1016/j.polymer.2007.11.019.

Taepaiboon, P., Rungsardthong, U., Supaphol, P., 2007. Vitamin-loaded electrospun cellulose acetate nanofiber mats as transdermal and dermal therapeutic agents of vitamin A acid and vitamin E. European Journal of Pharmaceutics and Biopharmaceutics 67, 387–397. https://doi.org/10.1016/j.ejpb.2007.03.018.

Taha, A.A., Wu, Y.N., Wang, H., Li, F., 2012. Preparation and application of functionalized cellulose acetate/silica composite nanofibrous membrane via electrospinning for Cr (VI) ion removal from aqueous solution. Journal of Environmental Management 112, 10–16. https://doi.org/10.1016/j.jenvman.2012.05.031.

Teo, W.E., Ramakrishna, S., 2006. A review on electrospinning design and nanofibre assemblies. Nanotechnology 17. https://doi.org/10.1088/0957-4484/17/14/R01.

Thie, A., 2012. e-Beam Lithography, EBL. https://doi.org/10.1007/978-90-481-9751-4.

Tungprapa, S., Jangchud, I., Supaphol, P., 2007a. Release characteristics of four model drugs from drug-loaded electrospun cellulose acetate fiber mats. Polymer 48, 5030–5041. https://doi.org/10.1016/j.polymer.2007.06.061.

Tungprapa, S., Puangparn, T., Weerasombut, M., Jangchud, I., Fakum, P., Semongkhol, S., Meechaisue, C., Supaphol, P., 2007b. Electrospun cellulose acetate fibers: effect of solvent system on morphology and fiber diameter. Cellulose 14, 563–575. https://doi.org/10.1007/s10570-007-9113-4.

Vallejos, M.E., Peresin, M.S., Rojas, O.J., 2012. All-cellulose composite fibers obtained by electrospinning dispersions of cellulose acetate and cellulose nanocrystals. Journal of Polymers and the Environment 20, 1075–1083. https://doi.org/10.1007/s10924-012-0499-1.

Vetrivel, S., Saraswathi, M.S.A., Rana, D., Divya, K., Nagendran, A., 2018. PT #pagerange#. https://doi.org/10.1016/j.ijbiomac.2018.04.091.

Wang, Y.K., Yong, A.T., Ramakrishna, S., 2005. Nanofibres and their influence on cells for tissue regeneration. Australian Journal of Chemistry 58, 704–712. https://doi.org/10.1071/CH05222.

Weitao, Z., Jianxin, H., Shan, D., Shizhong, C., Weidong, G., 2011. Electrospun silk fibroin/cellulose acetate blend nanofibres: structure and properties. Iranian Polymer Journal 20, 389–397.

Wu, S., Qin, X., Li, M., 2014. The structure and properties of cellulose acetate materials: a comparative study on electrospun membranes and casted films. Journal of Industrial Textiles 44, 85–98. https://doi.org/10.1177/1528083713477443.

Xu, C.Y., Inai, R., Kotaki, M., Ramakrishna, S., 2004. Aligned biodegradable nanofibrous structure: a potential scaffold for blood vessel engineering. Biomaterials 25, 877–886. https://doi.org/10.1016/S0142-9612(03)00593-3.

Zhong, S., Teo, W.E., Zhu, X., Beuerman, R.W., Ramakrishna, S., Yue, L., Yung, L., 2006. An aligned nanofibrous collagen scaffold by electrospinning and its effects on in vitro fibroblast culture. Journal of Biomedical Materials Research 79A (3), 456–463. https://doi.org/10.1002/jbm.a.

CHAPTER 10

Medical Implementations of Biopolymers

AMANI M. AL-GHRAIBAH • MAHA AL-QUDAH • FARIS M. AL-OQLA

1 CROSS-LINKING BIOPOLYMERS FOR MEDICAL APPLICATIONS

1.1 Biomaterials, Cross-linkers, and the Need for Cross-Linking

Biopolymer-based composites became a substantial alternative to synthetic materials in an emerging number of industries including food, packaging, tissue engineering, medical, and drug applications. They have several advantages and credentials over synthetic ones. This in fact has encouraged researchers to explore more valuable characteristics of such biocomposite materials (Al-Oqla and El-Shekeil, 2019; Alaaeddin et al., 2019b). In fact, a good number of important factors had led to more investigations for biopolymer-based composites including the high cost of synthetic materials, hazards of the industrial synthetic fibers, and environmental issues. Thus, efforts have been put to find out better solutions for the future by means of biopolymer-based composites with more desirable and functional properties like high mechanical properties, renewability, durability and biodegradability with low cost, and low density (AL-Oqla et al., 2019; Al-Oqla and Sapuan, 2018; Alaaeddin et al., 2019e; Sadrmanesh et al., 2019).

Moreover, more than a few of biopolymers are used to fabricate biomaterials; some of them are preferable more than others because they are more compatible with the human body. Collagen, proteins (like albumen), and silk are examples of these biopolymers (Butcher et al., 2014). Some examples of scaffolds that are used for tissue engineering and controlled drug delivery devices are made from biopolymers and are used in in vitro and in vivo applications such as hydrogels, films, fibers, and micro- and nanoparticles, as well as 2D and 3D structure scaffolds as illustrated in Fig. 10.1.

However, there are some restrictions that limit the usage of biopolymers in biomedical applications (regardless of the well-known advantages and the broad applicability of biomaterials) (AL-Oqla et al., 2018; Alaaeddin et al., 2018, 2019a, c, d). For example, materials made from biopolymers have a limited mechanical characteristics, and mostly the stability in physiological and aqueous situations required for medical uses is low (Jiang et al., 2010). Biomaterials are usually extracted from several biological species like silk, starch, cellulose, poly(lactic acid), collagen, and chitosan. These biopolymers have many advantages such as cytocompatibility and capability to dissolve in the human body with no result of any toxic or harmful materials.

1.2 Cross-Linking Biopolymers to Form Films

In medical applications like that of tissue engineering and controlled release, the biomaterials are fabricated as films. Films are considered mostly as the easiest structure of biomaterials, and they are made from biopolymers in general, which includes collagen. In the case of using collagen, cross-linking is required in order to develop its mechanical properties and decrease its ability of degradation, where the collagen has poor mechanical characteristics and it is quickly dissolved in water or any watery solutions (Mitra et al., 2011).

Cross-linking of collagen films was done using a mixture of UV irradiation and glucose, which provide a synergetic effect where UV light produced free radicals that usually show a response to a stimulus and form linear glucose molecules that lead to improve cross-linking. The result of using this process is improving the mechanical properties and decreasing the ability of degradation (Sionkowska, 2011). Another example of cross-links is the natural cross-linker originating from grape seeds, which are called proanthocyanidin. It can increase the thermal and enzymatic degradation resistances of collagen films (Sionkowska, 2011).

Advanced Processing, Properties, and Applications of Starch and Other Bio-Based Polymers. https://doi.org/10.1016/B978-0-12-819661-8.00010-X

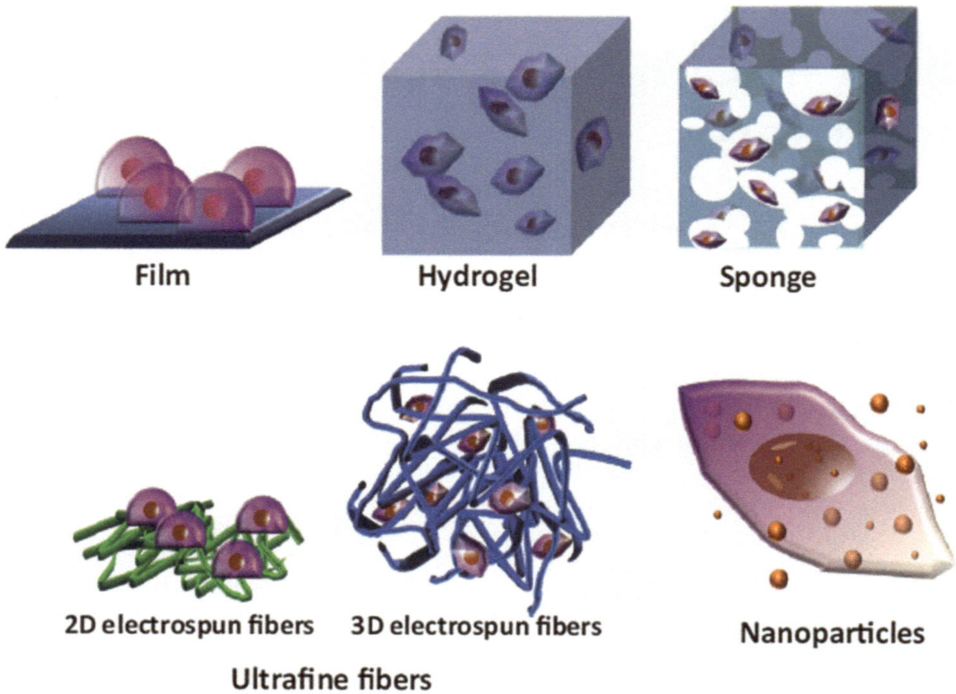

FIG. 10.1 Schematic Illustration of Commonly Used Forms of Scaffolds in Tissue Engineering.

1.3 Porous Structures and Cross-Linking of Biomaterials

There are many medical applications that use porous or sponge materials such as tooth tissue engineering, scaffolds in osteoblasts culturing for bone formation, and others (Sionkowska, 2011). The sponge structure of biomaterials makes their mechanical properties very weak and make their degradation process very fast in watery environment; therefore, cross-linking is required in this case (Sionkowska, 2011). Also, mixing different types of polymers can enhance the desired polymers properties, for example, enhancing stability as well as mechanical properties of the porous collagen-chitosan scaffolds can be reached by treating it with glutaraldehyde (Ma et al., 2003).

The original structure of collagen can be preserved while establishing the hybrid cross-linking systems, and controlling the mechanical characteristics and the degradation properties can be done with the presence of PCL, which makes sponges appropriate for increasing wound coverings and periodontal membranes (Ma et al., 2003).

There are some limitations in using the sponge's biopolymers, dissimilar than the porous ceramic's structures, which are because of their lower stability and poorer mechanical properties.

1.4 Cross-linking of Biopolymeric Hydrogels

Recently, hydrogels became very common in biomedical applications because their unique possessions such as high water content, flexibility, softness, and biocompatibility. Moreover, hydrogels are used in tissue engineering and in drug delivery devices because of the previous properties (Ma et al., 2003). However, hydrogels must be cross-linked to prevent degradation and improve it to be used in in vivo and in vitro applications (Hennink and van Nostrum, 2012). One example of a cross-linked hydrogel is the gelatin gel, which was prepared into hydrogels and cross-linked with dextran dialdehyde; this will increase the storage modulus than the cross-linked gels (Ma et al., 2003).

The main advantage of cross-linking the collagen is to decrease the degradation, but collagen is still not good enough to be used in tissue engineering applications, so collagen gels have been mixed with some other biopolymers, which will help in improving the stability and mechanical properties. The hydrogels resulted from mixing collagen with gelatin and then cross-linking with some amount of glutaraldehyde can also increase the viscoelastic characteristics and the breaking modulus of the previous gel without any toxicity to the living cells (Hennink and van Nostrum, 2012).

1.5 Cross-Linking of Coarse (Regular) Fibers

Different diameters of fibers usually some 100 μm have been established for some medical products and applications beside the usage of films, sponges, and hydrogels. Most of the time films are used as medical sutures and scaffolds. The approach of producing these microdiameter fibers is the traditional fiber spinning approach, like melt and wet spinning, subject to the kind of biopolymer utilized in the process. A direct method used to develop fibers is usually to melt and extrude the polymers as fibers using a syringe or an extruder.

Collagen, plant proteins like wheat gluten, and soy proteins are usually converted into filaments or fibers for possible utilization as drug delivery scaffolds and in tissue engineering applications (Ma et al., 2003). The previous fibers, under dry conditions, have worthy mechanical characteristics, but under wet circumstances have essentially low stability. Therefore, they are usually cross-linked with carboxylic acids as well as other cross-linkers (Ma et al., 2003).

1.6 Cross-Linking of Ultrafine Fibers

Ultrafine fibers have several novel characteristics, in comparison to the regular fibers. They have been extensively studied and approved for many medical uses. One of the known methods of producing the ultrafine fibers is the electrospinning method. Electrospun fibers are made from biopolymers where extracellular matrices (ECMs) usually resemble the ultrafine fibrous. Biopolymer electrospun fibers are usually proteins with low stability in water because of their smooth structure as well as the large surface area (Ma et al., 2003). Thus, cross-linking of kinds of fibers is necessary and needs a particular cross-linking methods and techniques.

Beside the electrospun protein fibers, some other polysaccharides like starch, and chitosan, are usually utilized to make electrospun structures, which are also cross-linked to enhance their stability and strength. Another method, in addition to the electrospinning method, is used to fabricate fibrous structure from biopolymers, which is called the phase separation method. Structures made from phase separation method have randomly oriented fibers, not like the layer-by-layer structures made from electrospun materials.

1.7 Cross-Linking Micro- and Nanoparticles

Biopolymer nanoparticles have the ability to load high quantities of drugs and gather in tumors and other organs. They can also offer effective transport of payloads. Micro- and nanobiopolymeric particles are better than metallic and artificial polymers. They are also utilized in in vivo distribution of drugs and other pharmaceuticals.

As any other biopolymers, the nanoparticle biopolymers have some limitations like weak stability, consequential accumulation that usually makes the size of the particle to increase, and fairly fast degradation in comparison to the metallic and artificial polymeric nanoparticles. As a result to these limitations, significant physical and chemical adjustments that include cross-linking were performed to enhance the act of the polymer-based nanoparticles (Almagableh et al., 2017). Even though, the development of cross-linking in biopolymer-based nanoparticles as well as the essential behavior of nanoparticles is still inadequate for the current biomedical applications.

2 BIOPOLYMERS APPLICATIONS FOR BONE REGENERATION

Many biomaterials are used in bone fracture treatment; the main role is to be an alternative substitute of bone grafts, as the number of donors decrease with time (AL-Oqla and Omari, 2017; AL-Oqla et al., 2015a). Examples of the materials that offer significant advantages to the development of bone scaffold are metals, polymers, and ceramics; they have many properties that make them important to be studied; biocompatibility is one of the most important properties. Although materials are generally used individually, biomaterial research focuses on mixing many materials to reach their maximum strength (AL-Oqla, 2017; AL-Oqla et al., 2017; AL-Oqla et al., 2014a). Moreover, metal alloys are classified into three classes: stainless steels, pure titanium- and titanium-based alloys, and cobalt-based alloys. All of these metal alloys are biocompatible, quit cheap, and have good mechanical characteristics. The second type, polymers, is greatly known as a part of tissue engineering processes and classified as natural polymers and synthetic polymers. They have good mechanical characteristics, compatible with the human body, and sometimes biodegradable, which is needed for some applications in tissue engineering processes. However, ceramics are nonmetallic compounds that include bioglass, glass ceramics, and carbon.

In fact, such materials are composites materials, which are classified depending on the shape of the reinforcing materials, such as 1. fibers are integrated in the matrix, called fibrous, and 2. if particles are integrated instead, called particulate. The formation of tissues and cellular functions is directly affected by the structure of the scaffolds. Scaffold structure can be porous; the interconnectivity and the size of the pores are significant for nutrient's distribution and for determining the type of molecules that travel between pores, respectively.

One of the important mechanical properies used to study the consequences of implantation on the scaffold is the compression, measuring the Young's (elastic) modulus which gives information about the amount of force needed to compress the material, which is also known as the "stiffness" of the material (AL-Oqla and Salit, 2017; AL-Oqla et al., 2014b; AL-Oqla et al., 2016). Researchers can compare between biopolymers using their unique value of Young's modulus, and it has a wider range for composites made from biopolymers. The rate of degradation and absorption is controlled in scaffolds in order to equalize the tissue and cell growth rates in vivo and in vitro (Cen et al., 2004).

An integration of two or more biomaterials will form a composite; materials used usually come from ceramics, metal, or polymers, which include natural and/or synthetics polymers, based upon the required possessions. If a composite contains of two polymers or more, then it is called "polymer blend." Biocomposites fabricated from natural polymers are still found little in the literature compared with synthetic polymer blend biocomposites (AL-Oqla et al., 2015b, c).

3 APPLICATIONS OF BIOPOLYMERS AND CALCIUM PHOSPHATE SCAFFOLD FOR BONE TISSUE ENGINEERING

During the last 30 years, tissue engineering has provided a good contribution in finding solutions for repairing and regenerating tissues. Scaffolds besides cells and growth factors are considered to be the key components of tissue engineering development. Recently, there are many studies that focus on materials used to design and fabricate the scaffold. However, significant challenges are faced when someone needs to develop bone tissue engineering (BTE) scaffolds successfully (Okamoto and John, 2013). For example, in bone regenerating, balancing between the need for a porous structured material and a suitable mechanical support is a real challenge. Both processing methods and the main components of the materials play a vital role in finding the scaffold properties. For example, artificial and biopolymers are broadly utilized because of their biocompatibility, flexibility to be shaped into various structures, and their fixed rates of degradation. Moreover, their mechanical and chemical characteristics are much desired in various applications (AL-Oqla and Sapuan, 2015, 2017; AL-Oqla et al., 2015d). On the other hand, these polymers have some limitations, including low mechanical strength. Biopolymers are also difficult to handle and have weak mechanical properties. Polymer-inorganic composites as a combination of inorganic polymers with some inorganic materials like calcium phosphate (CaP) are utilized to produce the so-called bioactive polymer/ceramic composite scaffolds. They are close to the bone nature and have many advantages like their delivery of agents and their interaction with stem cells (Okamoto and John, 2013).

3.1 Natural Biopolymers Uses in BTE

Natural biopolymers are exclusively similar to biological macromolecules, and they are created by normal processes and are extracted from bacteria, plants, and animals (Okamoto and John, 2013). Many biopolymer production have been approved by Food and Drug Administration (FDA). Depending on the chemical structure, natural biopolymers are divided into three classes: polysaccharides, proteins, and polyesters. Table 10.1 below summarizes many of the popular biopolymers applied in BTE and presents their processing, antigenicity (i.e., the capacity of a chemical structure to bind specifically with a set of certain products that have adaptive immunity), and cross-linking agent, which helps to make cross-linking bonds, which increase the material's mechanical properties and provide stability. Also, the table includes the biopolymer material ability for in vitro forming.

3.2 Scaffolds Including Calcium Phosphate

ECM of the bone is composed essentially of fibrous collagen and hydroxyapatite (HAp) (Li et al., 2014). One of the popular methods used to produce scaffolds for BTE is to imitate the natural organic-inorganic composite, using an integration of CaP into polymer matrices (Li et al., 2014). The integration process will improve the bone bonding behavior and play an important role in enhancing cell adhesion. Some of the combination ways used in scaffold synthesis to create natural biopolymer-based CaP materials are listed below beside some examples of the biopolymer's composite produced:

1. Physical mixture: collagen/HAp, chitosan/nHAP, silk fibroin/HAp with low crystallinity, and alginate/HAp partied.
2. Chemical deposition: Collagen/HAp, chitosan/nHAp, and silk fibroin/calcium-deficient HAp.
3. Biomimetic mineralization: Collagen/CaP, chitosan/HAp, and bacterial cellulose/calcuim-deficient HAp.

To generate a "living" scaffold in tissue engineering, scaffolds are usually mixed with stem cells in one

TABLE 10.1
Common Biopolymers Applied in BTE (Li et al., 2014).

Biopolymer		Origin	Major antigenicity	Cross-linking agent	Dominating nucleation sites
Protein	Collagen	Animal tissue	Pathogen, terminal telopeptides	Dehydrothermal treatment, UV light, carbodiimide, aldehydes, transglutaminase Riboflavin	Charged amino acid
	Silk fibroin	Silkworm, bombyx mori	Sericin	Glutaraldehyde, carbodiimide	Electronegative amino acid sequences
Polysaccharides	Chitosan	Crab shells, shrimps, fungal fermentation	Low toxicity	Glutaraldehyde, genipin, epoxy compound, sodium tripolyphosphate	Cationic amine groups
	Starch	Plant	Nontoxicity	Sodium trimetaphosphate, malonic acid, formaldehyde, anhydride	OH^- groups
	Cellulose	Plant, bacteria	Low toxicity	Aldehydes, carbodiimides, carboxylic acids, irradiation	OH^- groups

method, and in other methods, they are mixed with bioactive molecules (Li et al., 2014). In all cases, the scaffold proposes a suitable environment for the cell's growth and a delivery factor for cell-based therapies (Li et al., 2014). One of the most commonly used cells in scaffold's production are the mesenchymal stem cells, which are derived from mature human tissues. Alternatively, native protein plays a vital role in bone formation and regulation of cells (Li et al., 2014).

In general, a new drug application should be submitted to FDA or the European Medicines Agency (EMEA), and then followed by clinical trials, before it is released as a tissue engineered product. They use different animal models (like goat, mouse, rat, and rabbit), for bone tissue study, where they provide scaffolds with best performance.

4 BIOPOLYMERS AND SUPRAMOLECULAR POLYMERS APPLICATIONS

The mechanical support of the tissue organs is provided by the ECM. The ECM consisted of two types of proteins: proteoglycans and fibrous proteins. In engineering applications and constructing scaffold for regeneration medicine, proteins that mimic the function and structure of ECM mainly fibrillar, collagen, and silk protein are of particular interest. Those proteins have another distinctive capability of constructing long uniform structures from smaller subunit. It became stable using noncovalent interactions (Charras and Sahai, 2014). Tissue engineering researchers assembled artificial supramolecular architecture. It was stabilized by noncovalent intermolecular interactions, such as hydrophobic interactions and metal-ligand coordination (Charras and Sahai, 2014).

The most common supramolecular polymers are peptide based; the peptide sequence is the main biomechanical and biochemical signals in the ECM as well as it can synthesize quickly.

4.1 Structure and Organization of Protein Biopolymers

Amino acids are the main element of forming protein. The arrangements of the amino acid determine the complexity of the biomolecule. Collagen is the most researched families of protein biopolymers and the main component of the ECM. Collagen provides stiffness and strength to the most tissues in the human being (Gautieri et al., 2011). Human body is rich of collagen type I. It consist of three polypeptide called α-chain and a sequence of any amino acid. Then three α-chains bring together a triple helix tertiary structure. If it was stabilized by hydrogen bonding, it assembles into fibrils and then into supramolecular complexes (Gautieri et al., 2011). Collagen fibril's diameter is variable; it can be small as in cornea (20 nm) and can be large as in tendons (500 nm). Fig. 10.2 shows the main steps of polymer's assembly, i.e., the self-assembly biopolymers and the supramolecular polymers.

FN is another ECM protein; it is assembled from three repeated β-sheets. When two FNs combined, it can create various ECM components. The combination of FN with collagen forms a stabilized ECM and creates scaffold. FN fibril is very dynamic. Once it is coupled with cells, it will be ready for rearrangement, remolding, and recycling.

Silk is a different form of proteins that is produced outside the physical activity of the creatures like spiders and silkworms. Fibroin is the main component of silk. Fibrils are created by spinning fibroins up to 25 μm diameter. Sericins, which is a glue type protein, is used to coat and cement the fibroin together. This forms a protective shield environment for silkworm (Freeman et al., 2015). Fig. 10.3 presents an example of silk formation.

The common property of all ECM proteins is that they are built from repetitive sequence of main elements. This allows proteins to structure hierarchical structures with different arrangements and make them distinctive in responding to different stimuli from outside environment.

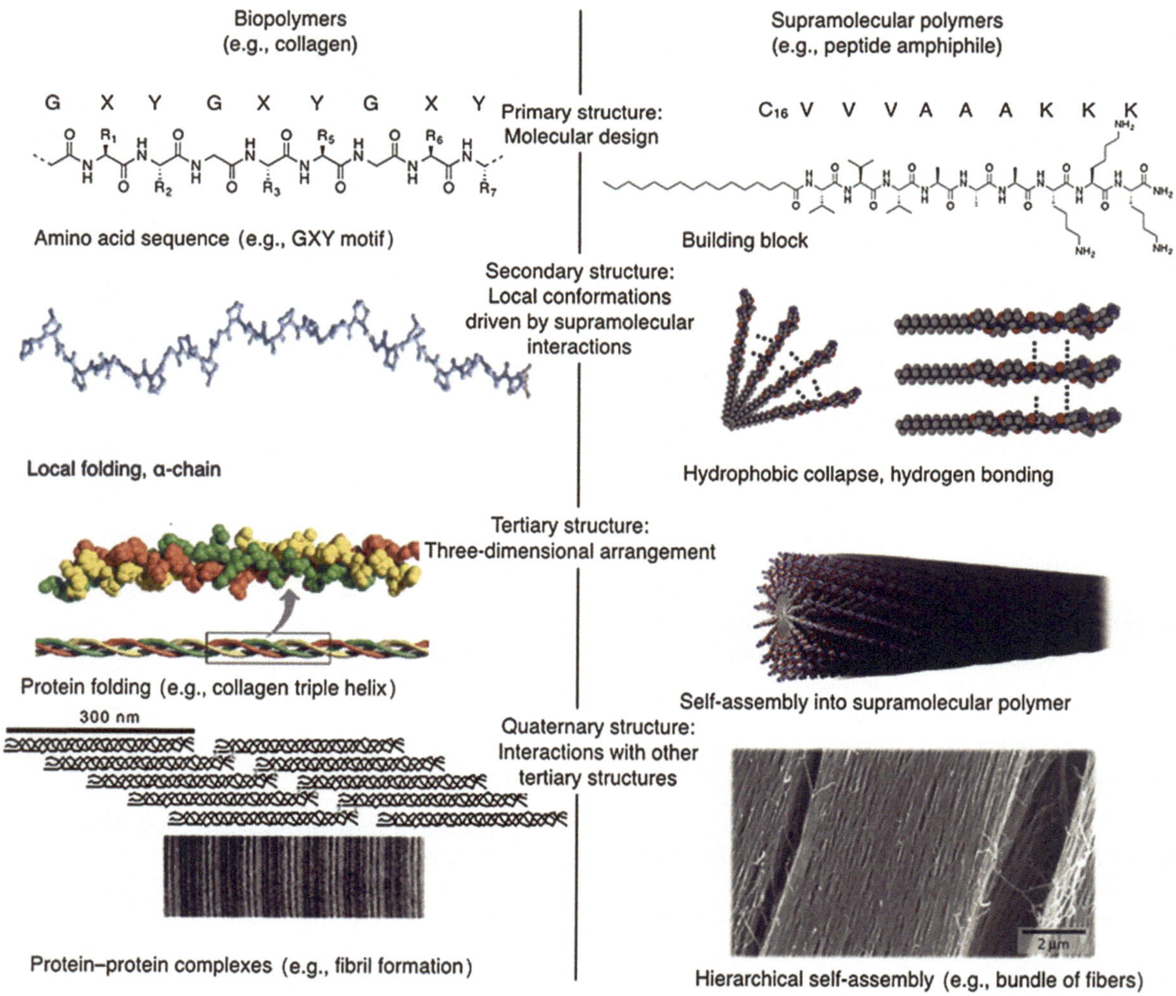

FIG. 10.2 Self-Assembly of Biopolymers (Left) and Supramolecular Polymers (Right) (Freeman et al., 2015).

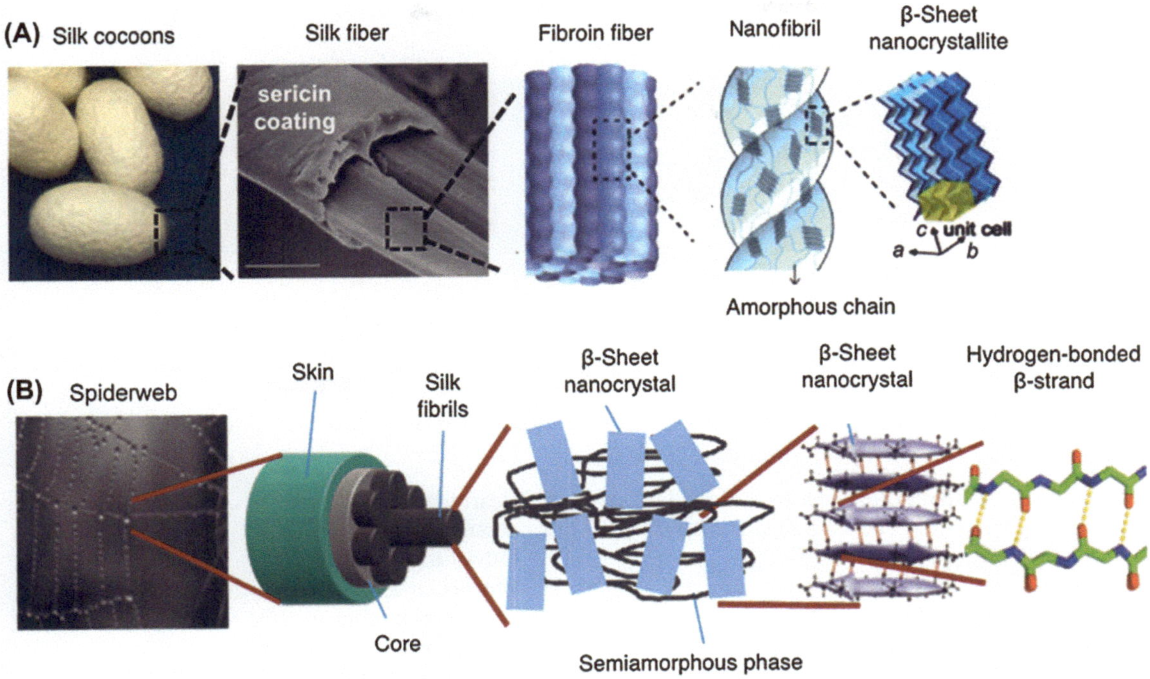

FIG. 10.3 Ordered Organization of Silk Fibers: **(A)** Structure of Silk Fibers Comprising Silk Cocoons Produced by the Silkworm and **(B)** Structure of Spider Dragline Silk (Freeman et al., 2015).

4.2 Bioinspired Supramolecular Polymers

The hierarchical structure and assembly form an inspiration engineering to find a new way of molecule assemble and to create an engineered material that can achieve a specific properties. Like the formation of underivatized peptides, Fmoc dipeptides, and peptides amphiphiles, protein biopolymers and supramolecular polymers are considered to be biomaterials for regenerative medicine.

Regenerative medicine aims to restore the function of impaired tissue or organ that can be detected by disease or age by building scaffolds that interact with outside environment. Cells are constantly stimulated by surrounding factors. Characteristics of protein biopolymers as well as bioinspired supramolecular polymers, including mechanical possessions, biocompatibility, and degradability, make them perfect alternatives for regenerative medicine. They are engineered to create growth factor domain that can enable the vascular endothelial growth factor. On other hand, the capsulate and immobilize factor that is available in silk protein make them a potential material to excite the nervous system, so enabling the treatment of outer nerve injury (Fig. 10.4)

There are various clinical applications of these proteins; however, it is limited because of the short degradation half-lives and high costs in addition to immune-related side effects. Moreover the biomaterials for regenerative medicine should be able to do different mechanisms such as adhesion, proliferation, migration, and differentiation. Besides, the scaffold must integrate with outer environment with no innervation of immunological response.

5 BIOPOLYMERS APPLICATIONS FOR DISEASES THERAPY

This section presents some applications of polymeric biomaterials in some medical diseases therapy like ophthalmology, orthopedics, cardiovascular diseases, wound closure, and nerve regeneration.

5.1 Polymeric Biomaterials in Ophthalmology

In general, ophthalmology is the study and treatment of any disorders and diseases in the eye. The history of using biomaterials in ophthalmology returns to the mid of the 19th century. Meanwhile, a wide variety of biomaterials used in ophthalmological applications are now established and some are having massive success in medical applications. Such utilization of

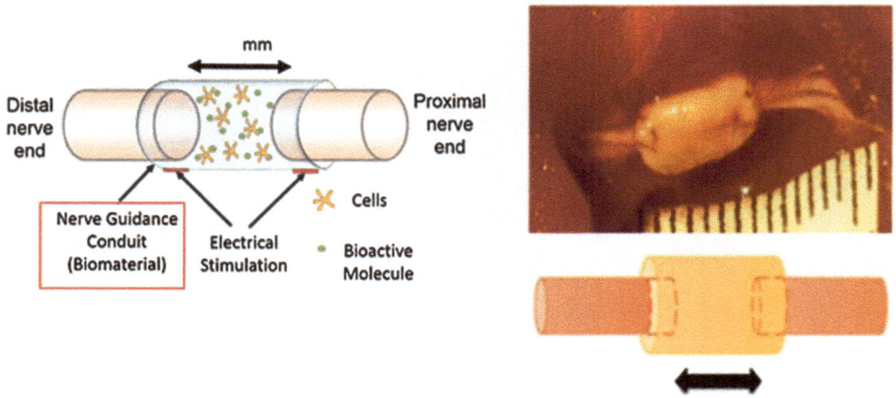

FIG. 10.4 Illustrative Diagram for the Applications of Biopolymers-Based Materials for Nerve Repairing.

biomaterials in ophthalmology can be summarized in the following (Modjarrad and Ebnesajjad, 2013):

1. Contact lenses
2. Intraocular lenses
3. Artificial orbital walls
4. Artificial corneas
5. Artificial lacrimal ducts
6. Glaucoma filtration implants
7. Viscoelastic replacements
8. Drug delivery systems
9. Scleral buckles
10. Retinal tacks and adhesives
11. Ocular endotamponades

5.2 Polymeric Biomaterials in Orthopedics

Usually, materials used in orthopedic applications are made mainly from metals, mostly because of its similarity to bone tissue properties, for example, fracture toughness, high strength, and hardness. Over the years, polymers have been used in orthopedics application, and they are always getting an increasing interest in BTE. In the history, polymer applications in orthopedics were mostly used in parts that perform well in bone fixation and can perform well under the cyclic load, for example, at the knee and hip joints. Regardless of the numerous application of orthopedics existing in market, they are dominated by few polymers, such as poly(methyl methacrylate) and ultrahigh-molecular weight polyethylene (Modjarrad and Ebnesajjad, 2013).

5.3 Polymeric Biomaterials for Cardiovascular Diseases Therapy

Biomaterials have occupied a very important role in treatment procedures of cardiovascular diseases; some examples of these applications are

1. Heart valve prostheses
2. Vascular grafts and stents
3. Indwelling catheters
4. Ventricular assist devices
5. Total implantable artificial hearts
6. Pacemakers
7. Automatic internal cardioverter defibrillators
8. Intraaortic balloon pumps

The main necessity for biomaterials of cardiovascular applications, mainly devices that are in contact with blood, is to be compatible with the blood (i.e., nonthrombogenic). Other requirements than biocompatibility are the mechanical and surface characteristics that have specific application. Researchers studied the types of polymers used in cardiovascular applications, and they found that polyurethanes, expanded PTFE, and polyethylene terephthalate (PET) are mostly used (Modjarrad and Ebnesajjad, 2013).

5.4 Polymeric Biomaterials for Wound Closure

There are many surgical methods used to close the operating wounds, for example, physicians use sutures, adhesives, tapes, staples (Modjarrad and Ebnesajjad, 2013), and recently they use laser tissue repairing. However, sutures are mostly and frequently used than the other methods. Sutures are sterilized filaments that are used to hold tissues together until the wound is healed and can withstand the mechanical stresses applied to it. We can classify sutures dependin on the type of the material into normal or synthetic, and depending on the performance of the material, it can be classified as absorbable or nonabsorbable sutures. Also, they are classified as monofilament, multifilament, braided, or twisted shape depending on the physical arrangements of the filaments.

Generally, when forming sutures, polymers should be selected to be minimally adverse biological reactions, and taking fiber-forming rheological characteristics. Sutures must be conforming to have minimum tissue slog as well as strength holding and knot security. In order to enhance the lubricity and decrease the tissue slog, they used coatings with polymers, which are applied to sutures normally like silicones and tetrafluoroethylene.

5.5 Polymeric Biomaterials for Nerve Regeneration

As known, nervous system is one of the complicated physiological systems in the human body, so repairing their damaged parts is a very hard challenge for physicians. However, many progresses have been done during the last decades, but still it is not easy to repair the nerve's damage totally, so restoring the function of the nerves can be lost (Modjarrad and Ebnesajjad, 2013).

The main classifications of the nervous system in general are the peripheral nervous system (PNS) and the central nervous system (CNS). Many methods have been discovered for nerve repair, together in the CNS and the PNS, including direction channels, cell transplantation with scaffolds, and transfer of therapeutics.

6 BIODEGRADABLE POLYMERS

Biopolymers are polymers produced in nature while the organisms are grown; therefore, they are also stated as natural polymers. Their production in general involves enzymes catalyzed, where chain grows with polymerization reactions of the activated monomers, and they are typically produced inside cells using complex metabolic procedures. The main two classes of the natural biopolymers depending on their production method are biopolymers that are directly extracted from biomass and biopolymers that are formed directly by natural or genetically adapted organism.

In different fields, the biodegradable polymers seem to have many properties that allow them to be in competition with the nonbiodegradable thermoplastics, such as using them in packaging, textile, biomedical applications, and others. From these biopolyesters, PLA presents as the most promising and favorable biodegradable polymers. During the last decade, PLA has been the main topic in many studies in literature as in several book chapters and reviews (Averous, 2012). Also, PLA is commercially available in a wide range of grades with a reasonable price and can achieve many applications. Various biomedical applications of biopolymers are illustrated in Fig. 10.5. Such applications include dental applications such as endosteal root form dental implant, ophthalmologic applications like artificial cornea hip, orthopedic applications such as artificial knee, cardiovascular applications like heart valve as well as several other applications like drug delivery, development of synthetic organs like artificial skin, synthetic kidney (hemodialyzer), heart lung machine, and organ implantations like breast. To enhance the duration of orally administered drugs in drug delivery applications, the products are designed to be consisted of small spheres with a soluble coating with varying thicknesses so that dissolution times could be varied.

6.1 Polylactide

6.1.1 PLA formation

The formation of PLA includes many steps that begin with the synthesis of the lactic acid and end with the lactic acid polymerization, where the intermediate step is the creation of the lactide. The following steps summarize the synthesis of PLA which includes three major tracks:

1. The first track is the condensation polymerization of the lactic acid, which produces brittle and low molecular weight polymer that is, for the most part, impracticable, unless coupling external agents are worked to improve the length of its chain.
2. The second track is the azeotropic dehydrative condensation of lactic acid. It can produce high molecular weight PLA with no chain extenders or any exceptional adjuvants.
3. The third and core procedure is the ring-opening polymerization of lactide; this process is mainly used to produce a high molecular weight PLA.
4. At the end, lactic acid units will be a portion of some complex macromolecular chain as in copolymers.

6.1.2 Properties of PLA

In this section, the most important properties of PLA is summarized and explained briefly:

1. Crystallinity and thermal possessions:
   PLA orientation can be created by processing at low temperature to enhance the mechanical possessions of the biopolymer, then the resulting PLLA will have higher modulus without any increment in its crystallinity.
   Using the differential scanning calorimetry (DSC), you can determine the levels of PLA crystallinity. The value of PLA melt enthalpy at 100% cystallinity is equal to 93 $\frac{J}{g}$.

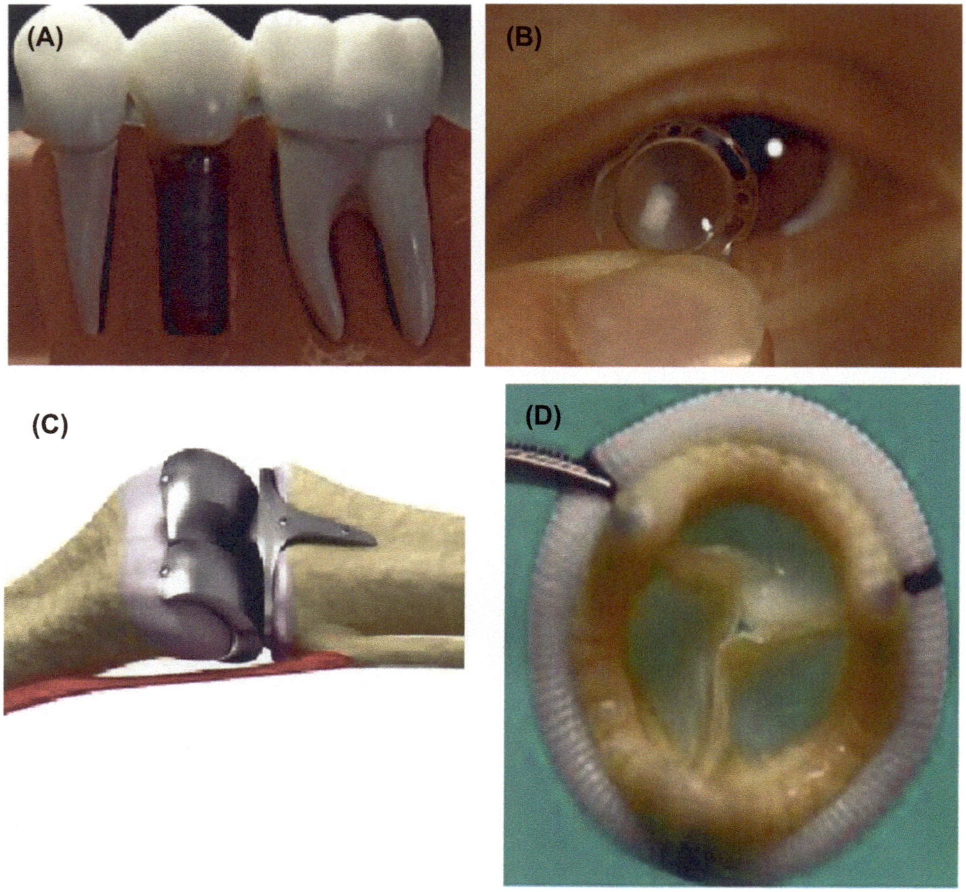

FIG. 10.5 Applications of biopolymers in medical applications. **(A)** Dental applications: endosteal root form dental implant; **(B)** ophthalmologic applications: artificial cornea hip; **(C)** orthopedic applications: artificial knee; and **(D)** Cardiovascular applications: heart valve.

Researchers found that the initiation of the crystallization process of the amorphous PLA, which is also thermally crystallizable, can be done by annealing it at a maximum temperature equal to 75 C and minimum temperature equal to the melting point.

The resulting copolymers of annealing crystallizable PLA are frequently yielded at two melting points, and different theories have been placed forward to clarify this feature (Averous, 2012). The melting temperature (Tm) of PLA ranges from 130 to 180 C, whereas the glass transition temperature (Tg) ranges from 50 to 80 C. The Tm, for semicrystalline PLA, can be controlled using altered processing parameters and the preliminary PLA structure.

If the crystallinity degree and the melting temperature (Tm) of the materials that are based on PLA need to be reduced, a random copolymerization with various comonomers (e.g., GA, CL, or others) should be done. The Tg of PLA can also be determined by the ratio of the various kinds of lactide to its macromolecular series (Averous, 2012).

2. Surface energy:

   Surface energy is very significant for many procedures (like printing, multilayering, etc.), and it has an effect on the interfacial tension. The surface energy of PLA which is made of 92% of L-lactide and 8% of meso-lactide was found to be 49 $\frac{mJ}{m^2}$, with dispersive and polar constituents of 37 and 11 $\frac{mJ}{m^2}$, respectively, which has a relatively hydrophobic organization when compared with other biopolyesters.

3. Solubility:

   Chloroform is considered as one of the good solvents for PLA and mostly for the consistent copolymers. Also, there are some other solvents like chlorinated or fluorinated organic compounds, dioxane, dioxolane, and furan. Poly(rac-lactide) and poly(meso-lactide) are soluble in several organic solvents, for example, acetone, pyridine, ethyl lactate, tetrahydrofuran, xylene, ethyl acetate, dimethyl formamide, and methyl ethyl ketone.

4. Barrier properties:
   Barrier properties (mainly to oxygen, water vapor, and carbon dioxide) of PLA are highly investigated, as it has a lot of applications in food packaging (Averous, 2012). One of the barrier properties that is studied for PLA polymers is the coefficients, where it is found that these coefficients are lower than those stated for crystalline polystyrene at 25 C and 0% related humidity (RH) and higher than the one for PET. As PET and PLA both are hydrophobic and their films absorb very low quantities of water, they have equivalent barrier properties. That was examined by looking at their water vapor permeability coefficient, which was found in the range from 10 to 37.8 C with RH in the interval of 40%–90%.
5. Mechanical properties:
   The mechanical properties of PLA can vary in a large range, beginning from elastic and soft materials and ending with a high strength materials (stiff materials), regarding some different factors, for example, 1. crystallinity, 2. polymer structure and molecular weight, 3. material formulation (blend, composites, plasticizers, etc.), and 4. processing (or orientation). The commercial PLA has modulus of about 2.1 GPa and about 9% of elongation at break. However, when the plasticization process occurred, the elongation at break usually increases to 200%, while its strength modulus reduces to about 0.7 MPa. This in order demonstrates the flexibility of modifications the mechanical properties of such polymer to fit various industrial applications (Averous, 2012).

6.2 Medical Applications of PLA

This section explains some of the medical applications of the polylactic acid (PLA), which is, as noted before, one of the famous biodegradable polymers. These days, PLA-based materials are basically referenced on diverse markets like biomedical (initial market), textile, and packaging (basically food, as short-term applications). PLA is widely used in biomedical applications because it has a good biocompatibility and bioresorbability properties in the human body. The main specified examples of using PLA in biomedical products are screws and sutures, which are used in fracture fixation, and microtitration plates, and some of the drug delivery devices are also made from PLA.

As known, bone faces a slowly increment in stress during the bone healing process, thus researchers found that the stress can be decreased if the plate loses stiffness in in vivo environment. So, they used a resorbable and degradable PLA, which they do not need to remove after healing using a second medical surgery operation; however, most of its mechanical properties are lost after few weeks of the implantation. So, to improve the mechanical properties of PLA, it has been reinforced with many of the nonresorbable materials, like polyamide fibers and carbon, to form carbon fiber/PLA composites.

The carbon fiber/PLA composites have high mechanical characteristics at the beginning, i.e., before the implantation process, but they start to lose them rapidly in vivo because of degradation. The effects of the resorbable materials or the slowly degraded fibers in the living tissues for long time usage are not totally known, and there are many concerns to be resolved.

PLA fibers are successfully used as resorbable sutures, although they were used in different textile applications. For example, one of the commercially existing fiber-formed bioresorbable medical production is dependent on copolymers of glycolide (GA) blended with lactide (Vicryl).

In medicine, PLA is motivating because of it has a low toxicity and hydrolytic degradability properties. Also, copolymers of GA and rac-lactide appear to be the greatest suitable mixture to be used in drug delivery mediums. Another product of PLA is the porous PLA scaffolds, which was established to be possible renovation matrices for injured organs and tissues.

6.3 PLA Packaging Applications

To provide better mechanical characteristics for PLA than polystyrene and have more or less comparable properties than PET, a commercially packaging technique is used for PLA. Also, in some studies of market, they found that PLA is economically good for packaging and very important in volume for bioresorbable packaging. Because of its high cost, the first use of it was like films, packaging papers, and beverage and food container.

There is a continuous increments of using PLA all over the world including, Europe, Japan, and the United States, where PLA is used for packaging of short shelf life food products like fruits and vegetables. Many packaging applications are also made from PLA, which includes containers, sundae, covers for sweets, plating films and drinking cups, and others.

7 BIOPOLYMER GREEN LUBRICANT FOR SUSTAINABLE MANUFACTURING

The idea of "biopolymer green lubricant," "dry coating methods," and "minimum quantity lubrication" was evolved to meet the 3Rs demands (the climate change conference, which was held in France, Rhodes, 2016, focused on "reducing," "reusing," and "recycling" resources and called them the 3Rs).

7.1 Green Lubricant

Green lubricant technologies are environmentally friendly development lubricants that contain harmless and contamination-free lubricating substances such as heavy metals and sulfides (Palacio and Bhushan, 2010; Shi and Lu, 2016). A preferred choice of green lubricates is nonpetrochemical natural oils such as vegetable oil. The addition of suitable chemicals makes the natural oil properties similar to petrochemical oils. Testing lubricants include replacing the original tissue fluids with solutions (e.g., biopolymers) or by mixing biopolymers with water. The toxicity of green lubricants was the hydroxypropyl methylcellulose (HPMC) (Shi and Lu, 2016).

It was found that the HPMC lubrication thickness reaches the max at 40 mL of water and 100 mL of alcohol (Shi and Lu, 2016). Although the lubrication thickness was not influenced by thickness, the lubrication duration increased continuously with the thickness. Fig. 10.6 shows SEM images of thin films at different thicknesses.

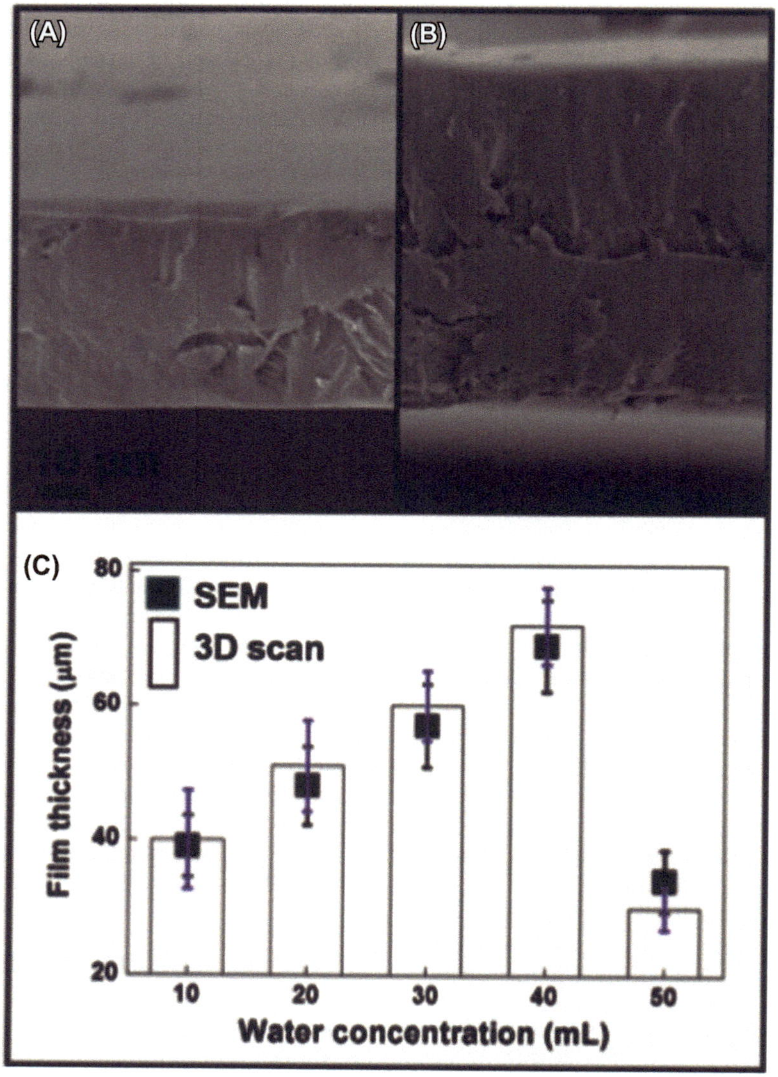

FIG. 10.6 SEM Images of Thin Film at Thicknesses of **(A)** 40 μm and **(B)** 70 μm; **(C)** the Thickness Difference Drawing Measured by SEM and 3D Scan (Shi and Lu, 2016).

7.2 Raman Spectroscopy and EDS Analysis

The assessment of film uniformity was performed using Raman spectrum at three major characteristic peaks (Fig. 10.7): 1110 cm^{-1}, 1360 cm^{-1}, and 1540 cm^{-1} used for symmetric C–O–C, COH bending, and CH_2 scissor, respectively (Shi and Lu, 2016). The results indicate that the HPMC tend to be uniform (Shi and Lu, 2016).

Mapping of the energy-dispersive spectroscopy for carbon (C), oxygen (O), and silicon (Si) signals from HPMC shows a circular wear mark on the test of specimen that underwent lubrication test. Si signals had been worn over a long period of time and the coefficients of friction (COFs) of the silicon remain low (Shi and Lu, 2016).

7.2.1 Tribology test

The tribological properties of coating of HPMC were analyzed using pin-on-disk tribo test (Shi and Lu, 2016). The transfer layer prevents the influence of COFs for different loads. When the lubrication was tested at different speeds, the COFs were marked high because of the lack of transfer layer.

7.2.2 Zebrafish embryo toxicity test

The zebrafish embryo toxicity test shows that the HPMC concentrations can go up to 0.5% in treatments with no change in zebrafish embryo development (Shi and Lu, 2016). However, if the concentration is 1%, then the chroine in zebrafish will change and the hatching will be delayed.

8 CONCLUSIONS

Natural biopolymers are exclusively similar to biological macromolecules; they are created by normal processes and are extracted from bacteria, plants, and animals. Many biopolymer productions have been approved by FDA. PLA-based materials are basically referenced on diverse markets like biomedical (initial market), textile, and packaging (basically food, as short-term applications). PLA is widely used in biomedical applications because it has a good biocompatibility with the human body. In medicine, PLA is motivating because of it has a low toxicity and hydrolytic degradability properties. Also, copolymers of GA and rac-lactide appear to be the greatest suitable

(A)

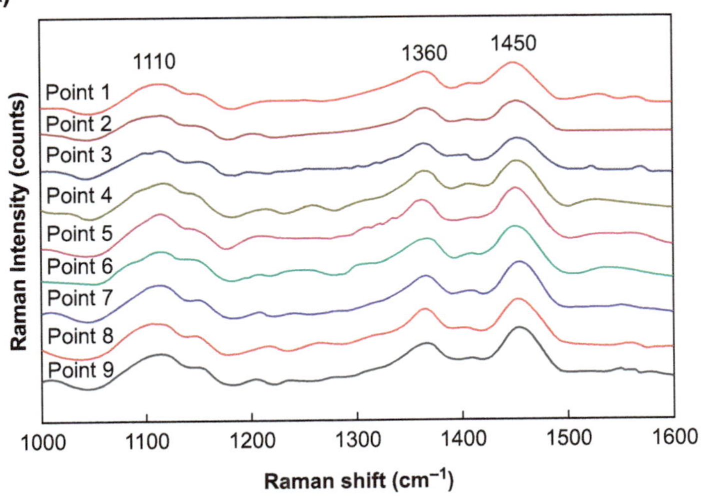

(B)

Peak (cm^{-1})	Point 1	Point 2	Point 3	Point 4	Point 5	Point 6	Point 7	Point 8	Point 9
1110	1.00	0.96	1.03	1.04	1.06	0.96	0.96	0.99	0.97
1360	1.00	0.96	1.01	1.04	1.06	1.02	0.96	0.99	0.95
1450	1.00	0.96	1.02	1.05	1.05	0.97	0.96	0.95	0.96

FIG. 10.7 Examining Film Uniformly Using Raman Spectrum: **(A)** the topmost values and **(B)** ratio of each point in details (Shi and Lu, 2016).

mixture to be used in drug delivery mediums. Another product of PLA is the porous PLA scaffolds, which was established to be possible renovation matrices for injured organs and tissues. On the other hand, protein biopolymers as well as bioinspired supramolecular polymer characteristics including mechanical possessions, biocompatibility, and degradability make them perfect alternatives for regenerative medicine. They are engineered to create growth factor domain that can enable the vascular endothelial growth factor. The potential of biopolymers in both macro- and nanoscales are very promising for the near future to develop sustainable green products in various applications, particularly the medical ones.

REFERENCES

AL-Oqla, F.M., 2017. Investigating the mechanical performance deterioration of Mediterranean cellulosic cypress and pine/polyethylene composites. Cellulose 24, 2523–2530.

AL-Oqla, F.M., Almagableh, A., Omari, M.A., 2017. Design and Fabrication of Green Biocomposites, Green Biocomposites. Springer, Cham, Switzerland, pp. 45–67.

AL-Oqla, F.M., Alothman, O.Y., Jawaid, M., Sapuan, S., Es-Saheb, M., 2014a. Processing and Properties of Date Palm Fibers and its Composites. Biomass Bioenergy. Springer, Cham, Switzerland, pp. 1–25.

Al-Oqla, F.M., El-Shekeil, Y., 2019. Investigating and predicting the performance deteriorations and trends of polyurethane bio-composites for more realistic sustainable design possibilities. Journal of Cleaner Production 222, 865–870.

AL-Oqla, F.M., Hayajneh, M.T., Fares, O., 2019. Investigating the mechanical thermal and polymer interfacial characteristics of Jordanian lignocellulosic fibers to demonstrate their capabilities for sustainable green materials. Journal of Cleaner Production 118256.

AL-Oqla, F.M., Omar, A.A., Fares, O., 2018. Evaluating sustainable energy harvesting systems for human implantable sensors. International Journal of Electronics 105, 504–517.

AL-Oqla, F.M., Omari, M.A., 2017. Sustainable biocomposites: challenges,Potential and barriers for development. In: Jawaid, M., Sapuan, S.M., Alothman, O.Y. (Eds.), Green Biocomposites: Manufacturing and Properties. Springer International Publishing (Verlag), cham, switzerland, pp. 13–29.

AL-Oqla, F.M., Sapuan, S.M., Ishak, M.R., Nuraini, A.A., 2015a. Selecting Natural Fibers for Industrial Applications. Postgraduate Symposium on Biocomposite Technology Serdang, Malaysia.

AL-Oqla, F.M., Salit, M.S., 2017. Materials Selection for Natural Fiber Composites. Woodhead Publishing, Elsevier, Cambridge, USA.

AL-Oqla, F.M., Sapuan, M.S., Ishak, M.R., Aziz, N.A., 2014b. Combined multi-criteria evaluation stage technique as an agro waste evaluation indicator for polymeric composites: date palm fibers as a case study. BioResources 9, 4608–4621.

AL-Oqla, F.M., Sapuan, M.S., Ishak, M.R., Nuraini, A.A., 2015b. Decision making model for optimal reinforcement condition of natural fiber composites. Fibers and Polymers 16, 153–163.

AL-Oqla, F.M., Sapuan, M.S., Ishak, M.R., Nuraini, A.A., 2015c. Selecting natural fibers for bio-based materials with conflicting criteria. American Journal of Applied Sciences 12, 64–71.

AL-Oqla, F.M., Sapuan, S., 2015. Polymer selection approach for commonly and uncommonly used natural fibers under uncertainty environments. Journal of Occupational Medicine 67, 2450–2463.

AL-Oqla, F.M., Sapuan, S., 2017. Investigating the inherent characteristic/performance deterioration interactions of natural fibers in bio-composites for better utilization of resources. Journal of Polymers and the Environment 1–7.

Al-Oqla, F.M., Sapuan, S., 2018. 1 natural fiber composites. In: Kenaf Fibers and Composites.

AL-Oqla, F.M., Sapuan, S., Ishak, M., Nuraini, A., 2015d. Predicting the potential of agro waste fibers for sustainable automotive industry using a decision making model. Computers and Electronics in Agriculture 113, 116–127.

AL-Oqla, F.M., Sapuan, S., Jawaid, M., 2016. Integrated mechanical-economic–environmental quality of performance for natural fibers for polymeric-based composite materials. Journal of Natural Fibers 13, 651–659.

Alaaeddin, M., Sapuan, S., Zuhri, M., Zainudin, E., AL-Oqla, F.M., 2018. Properties and common industrial applications of polyvinyl fluoride (PVF) and polyvinylidene fluoride (PVDF). In: IOP Conference Series: Materials Science and Engineering. IOP Publishing, p. 012021.

Alaaeddin, M., Sapuan, S., Zuhri, M., Zainudin, E., AL-Oqla, F.M., 2019a. Photovoltaic applications: status and manufacturing prospects. Renewable and Sustainable Energy Reviews 102, 318–332.

Alaaeddin, M., Sapuan, S., Zuhri, M., Zainudin, E., AL-Oqla, F.M., 2019b. Physical and mechanical properties of polyvinylidene fluoride-Short sugar palm fiber nanocomposites. Journal of Cleaner Production 235, 473–482.

Alaaeddin, M., Sapuan, S., Zuhri, M., Zainudin, E., AL-Oqla, F.M., 2019c. Polymer matrix materials selection for short sugar palm composites using integrated multi criteria evaluation method. Composites Part B: Engineering 107342.

Alaaeddin, M., Sapuan, S., Zuhri, M., Zainudin, E., AL-Oqla, F.M., 2019d. Polyvinyl fluoride (PVF); its properties, applications, and manufacturing prospects. In: IOP Conference Series: Materials Science and Engineering. IOP Publishing, p. 012010.

Alaaeddin, M., Sapuan, S., Zuhri, M., Zainudin, E., M AL-Oqla, F., 2019e. Lightweight and durable PVDF–SSPF composites for photovoltaics backsheet applications: thermal, optical and technical properties. Materials 12, 2104.

Almagableh, A., Al-Oqla, F.M., Omari, M.A., 2017. Predicting the effect of nano-structural parameters on the elastic properties of carbon nanotube-polymeric based composites. International Journal of Performability Engineering 13, 73.

Averous, L., 2012. Synthesis, properties, environmental and biomedical applications of polylactic acid. In: Handbook of Biopolymers and Biodegradable Plastics: Properties, Processing and Applications, pp. 171–188.

Butcher, A.L., Offeddu, G.S., Oyen, M.L., 2014. Nanofibrous hydrogel composites as mechanically robust tissue engineering scaffolds. Trends in Biotechnology 32, 564–570.

Cen, L., Neoh, K., Li, Y., Kang, E., 2004. Assessment of in vitro bioactivity of hyaluronic acid and sulfated hyaluronic acid functionalized electroactive polymer. Biomacromolecules 5, 2238–2246.

Charras, G., Sahai, E., 2014. Physical influences of the extracellular environment on cell migration. Nature Reviews Molecular Cell Biology 15, 813.

Freeman, R., Boekhoven, J., Dickerson, M.B., Naik, R.R., Stupp, S.I., 2015. Biopolymers and supramolecular polymers as biomaterials for biomedical applications. MRS Bulletin 40, 1089–1101.

Gautieri, A., Vesentini, S., Redaelli, A., Buehler, M.J., 2011. Hierarchical structure and nanomechanics of collagen microfibrils from the atomistic scale up. Nano Letters 11, 757–766.

Hennink, W.E., van Nostrum, C.F., 2012. Novel crosslinking methods to design hydrogels. Advanced Drug Delivery Reviews 64, 223–236.

Jiang, Q., Reddy, N., Yang, Y., 2010. Cytocompatible cross-linking of electrospun zein fibers for the development of water-stable tissue engineering scaffolds. Acta Biomaterialia 6, 4042–4051.

Li, J., Baker, B.A., Mou, X., Ren, N., Qiu, J., Boughton, R.I., Liu, H., 2014. Biopolymer/calcium phosphate scaffolds for bone tissue engineering. Advanced healthcare materials 3, 469–484.

Ma, L., Gao, C., Mao, Z., Zhou, J., Shen, J., Hu, X., Han, C., 2003. Collagen/chitosan porous scaffolds with improved biostability for skin tissue engineering. Biomaterials 24, 4833–4841.

Mitra, T., Sailakshmi, G., Gnanamani, A., Mandal, A.B., 2011. Cross-linking with acid chlorides improves thermal and mechanical properties of collagen based biopolymer material. Thermochimica Acta 525, 50–55.

Modjarrad, K., Ebnesajjad, S., 2013. Handbook of Polymer Applications in Medicine and Medical Devices. Elsevier.

Okamoto, M., John, B., 2013. Synthetic biopolymer nanocomposites for tissue engineering scaffolds. Progress in Polymer Science 38, 1487–1503.

Palacio, M., Bhushan, B., 2010. A review of ionic liquids for green molecular lubrication in nanotechnology. Tribology Letters 40, 247–268.

Rhodes, C.J., 2016. The 2015 Paris climate change conference: COP21. Science progress 99 (1), 97–104. https://doi.org/10.3184/003685016X14528569315192.

Sadrmanesh, V., Chen, Y., Rahman, M., AL-Oqla, F.M., 2019. Developing a decision making model to identify the most influential parameters affecting mechanical extraction of bast fibers. Journal of Cleaner Production 238, 117891.

Shi, S.-C., Lu, F.-I., 2016. Biopolymer green lubricant for sustainable manufacturing. Materials 9, 338.

Sionkowska, A., 2011. Current research on the blends of natural and synthetic polymers as new biomaterials. Progress in Polymer Science 36, 1254–1276.

CHAPTER 11

Modern Electrical Applications of Biopolymers

OSAMA O. FARES • FARIS M. AL-OQLA

1 INTRODUCTION

The everlasting challenges toward green technology and clean energy can be summarized by the objective of providing eco-friendly systems with excellent environment record of recyclability, degradability, and renewability at lower costs and lower power consumption levels while meeting pre-prescribed electrical, mechanical, and chemical standards (AL-Oqla et al., 2019; Fares et al., 2019). Biopolymers, both of structured and synthetic origins, are becoming essential in modern technologies to meet these challenges. When combined with other materials and particles, many properties of these composites can be altered to meet specific requirements leading to smart biocomposites (Al-Oqla and El-Shekeil, 2019).

As green technology and clean energy are becoming fundamental strategies in modern industry, many biopolymers have been verified for various industrial applications including transient electronic devices (Alaaeddin et al., 2019a,c,d). The term transient reflects the need of such electronic devices to be built out of specially engineered materials that are programmed to decay within precise period of time when put to function, whereas the term bioelectronics refers to that unlike the traditional solid state silicon-based electronics, the semiconducting materials used are fabricated out of organic materials of living origin. The main expected advantages of bioelectronics systems are to be nontoxic, biocompatible, and biodegradable. These added values make this new type of electronic systems very important in biomedical applications. Flexibility of the electronic devices in this case is of crucial importance to ensure matching with the structure of the subjected organ. Other fields of interest of transient electronics include, but not limited to, field-effect transistors (FETs), biosensing and actuating, robotics, energy harvesting, and manufacturing of supercapacitors (Arnal et al., 2019; Cho et al., 2017; Chu et al., 2016; Kim and Seo, 2002; La Notte et al., 2018; Rullyani et al., 2019).

Mechanical flexibility is a major privilege that modern electronic systems need to have to compile with requirements of the industry (AL-Oqla et al., 2017; AL-Oqla and Hayajneh, 2007; Alaaeddin et al., 2019b). Flexibility enables electronic systems to respond and tolerate any possible mechanical bending, stretching, or even twisting (AL-Oqla et al., 2018b).

In summary, biocomposite materials are expected to have the following features:

- Naturally abundant
- Lightweight
- Low production and processing expenses
- Biocompatible
- Degradable
- Enhanced electrical, thermal, and mechanical characteristics

In addition to the special features, a certain design may require flexibility, transparency, good electrical insulating, conducting, or even semiconducting properties, and prespecified acoustic properties (AL-Oqla, 2017; AL-Oqla et al., 2018a; AL-Oqla and Omari, 2017; AL-Oqla et al., 2015; AL-Oqla and Salit, 2017b; AL-Oqla and Sapuan, 2018a).

In this chapter, a brief introduction to some of the most important modern advanced electrical applications of biopolymers is presented. The chapter will mainly discuss the use of biopolymers in the fields of organic thin film transistors (OTFTs), flexible displays, biosensing and actuators, supercapacitors, photovoltaic (PV) solar cell, and optoelectronics.

2 ORGANIC THIN FILM TRANSISTORS

Since they were first invented back in 1970s, OTFTs received much attention from researchers as part of the organic electronics field (Arnal et al., 2019; Kumagai et al., 2018; Singh et al., 2018; Tortora et al., 2018).

Advanced Processing, Properties, and Applications of Starch and Other Bio-Based Polymers. https://doi.org/10.1016/B978-0-12-819661-8.00011-1

OTFTs are special type of field effect transistors (FETs) that utilize organic composites instead of the regular nonorganic semiconductor materials usually used. Generally, the physical structure of a thin film transistor (TFT) is very similar to that of FETs having a drain, a source, a gate, a channel, and a substrate layers (Lu et al., 2018). The substrate of the transistor is either n-type semiconductor for holes to be the current carriers or p-type semiconductor for electrons to be the current carriers. The drain and source are formed from thin layers of heavily doped n-type or p-type semiconductors opposite to the doping of the substrate. A thin layer of insulating dielectric material is placed in the region between the source and the drain to form the gate. The contact electrodes are formed by depositing metallic layers above the source, drain, gate, and substrate regions. It is the electric field imposed on the gate terminal that is responsible for the switching function of the transistor. Depending on the type of the transistor current carrier charges, a suitable electric potential is applied on the gate terminal relative to the source terminal to allow or stop the electric current to flow between the source and the drain terminals. If the substrate of the transistor is a p-type semiconductor, a positive potential should be applied across the gate terminal to attract the free electrons from the substrate and thus form an enhanced n-type channel just beneath the gate region. This will allow a current to flow from the drain to the substrate once an electric potential is applied between the drain and the source terminals. In this case, an n-type enhancement mode FET is formed. If, on the other hand, the substrate is made of an n-type semiconductor, both the source and the drain are made of p-type materials and a negative potential should be applied across the gate terminal to attract holes, thus forming a p-type channel beneath the gate. In this case, the current carrier charges are holes and a p-type transistor is formed. In case of TFTs, all these layers are made of thin films.

Fig. 11.1 below is an example of a multilayer OTFT reported by L. Tortora et al. (2018). The reported OTFT is constructed from PDIF-CN_2 as the organic semiconductor forming the channel between two 30 nm thick gold drain and source electrodes. The gate electrode was made from aluminum layer with thickness of 50 nm. These layers were deployed on a flexible substrate made of polyethylene naphthalate (PEN) with a thickness of 100 μm. Fig. 11.1A shows a schematic of the three-dimensional structure of the reported OTFT. Fig. 11.1B represents a photo of the manufactured PEN flexible wafer. The current-voltage (*i-v*) characteristic curves of the OTFT are presented in Fig. 11.1C. This set of *i-v* curves shows good field effect characteristics with a possible applied drain to source potential difference (V_{DS}) up to 30 V. The recorded current levels, however, were in the order of 1 μA. The gate to source potential difference (V_{GS}) was in the range of ±20 V.

The characteristics of OTFTs, such as the threshold voltage and the I_{on}/I_{off} ratio, depend mainly on the properties of the semiconducting material and the electrodes being used. However, these characteristics can be modified by using a proper transistor geometry as needed even with same materials and fabrication process. TFTs, in general, come with four different physical structures depending on the manufacturing process used and thus resulting in different performance characteristics. The gate layer can be placed either at the top or the bottom of the TFT resulting in a normal or inverted structure, respectively. If the source, drain, and gate are at the same side, the TFT is said to have a coplanar structure. Otherwise, it is staggered. Usually a top-gate staggered structure will result in lower parasitic capacitances and enhanced channel protection (Jiang et al., 2019). Other commonly used geometries are the interdigitated and the Corbino structures. Fig. 11.2A and B represents an illustration of the interdigitated geometry and the Corbino geometry, respectively (Arnal et al., 2019). Interdigitated geometry provides higher surface-to-

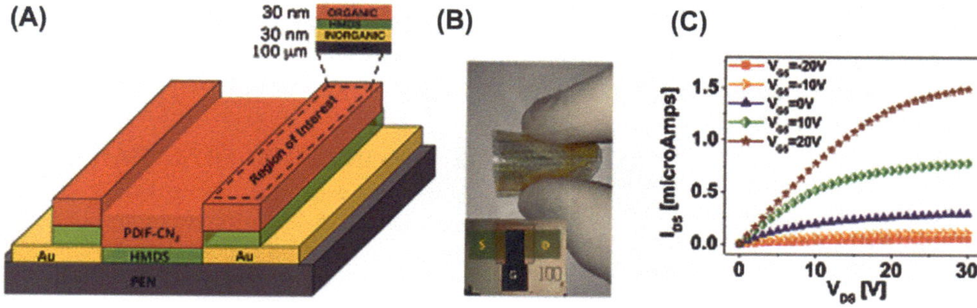

FIG. 11.1 (**A**) The three-dimensional structure. (**B**) A picture of flexible PEN wafer, (**C**) with its *i-v* characteristics of the OTFT investigated (Tortora et al., 2018).

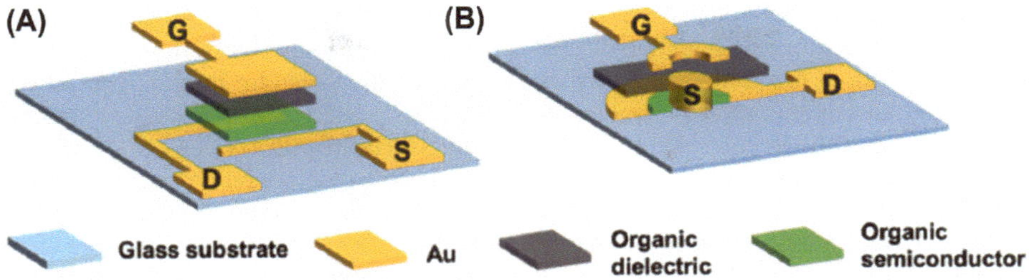

FIG. 11.2 Illustration of the 3D structure of (A) interdigitated geometry and (B) Corbino geometry of OTFT (Arnal et al., 2019).

width ratios of the transistor so as to compensate for the expected limited conductivity of the organic semiconductors. However, by having the channel of circular shape, Corbino geometry results in lower expected threshold voltages. Since the electrodes overlying areas in Corbino geometry are usually lower than that of conventional geometry, the parasitic capacitances are expected to be less (Arnal et al., 2019).

In silicon-based technology, crystalline silicon (c-Si), amorphous silicon (a-Si), and poly-Si are being used resulting in TFTs with different operation modes. Many other materials are being widely used as the active layers of TFTs. Zinc oxide (ZnO) is another widely used material in TFT technology. This material is highly transparent with large energy gap in the order of 3.3 ev at room temperature (300 K) making it very suitable for transparent electronics application. The basic ZnO-based TFT usually uses top-gate structure formed of conductive channel of thick active layer of ZnO deposited on glass substrate. Due to their high switching ratio (I_{on}/I_{off}) of the drain current, these transistors found many applications in mixed digital and analog electronic circuits.

OTFTs, on the other hand, make the most of organic materials as the active layer with plastic substrate. OTFT technology is less complex than traditional Si-based electronics, allowing the fabrication to be achieved at low temperature, while providing transparency and flexibility as needed (Arnal et al., 2019; Kumagai et al., 2018; Singh et al., 2018; Tortora et al., 2018). OTFTs enjoy many other advantages such as low cost, lightweight, and ruggedness. All these features made them suitable for electronic applications where flexibility is essential. Such applications include flexible displays, electronic papers, sensor arrays, and radio-frequency identification cards. OTFTs are currently being used in implementing not only basic digital blocks such as memory registers, flip-flops, logic functions, and pixel switching and sensing but also in the analog parts of electric circuits to implement different signal processing functions such as amplification, waveforms generation, and signal conditioning. To overcome the disadvantage of having relatively high threshold voltages in OTFTs, dual gate structures can be used.

3 ORGANIC LIGHT-EMITTING DIODES AND FLEXIBLE DISPLAYS

Organic light-emitting diodes (OLEDs) and flexible displays are direct applications of organic electronics that are used to convert electric signal into light. Conventional inorganic LED is fabricated from a forward-biased *pn* junction made from direct bandgap materials such as gallium arsenide (GaAs). The emitted light is basically resulting from the recombination process between electron-hole pairs within the semiconductor material. When a *pn* junction is subjected to forward biasing conditions, a minority carrier diffusion current will result due to diffusion process of the minority carriers at both sides of the depletion region toward the *p* and *n* regions. As the carriers cross the depletion region, they start recombining with the majority carriers. In direct band gap materials, this recombination will result in light emission phenomenon. The intensity of the emitted light is directly proportional to the forward current flowing through the diode which actually controls the number of electron-hole recombination.

In spite of the many advantages they have, inorganic-based LEDs suffer from poor electric current to light conversion efficiency, short life cycle, pitiable environmental record, and relatively high prices. Due to the natural electrooptical properties they are having, biopolymers present an excellent replacement in the field of LED industry. Such displays exploit the good conducting and transparency properties of biopolymers to fabricate organic-based light emitting displays (OLEDs). OLEDs are becoming widely used in fabricating not only point-source displays but more importantly surface-source displays. In order to get benefit out of the attractive features they have, more than one polymer are conjugated together to enhance the overall characteristics of specially engineered materials in the industry of OLEDs. The use of biocomposite materials in the field of LEDs added many attractive values to the overall performance and characteristics of the display systems. Among these are lower power consumption, higher degree of degradability, renewable, lighter weight, lower costs, and extended color quality (AL-Oqla and Salit, 2017a; AL-Oqla et al., 2016).

Carbon nanotube (CNT) is another widely used material in the fabrication of OLEDs. CNT-based OLEDs enjoy good mechanical flexibility with excellent light outputs and relatively acceptable lifetime. However, this type of LED suffers from reduced surface roughness making them less attractive than graphene-based LEDs.

Cellulose-based OLEDs are another emerging and very promising flexible OLED family. As a way to enhance the properties of CTE in OLEDs, nanocellulose composites are mainly used in fabrication of the OLED substrates to provide them with the needed flexibility. Starch is cheap and widely available sustainable biocomposite that is known to contain a large number of hydroxyl groups that are functionalized around the surface (Haque et al., 2017). This physical structure makes it possible to integrate them with carbon nanodots (CDs) resulting in more ecological CD-based and phosphors-based LEDs.

Fig. 11.3 below summarizes the fabrication steps of an example of economic fabrication process of flexible substrates integrated for OLEDs reported in Ref. Cho et al. (2017). In this fabrication procedure, glass is being used as the substrate holding the other layers of the OLED. The electrodes were deployed using screen printing associated with UV curing.

Flexible displays rely on the new emerging technology of flexible electronics. Due to the expected nature of their usual use, flexible displays should be elastic and bendable. This is also beside the need that the materials used in manufacturing these displays should be of lightweight and able to tolerate the mechanical deformation expected from the continuous bending and twisting which can affect the quality of the performance as well. In addition, flexible displays are expected to operate at frequencies in the order of MHz, which is much lower than what CMOS technology already has. All these factors make OTFT the right candidate for such applications.

Typical flexible OLEDs usually have a multilayer structure as illustrated in Fig. 11.4 (Li and Sapatnekar, 2018). The first layer is a flexible substrate that provides all other parts with the required mechanical support. This substrate is usually manufactured from materials such as polyimide, paper, or plastic for transparent applications. The second layer consists of OTFT matrix forming the active layer of the display associated with the controlling system. The flexible display is finally covered with a transparent protection layer. Fig. 11.5 shows an example of an OLED based on

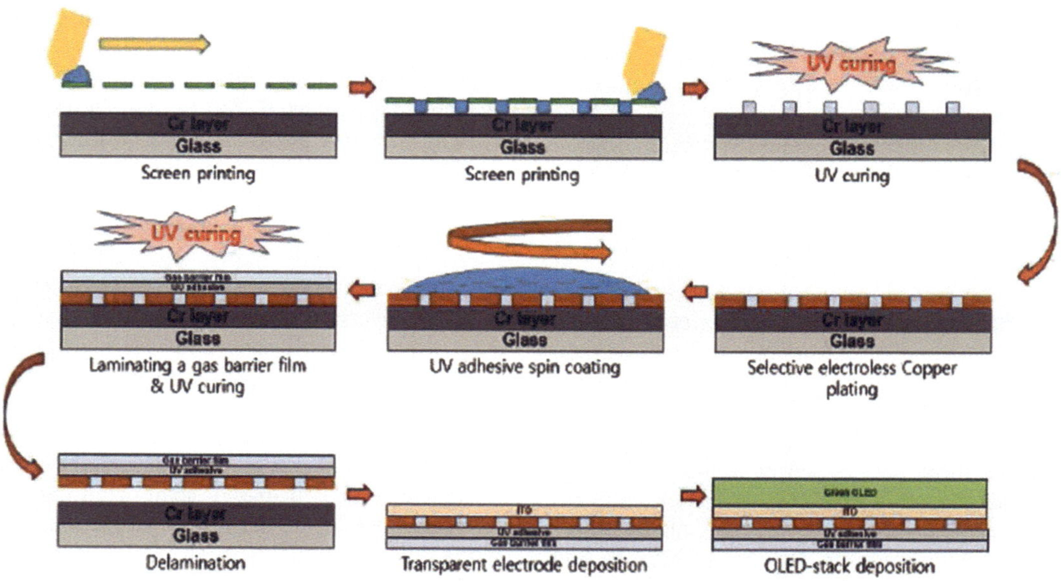

FIG. 11.3 Schematics of the steps of the fabrication process of the flexible OLED presented in Ref. Cho et al. (2017).

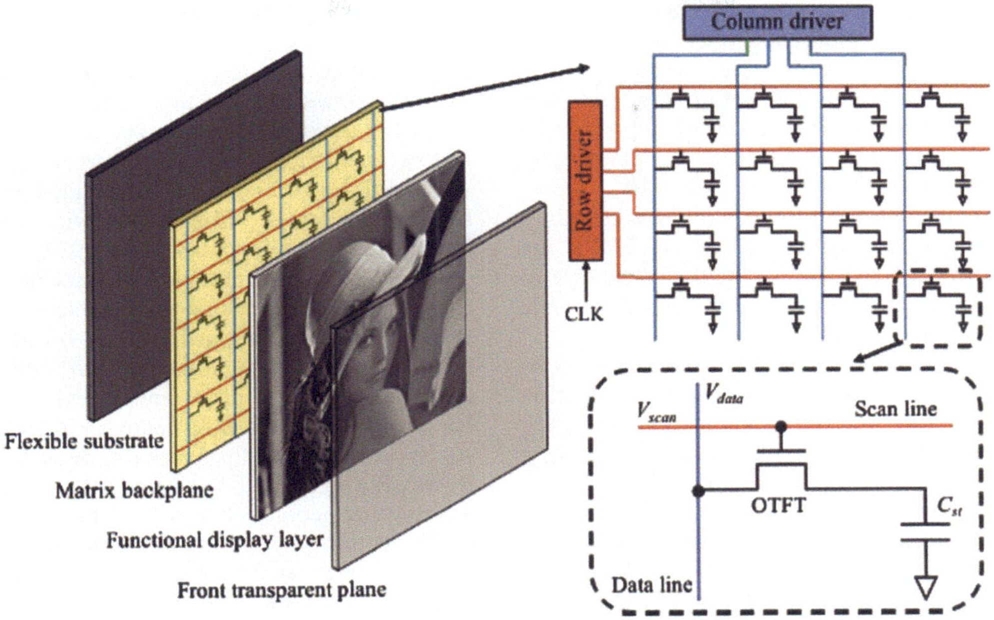

FIG. 11.4 Illustration of typical structure of OTFT-based flexible display (OTFT_Display) (Li and Sapatnekar, 2018).

bacterial cellulose nanocomposite proposed in reference (Ummartyotin et al., 2012).

4 BIOSENSORS AND ACTUATORS

Biosensors are systems that usually consist of two main parts integrated together, the receptors and the transducers. Receptors are responsible for interaction and recognition of the element being sensed. Many types of biorecognition elements are used for this purpose, including enzymes, antibodies, and nucleic acids. The main challenge in this field is to increase both the sensitivity and electivity of the receptor elements. In this regard, specially engineered biopolymers are having lot of attention from researchers. This is mainly due to the many advantages they have, making them suitable for both medical applications and ecological screening. The transducer part of the biosensor is responsible for

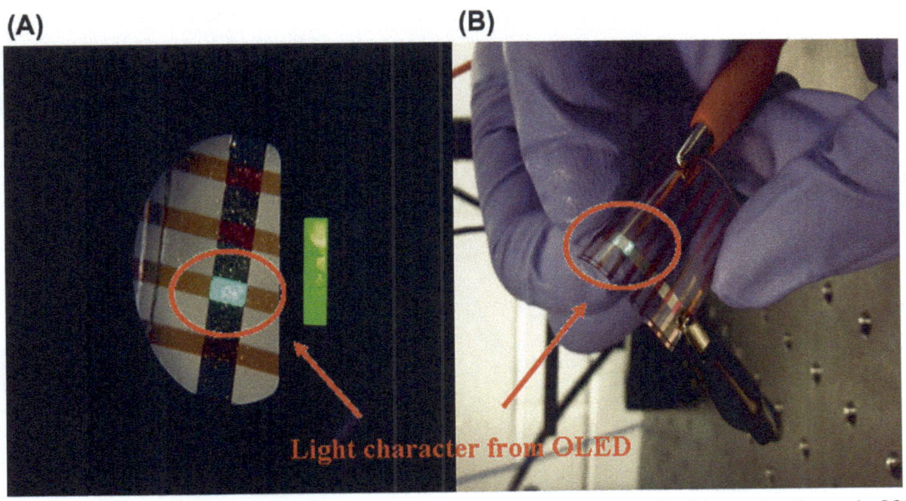

FIG. 11.5 Photo of an OLED based on bacterial cellulose nanocomposite (Ummartyotin et al., 2012).

converting the biological signal produced into electronic signals, which can be amplified and processed. Depending on the field of application of the biosensors, transducers could be thermal, optoelectric, piezoelectric, or electrochemical. Fig. 11.6 below represents an illustration of the physical structure of two examples of organic-based biosensors. The first sensor (Fig. 11.6A) is designed to detect the concentrations of mercuric ion (Hg^{2+}) (Rullyani et al., 2019). The sensor shown in Fig. 11.6B is designed to be able to monitor several environmental parameters at the same time (Surya et al., 2018).

To ensure responsive tapping of the sensed variations, one main challenge in the design and fabrication of biosensors is to have the receptor elements tightly held to the transducer element. FET-based biosensors have the ability to directly convert the biological signal into electric current. In such sensors, the bioreceptors are held on the dielectric material forming the gate. Any changes within the gate potential will be directly converted into a change in the drain-source current.

Electroactive paper (EAPap) is a cellulose-based electronic paper that has been recently invented to have part in sensing applications (Kim and Seo, 2002). EAPap enjoys the many attractive features cellulose-based composites have, including low cost, lightweight, sustainability, biodegradability, and renewability. This is beside the ability to operate at lower voltage levels and lower power consumption.

Actuators are specially engineered materials that are designed to be able to sense and react to any possible change within its ambience. Many smart materials have been used in the fabrication of actuators, including piezoelectric ceramics, piezoelectric polymers, electroactive polymers (EAPs), and many others. Among these, EAPs show high potential in the field of biosensing and actuating coming in two main categories, ionic EAPs and electronic EAPs. One very important merit of ionic EAPs over the electronic counterpart is the relatively low operating voltages they need. However, ionic EAPs suffer from the relatively slow response, and they have to preserve specific degree of humidity. Cellulose-based EAPap has recently emerged as a very promising smart material to be used in actuators (Kim et al., 2018). EAPap based actuators enjoy the many advantages cellulose-based materials have, including, but not limited to low cost, simple processing, compatibility with biosystems, biodegradable, and abundant (Kim, 2017).

One very important factor that should be taken into consideration when designing an EAPap is the permittivity of the material used. This is because in one hand permittivity will affect the maximum possible resulting mechanical displacement and on the other hand it affects the power dissipation within the EAPap. As the permittivity of the material increased, the power losses will also increase. Designers should also take into account that permittivity is usually heavily dependent on the frequency of the applied electric field as well as the environmental conditions at which the EAPap is supposed to act, i.e., ambient temperature, humidity, pH, etc. Fig. 11.7 below shows an illustration example of the fabrication process of a chitin-based piezoelectric sensor (Kim et al., 2018). The final step in the sequence shows its possible decaying when subjected to proper degradation conditions.

5 SUPERCAPACITORS

The need for sustainable and clean energy forced the industry to search and develop not only new eco-friendly energy sources but also new systems to store the generated energy. One of the main electrical energy storage devices is capacitors. A capacitor is defined as passive electric element that can store electric energy in the form of accumulated charges. The simplest form of a capacitor is two conducting plates separated by an insulation dielectric material. The capacitance depends on the physical geometry of the capacitor and the dielectric

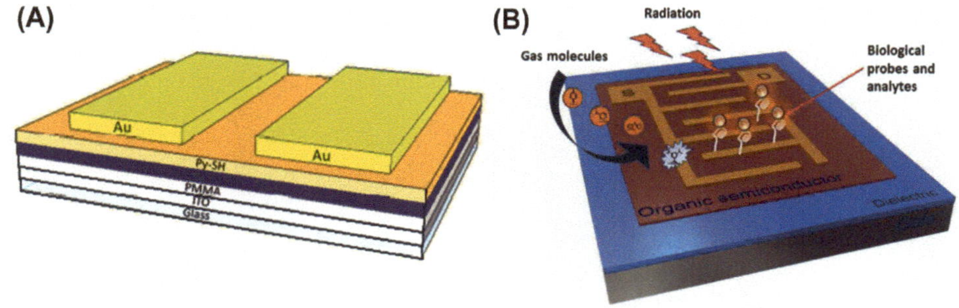

FIG. 11.6 **(A)** 3D schematic illustration of an organic based sensor for the detection of Hg^{2+} ions (Rullyani et al., 2019). **(B)** A multipurpose sensor for environment monitoring (Surya et al., 2018).

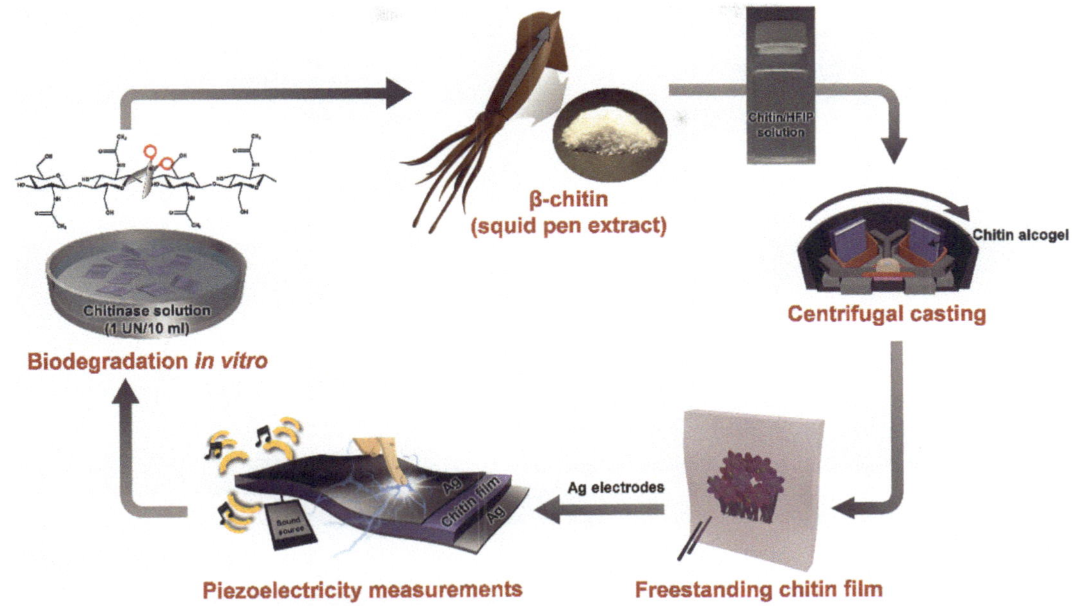

FIG. 11.7 Illustration of the fabrication sequence of a piezoelectric sensor utilizing chitin biopolymers as the active material (Kim et al., 2018).

properties of the separation material. For a simple parallel plate capacitor, the capacitance (C) can be expressed as

$$C = \varepsilon_o \varepsilon_r \frac{A}{d} \tag{11.1}$$

where ε_o is the free space permittivity and is given as 8.85×10^{-12} F/m, ε_r is the relative dielectric constant (permittivity) of the separation material, d is the separation distance between the conducting plates (equal to the thickness of the dielectric material), and A is the conducting plates surface area. The amount of energy (E) that can be stored within the capacitor directly depends on the dielectric constant of the separation dielectric material used as can be seen from Eq. (11.2):

$$E = \frac{1}{2} CV^2 = \frac{1}{2} \varepsilon_o \varepsilon_r \frac{A}{d} V^2 \tag{11.2}$$

where V is the voltage difference between the two conducting plates. As the dielectric constant increases, the amount of energy that can be stored within the capacitor is also increased. According to Eq. (11.2), the stored energy also depends on the square of the electric field applied across the plates of the capacitor. However, increasing the applied electric field is not a good solution since it will affect the physical properties of the dielectric material, forcing it to breakdown. Another possible way to enhance the storage capabilities of a capacitor is by increasing the surface area of the conducting electrodes by forming a set of capacitors parallel to each other.

One research trend toward increasing the capacity of capacitors has been focusing on manufacturing specially engineered materials that have a maximum possible dielectric constant while maintaining a relatively acceptable breakdown voltage. Another very important research trend toward maximizing the capacity of storage is to increase the area of the electrode material being used. To ensure high performance and thus high storage capabilities, electrode materials to be chosen should have large surface area while maintaining excellent electrical conductivity. These research trends led to what is known as supercapacitors. Supercapacitors can be defined as capacitors with specially engineered materials that allow them to store electric energy levels much higher than that of ordinary capacitors. To enhance the environmental profile of supercapacitors industry, many new materials such as cellulose, CNTs, and hydroxyapatite have been studied as additive materials to enhance both the conductivity and surface are of the electrodes (Okonkwo et al., 2017).

Fig. 11.8 below shows an illustration of the sequence of the preparation process of cellulose/ multiwalled carbon nanotube (MWCNT)/rGO/Co_3O_4/SnO_2 hybrid nanocomposite to be used for supercapacitor applications (Ramesh et al., 2018).

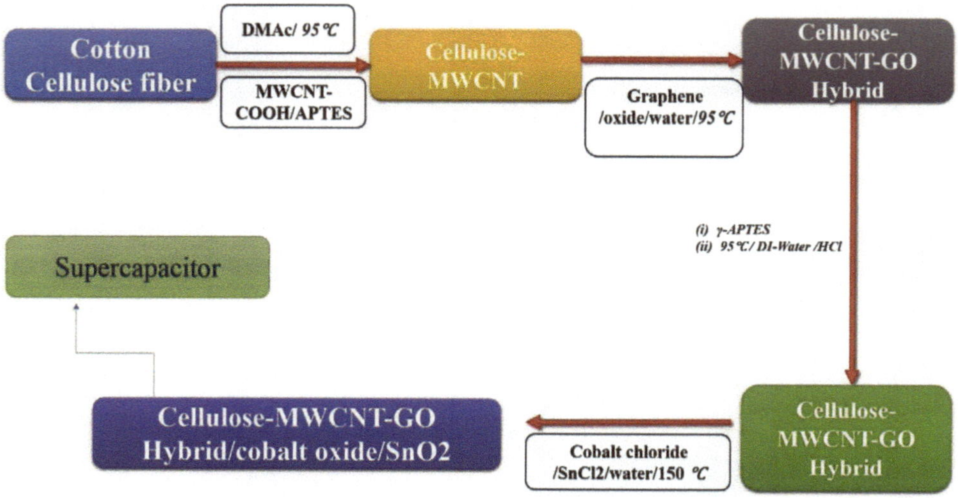

FIG. 11.8 Schematic of the fabrication sequence of cellulose biocomposites in supercapacitors (Ramesh et al., 2018).

6 PHOTODIODES, PHOTOTRANSISTORS, AND PHOTOVOLTAIC SOLAR CELLS

Optoelectronics is another area in which biopolymers found lot of attention. In optoelectronics systems, electronic circuits utilize both electronics and optics not only for converting electronic signals into light and vice versa but also for further processing, storing, and transmission. Optoelectronics found applications in diverse areas including communications, medicine, and sensing and detection. In this section, photodiodes, phototransistors (PTs), and PV solar cells are briefly discussed.

Photodiodes and PTs are semiconductor electronic devices that are able to sense and convert light into electrical signals. Similar to ordinary diodes, photodiodes are constructed from *pn* junctions. However, these *pn* junctions are biased to work in the reverse operation mode. When the photons impact the reverse junction, some of the covalent bonds break resulting in a generation of electron-hole pairs within the depletion layer. The number of these electron-hole pairs will depends on the intensity level of the incident light. The potential difference across the depletion region will force the generated electrons to sweep to the *n* side of the diode. Due to the same effect, the generated holes will be swept to the *p* side of the diode. This will result in a net drift current known as the photocurrent. It is this reverse bias operation that differentiates, in principle, between photodiodes and the PV solar cells. Inorganic photodiodes are usually constructed from a combination of elements rather than from just silicon. The most popular inorganic photodiodes are fabricated from GaAs. Photodiodes found lot of applications in many areas both of civilian and military nature.

Promising studies on using DNA to built UV photodetectors have also been reported. To overcome the poor conductivity they have, DNA composite films combined with mobility enhancers could be used in these applications. Fig. 11.9 below presents the architecture and the photo detection mechanism of an example of a bilayer PT based on SWNT/C60 reported in Ref. Park et al. (2015).

Fig. 11.10 below presents an example of another organic phototransistor (OPT) fabricated using the polar biomaterial polylactide as the dielectric material under the gate electrode with a silicon-based substrate (Chu et al., 2016). Fig. 11.10A shows a schematic representation of the top-contact geometry used in constructing the reported OPT. Fig. 11.10B illustrates the chemical structure of the treated gate dielectric material. Fig. 11.10C and D presents schematic of the resulting OPT array and a photo of the free standing device which was highly flexible with thickness of less than 4 μm.

Dye-sensitized solar cell (DSSC) is one type of organic-based PV solar cells. PV cells of the heterojunction structure, on the other hand, are known to provide conversion efficiency higher than that of the other solar cell types. This is mainly because such structure can provide larger area of interface between the electroactive donor and acceptor biopolymers. One example of the basic structure and principle of operation of DSSCs is illustrated in Fig. 11.11 (Mohiuddin et al., 2017). When the photosensitive dye is exposed to sunlight, some of the valence band electrons in the p-type

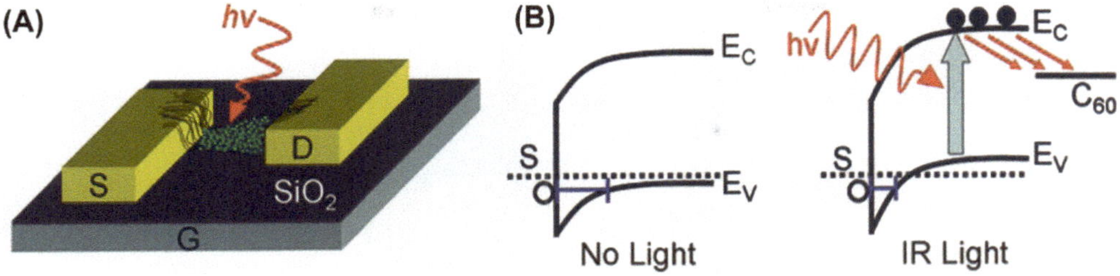

FIG. 11.9 (A) 3D illustration of the SWNT/C60 PT. (B) Energy band diagram under dark condition and when subjected to light (Park et al., 2015).

material will have sufficient energy to overcome the material's energy gap and jump to the conduction band to be free to move. Once an electric load is connected across the cell, an electric current resulting from the movement of these conduction electrons will be established.

In spite of the many advantages they enjoy, biopolymer-based PV solar cells still have some weakness points that form real challenges to this industry. Among these weaknesses are poor light-to-electric energy conversion rates, instability in the overall performance, and relatively short life cycle. It is believed that since making biopolymer composites made by combining different materials with each other somehow can improve the characteristics of the overall resulting composites, specific properties of PV solar cells can be improved by using specially engineered composites. Due to the many merits nanocellulose has, it is considered as one of the most promising materials in this regard. Among these advantages are the outstanding mechanical properties, elevated aspect ratio, low density, and ease of adaptation with diverse chemical groups.

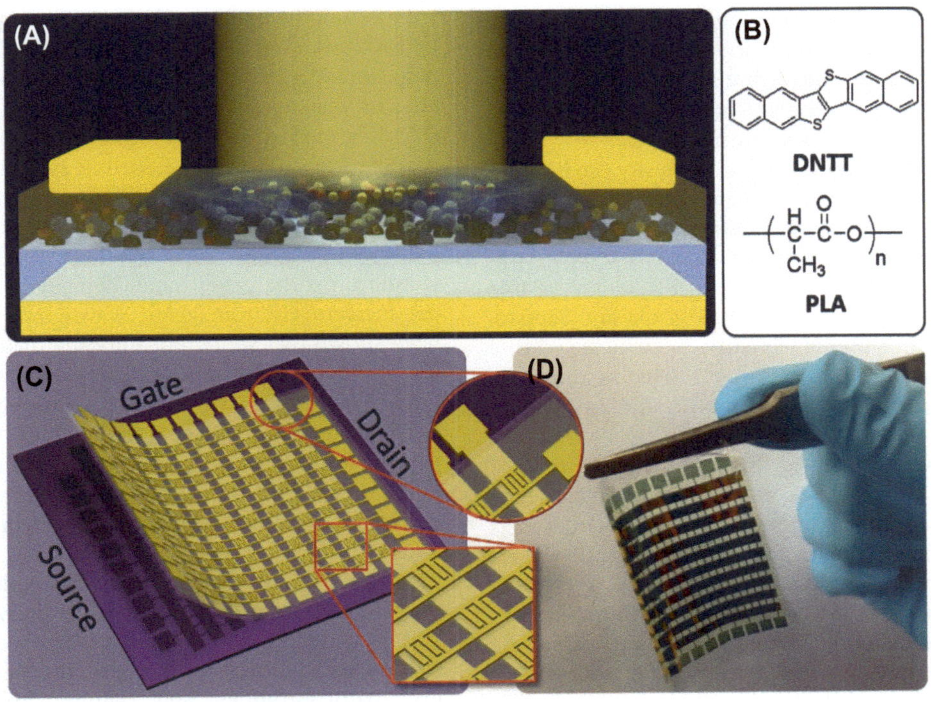

FIG. 11.10 (A) 3D representation of the top-contact structure of the organic phototransistor reported in Ref. Chu et al. (2016). (B) The chemical structure of the organic semiconductor (DNTT) and the dielectric (PLA) materials. (C) Schematic of the resulting pattern. (D) A photo of the final organic transistor array (Chu et al., 2016).

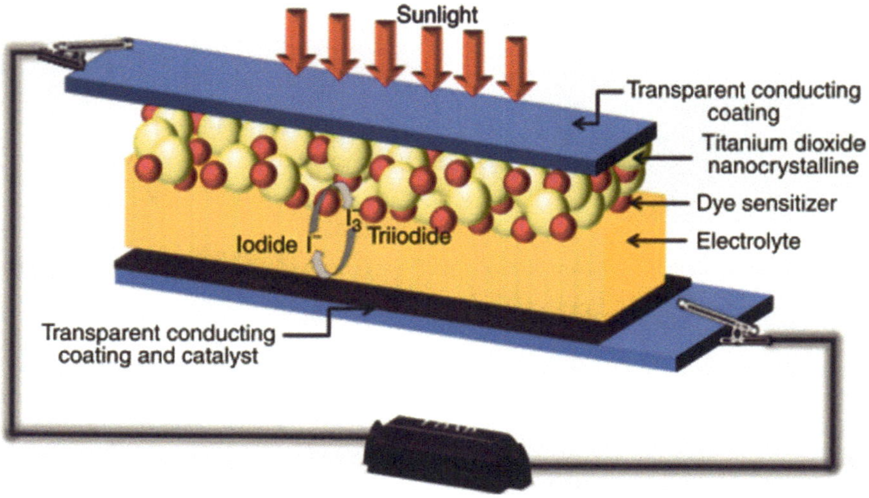

FIG. 11.11 Basic structure of a typical DSSC (Mohiuddin et al., 2017).

7 OTHER ELECTRICAL APPLICATIONS OF BIOPOLYMERS

Biopolymers have found very important role as fundamental components in many other modern and advanced applications in the field of electric and electronic systems. Examples of such applications include, but not limited to, fuel cell fabrication, microwave absorbers, electromagnetic interference shielding, and specially engineered piezoelectric and thermoelectric materials.

The main challenge in fuel cell industry can be summarized in having high efficiency, stable materials at wide range of temperature of operation, and extended lifetime, while maintaining commercially accepted prices using more eco-friendly, sustainable, and recyclable materials (AL-Oqla et al., 2014; AL-Oqla and Sapuan, 2014, 2018b). Biopolymers such as cellulose, chitin, chitosan, and starch are being considered as good candidates to be used in more economic and ech-friendly fuel cells. Specifically, it is chitosan which is being the most widely studied biopolymer for fuel cell application. Chitosan and its composites are used in fabricating the membrane part of modern fuel cells such as hydrogen polymer-electrolyte fuel cells, direct methanol fuel cells, and alkaline fuel cells.

As the advances in electronics and communication industry require multisystems to operate within the same limited area without affecting the performance and characteristics of each other, the need for effective electromagnetic shielding is becoming more argent and a strict requirement. Electromagnetic compatibility of an electronic device or a system usually needs to satisfy preassigned levels of effective shielding. High shielding efficiency within wide bandwidths of frequency at an unfailing behavior while being robust and eco-friendly with commercially acceptable costs are forming the main requirements for any material to be successful and attractive in this field. Shielding by reflection depends heavily on the use of metals. However, and due to reflected electromagnetic waves, this shielding mechanism could result in more interference problems to surrounding electronic systems. This is besides the many drawbacks the usage of metals could produce, including overall heavy weight, higher cost, inefficient flexibility, and corrosions. Shielding by absorption, on the other hand, depends on converting the incident electromagnetic waves into heat. It has been shown that some biocomposite materials can be used as absorbents to shield different frequency bands. For instance, the CNT/cellulose composite paper shows a relatively wide shielding bandwidth from 15 to 40 GHz with shielding efficiency in the range of 20 dB. On the other hand, it is believed that shielding efficiency can be improved by utilizing hybrid combinations of dielectric/magnetic biocomposites. Many other biopolymers have been extensively studied and reported in this regard, including polyaniline/polypyrrole and wood particles, nickel-plated wooden particles, copper-coated birch veneer substrate, flaky carbonyl iron particles/rubber, and cellulose triacetate and MWCNT composite films.

Energy harvesting by utilizing the mechanical stress (piezoelectric principle) and the temperature variations (thermoelectric principle) is another very important field for the utilization of the extraordinary properties of biopolymers(AL-Oqla et al., 2018a). The excellent flexibility biopolymers enjoy make them competitive candidates as energy harvesting materials. Such biocomposites have the ability to convert any applied mechanical stress regardless of its direction into free electric charges. Polyamides, liquid crystal polymers, and parylene C are reported for their piezo-electricity, whereas examples of thermoelectric materials are polyaniline, polypyrrole, and polythiophene and its derivatives.

8 CONCLUSIONS

Biopolymers possess many important natural electrical, chemical, mechanical, and optical properties that made them very attractive to be used in modern advanced electrical systems. The invention of the cellulosic EAPap was a breakthrough in the industry of biosensing and actuating. When reinforced with specific nanoparticles, the actuation force and frequency of EAPaps improve extensively while still enjoying the many merits biopolymers could provide. Cellulose biopolymers, when working as reinforcement agents, made it possible also to fabricate the OLEDs not only for point sources but also for panel displays that are flexible, biodegradable, and economic.

Starch, in addition, is another naturally available biocomposite with huge renewable resources. Lots of starch-dependent biocomposites have been investigated and fabricated in the field of OLEDs showing competitive properties in terms of quality, price, and environmental regulations, as another alternative for inorganic materials. Starch has been also successfully utilized to manufacture piezoelectric and thermoelectric materials with excellent conversion efficiencies to be used within self-powered systems.

Chitin is another biopolymer that is widely available in nature. This biocomposite has been extensively studied and used in eco-friendly industries. Special interest has been reported in the literature in using chitin as fillers in batteries and electrolytes fabrication as a solution to the high level of possible inflammation occurring in rechargeable lithium batteries. This is in addition to the expected economic and environmental enhancements that could be achieved.

Graphene represents another very attractive material that is being used to replace indium tin oxide as a conducting material in transparent electronics. This is mainly because of the unique structure it is having in addition to its high degree of transparency.

REFERENCES

AL-Oqla, F.M., 2017. Investigating the mechanical performance deterioration of Mediterranean cellulosic cypress and pine/polyethylene composites. Cellulose 24, 2523–2530.

AL-Oqla, F.M., Almagableh, A., Omari, M.A., 2017. Design and Fabrication of Green Biocomposites, Green Biocomposites. Springer, Cham, Switzerland, pp. 45–67.

Al-Oqla, F.M., El-Shekeil, Y., 2019. Investigating and predicting the performance deteriorations and trends of polyurethane bio-composites for more realistic sustainable design possibilities. Journal of Cleaner Production 222, 865–870.

AL-Oqla, F.M., Hayajneh, M.T., 2007. A Design Decision-Making Support Model for Selecting Suitable Product Color to Increase Probability, Design Challenge Conference: Managing Creativity, Innovation, and Entrepreneurship. Yarmouk University, Amman, Jordan.

AL-Oqla, F.M., Hayajneh, M.T., Fares, O., 2019. Investigating the mechanical thermal and polymer interfacial characteristics of Jordanian lignocellulosic fibers to demonstrate their capabilities for sustainable green materials. Journal of Cleaner Production 118256.

AL-Oqla, F.M., Omar, A.A., Fares, O., 2018a. Evaluating sustainable energy harvesting systems for human implantable sensors. International Journal of Electronics 105, 504–517.

AL-Oqla, F.M., Omari, M.A., 2017. Sustainable Biocomposites: Challenges, Potential and Barriers for Development, Green Biocomposites. Springer, pp. 13–29.

AL-Oqla, F.M., Sapuan, S.M., Ishak, M.R., N, A.A., 2015. Selecting Natural Fibers for Industrial Applications. Postgraduate Symposium on Biocomposite Technology Serdang, Malaysia.

AL-Oqla, F.M., Salit, M.S., 2017a. Material Selection of Natural Fiber Composites Using the Analytical Hierarchy Process, Materials Selection for Natural Fiber Composites. Woodhead Publishing, Elsevier, Cambridge, USA, pp. 169–234.

AL-Oqla, F.M., Salit, M.S., 2017b. Materials Selection for Natural Fiber Composites. Woodhead Publishing, Elsevier, Cambridge, USA.

AL-Oqla, F.M., Sapuan, S., 2018a. Investigating the inherent characteristic/performance deterioration interactions of natural fibers in bio-composites for better utilization of resources. Journal of Polymers and the Environment 26, 1290–1296.

Al-Oqla, F.M., Sapuan, S., 2018b. Natural fiber composites. In: Sapuan, S.M., Ishak, M.R., Sahari, J., Sanyang, M.L. (Eds.), Kenaf Fibers and Composites, Vol. 1. CRC Press.

AL-Oqla, F.M., Sapuan, S., Fares, O., 2018b. Electrical–Based Applications of Natural Fiber Vinyl Polymer Composites, Natural Fibre Reinforced Vinyl Ester and Vinyl Polymer Composites. Elsevier, pp. 349–367.

AL-Oqla, F.M., Sapuan, S., Ishak, M.R., Nuraini, A.A., 2014. A novel evaluation tool for enhancing the selection of natural fibers for polymeric composites based on fiber moisture content criterion. BioResources 10, 299–312.

AL-Oqla, F.M., Sapuan, S., Jawaid, M., 2016. Integrated mechanical-economic–environmental quality of performance for natural fibers for polymeric-based composite materials. Journal of Natural Fibers 13, 651–659.

AL-Oqla, F.M., Sapuan, S.M., 2014. In: Enhancement Selecting Proper Natural Fiber Composites for Industrial Applications, Postgraduate Symposium on Composites Science and Technology 2014 & 4th Postgraduate Seminar on Natural Fibre Composites 2014, 28/01/2014, Putrajaya, Selangor, Malaysia.

Alaaeddin, M., Sapuan, S., Zuhri, M., Zainudin, E., AL-Oqla, F.M., 2019a. Photovoltaic applications: status and manufacturing prospects. Renewable and Sustainable Energy Reviews 102, 318–332.

Alaaeddin, M., Sapuan, S., Zuhri, M., Zainudin, E., AL-Oqla, F.M., 2019b. Physical and mechanical properties of polyvinylidene fluoride-short sugar palm fiber nanocomposites. Journal of Cleaner Production 235, 473–482.

Alaaeddin, M., Sapuan, S., Zuhri, M., Zainudin, E., AL-Oqla, F.M., 2019c. Polyvinyl Fluoride (PVF); its Properties, Applications, and Manufacturing Prospects, IOP Conference Series: Materials Science and Engineering. IOP Publishing, p. 012010.

Alaaeddin, M., Sapuan, S., Zuhri, M., Zainudin, E., M AL-Oqla, F., 2019d. Lightweight and durable PVDF–SSPF composites for photovoltaics backsheet applications: thermal, optical and technical properties. Materials 12, 2104.

Arnal, A., Martínez-Domingo, C., Ogier, S., Terés, L., Ramon, E., 2019. Monotype organic dual threshold voltage using different OTFT geometries. Crystals 9, 333.

Cho, D.-H., Kwon, O.E., Park, Y.-S., Yu, B.G., Lee, J., Moon, J., Cho, H., Lee, H., Cho, N.S., 2017. Flexible integrated OLED substrates prepared by printing and plating process. Organic Electronics 50, 170–176.

Chu, Y., Wu, X., Lu, J., Liu, D., Du, J., Zhang, G., Huang, J., 2016. Photosensitive and flexible organic field-effect transistors based on interface trapping effect and their application in 2D imaging array. Advanced Science 3, 1500435.

Fares, O., AL-Oqla, F.M., Hayajneh, M.T., 2019. Dielectric relaxation of mediterranean lignocellulosic fibers for sustainable functional biomaterials. Materials Chemistry and Physics.

Haque, S., Shah, M.S., Rahman, M., Mohiuddin, M., 2017. Biopolymer Composites in Light Emitting Diodes, Biopolymer Composites in Electronics. Elsevier, pp. 277–310.

Jiang, A., Yuan, Y., Liu, N., Han, L., Xiong, M., Sheng, Y., Ye, Z., Liu, Y., 2019. Transparent capacitive-type fingerprint sensing based on zinc oxide thin-film transistors. IEEE Electron Device Letters 40, 403–406.

Kim, J., 2017. Multifunctional Smart Biopolymer Composites as Actuators, Biopolymer Composites in Electronics. Elsevier, pp. 311–331.

Kim, J., Seo, Y.B., 2002. Electro-active paper actuators. Smart Materials and Structures 11, 355.

Kim, K., Ha, M., Choi, B., Joo, S.H., Kang, H.S., Park, J.H., Gu, B., Park, C., Park, C., Kim, J., 2018. Biodegradable, electro-active chitin nanofiber films for flexible piezoelectric transducers. Nanomaterials and Energy 48, 275–283.

Kumagai, S., Nakano, M., Takimiya, K., Takeya, J., 2018. Solution-crystallized n-type organic thin-film transistors: an impact of branched alkyl chain on high electron mobility and thermal durability. Organic Electronics 62, 548–553.

La Notte, L., Cataldi, P., Ceseracciu, L., Bayer, I.S., Athanassiou, A., Marras, S., Villari, E., Brunetti, F., Reale, A., 2018. Fully-sprayed flexible polymer solar cells with a cellulose-graphene electrode. Materials Today Energy 7, 105–112.

Li, T., Sapatnekar, S.S., 2018. In: Strain-aware Performance Evaluation and Correction for OTFT-Based Flexible Displays, 2018 IEEE/ACM International Conference on Computer-Aided Design (ICCAD). IEEE, pp. 1–7.

Lu, N., Jiang, W., Wu, Q., Geng, D., Li, L., Liu, M., 2018. A review for compact model of thin-film transistors (TFTs). Micromachines 9, 599.

Mohiuddin, M., Kumar, B., Haque, S., 2017. Biopolymer Composites in Photovoltaics and Photodetectors, Biopolymer Composites in Electronics. Elsevier, pp. 459–486.

Okonkwo, P., Collins, E., Okonkwo, E., 2017. Application of Biopolymer Composites in Super Capacitor, Biopolymer Composites in Electronics. Elsevier, pp. 487–503.

Park, S., Kim, S.J., Nam, J.H., Pitner, G., Lee, T.H., Ayzner, A.L., Wang, H., Fong, S.W., Vosgueritchian, M., Park, Y.J., 2015. Significant enhancement of infrared photodetector sensitivity using a semiconducting single-walled carbon nanotube/C60 phototransistor. Advanced Materials 27, 759–765.

Ramesh, S., Khandelwal, S., Rhee, K.Y., Hui, D., 2018. Synergistic effect of reduced graphene oxide, CNT and metal oxides on cellulose matrix for supercapacitor applications. Composites Part B: Engineering 138, 45–54.

Rullyani, C., Shellaiah, M., Ramesh, M., Lin, H.-C., Chu, C.-W., 2019. Pyrene-SH functionalized OTFT for detection of Hg^{2+} ions in aquatic environments. Organic Electronics 69, 275–280.

Singh, V.K., Srivastava, R., Pal, S., Baliga, A.K., Kumar, B., 2018. In: Impact of Different Organic Semiconductor Materials on Performance of Dual Gate OTFT, 2018 5th IEEE Uttar Pradesh Section International Conference on Electrical, Electronics and Computer Engineering (UPCON). IEEE, pp. 1–5.

Surya, S.G., Raval, H.N., Ahmad, R., Sonar, P., Salama, K.N., Rao, V.R., 2018. Organic field effect transistors (OFETs) in environmental sensing and health monitoring: a review. TRAC Trends in Analytical Chemistry.

Tortora, L., Urbini, M., Fabbri, A., Branchini, P., Mariucci, L., Rapisarda, M., Barra, M., Chiarella, F., Cassinese, A., Di Capua, F., 2018. Three-dimensional characterization of OTFT on modified hydrophobic flexible polymeric substrate by low energy Cs^+ ion sputtering. Applied Surface Science 448, 628–635.

Ummartyotin, S., Juntaro, J., Sain, M., Manuspiya, H., 2012. Development of transparent bacterial cellulose nanocomposite film as substrate for flexible organic light emitting diode (OLED) display. Industrial Crops and Products 35, 92–97.

CHAPTER 12

Biopolymers in Building Materials

JARIN TUSNIM • MD ENAMUL HOQUE • MANIK CHANDRA BISWAS

1 INTRODUCTION

The geotechnical construction and building industry consumed not less than 36% energy of total global energy consumption and emitted approximately 39% energy-related carbon dioxide (CO_2) using 45% global resources (Global Status Report 2017). In 2040, the global population will be 7 to 9 billion, resulting in more demand for food, house, building, geotechnical construction, energy, water, and other resources. Besides, the environment will have limitations to meet those demands due to climatic change and other unavoidable reasons (da Cruz, RMS. 2019; Rangari et al., 2019). Biopolymers are polymers generated by living microorganisms and are used in various fields (Wahab et al., 2018; Hoque et al., 2019). They have gained immense interest in research for their excellent attributes. Applications of biopolymers as building material are widespread and diverse nowadays. Bio-based polymer products derived from biomass and agricultural feedstock can compete with commercially dominated petroleum-based products. Biopolymer-based composites or bioadmixtures are a potential alternative to petroleum-based reinforced composites. Bioadmixtures are functional molecules incorporated in building materials to optimize structural properties. The term "bioadmixture" comprises biopolymers such as natural or modified biopolymers and products such as biotechnological and biodegradable products, which are produced by biotechnological processes (Johann Plank, 2004). Two main application sectors of bioadmixtures are in concrete and dry-mix mortars (e.g., plasters/adhesives). Examples of bioproducts used in concrete are lignosulfonate, sodium gluconate, pine root extract, protein hydrolyzates, welan gum and methyl hydroxypropyl cellulose, tartaric acid, hydroxypropyl starch, casein, guar gum, succinoglycan, and xanthan gum, which are used in dry-mix mortar. Bioadmixtures are potential candidates as an alternative to synthetic admixtures in different applications due to their promising characteristics. The advantage of applying admixtures in construction is to enhance material properties such as an organic adhesive and water resistantance in mixtures with clay and straw. Proteins also have been used for the set impedance of gypsum and dried blood for air-entrapment to enhance the properties of building materials (Johann Plank, 2004). Nowadays, admixtures offering additional attributes, such as ductility, shrinkage reduction, water preservation, adhesion, etc., are available to the construction materials. Currently available bioproducts used in various industries will be reviewed for consideration in building materials. Product improvement and innovating utterly new technology will occur, such as the introduction of diutan, a novel biopolymer produced by *Sphingomonas* bacteria that achieves ultra-high zero shear rate rheology in cement-based grouts and drilling fluids (Navarrete et al., 2001), and the use of hydroxypropyl guar as a new class of water retention agents for dry-mix mortars (Johann Plank, 2004). This chapter deals with various biopolymers that contribute to geotechnical construction as an alternative to traditional building materials.

2 POLYMER CONCRETE

The most commonly used building material is the Portland cement concrete and other aggregates such as steel or aluminum, due to its low cost, application convenience, and high compressive strength. Portland cement and aggregates also possess some disadvantages, such as poor flexural strength, low chemical inertness, barrier properties, and high emission of greenhouse gas (Herczeg et al., 2014). To overcome those drawbacks, researchers tried to develop eco-friendly and lightweight building materials. Nowadays, the construction field starts using polymer concrete and aggregate materials with high structural properties and quick repairability to infrastructure. Polymer concrete is a kind of concrete-polymer composite where cement hydrate binders of conventional cement concrete are completely

Advanced Processing, Properties, and Applications of Starch and Other Bio-Based Polymers. https://doi.org/10.1016/B978-0-12-819661-8.00012-3

replaced by polymer binder or liquid resins (Bulut and Şahin, 2017). The production process of polymer concrete (as shown in Fig. 12.1) is like cement concrete, but here, liquid resins are used as polymeric binders and they polymerized in place. Different types of liquid resins can be used in preparing polymer concrete such as epoxy, acrylic, vinyl ester, polyurethane, phenol, methyl methacrylate, polyester styrene, furan, urea, etc. (Fowler, 1999). Besides liquid resin, hardeners or initiators are used for hardening. By selecting an appropriate type of liquid resin and hardener at specific ratios, working life and hardening time can be modified and controlled. After selection, they are dried as moisture hurts hardening. The strength of the polymer concrete depends on the strength of the aggregates. In polymer concrete production, aggregates and binders are used instead of cement and water for traditional concrete production. To avoid difficulties, aggregate with the lowest possible voids should be selected. At the time of mixing, silane coupling agents are added to ameliorate the bond between aggregates and polymeric binders. Nowadays, various industrial wastes got much attention as an alternative of aggregates for polymer concrete. Wastes are used by recycling biomass and plastic commodities, to form polymer concrete. For example, for making polyester concrete, the EPS solution is prepared by dissolving EPS in vinyl monomers, which works as a liquid resin and helps reducing shrinkage.

Although Portland cement concretes are used widely, they have some limitations that cannot be overlooked easily. Both internal and processing limitations of Portland cement concrete have bound scientists to invent polymer concrete, which has more strength, water tightness, chemical resistance, abrasion resistance, and durability. But polymer concretes show low thermal resistance, high cost, toxicity, flammability, and low-temperature resistance. To overcome this, the glass transition temperature of the polymer matrix phases should be noted from the very beginning point of showing undesirable thermal properties. Today, polymer concrete is used worldwide. In the United States, they use polymer concrete in highway, dam, electrical insulator, underground tunnels, coating materials, piping, etc. In Japan, it is used in product manufacturing such as telephone cable ducts, tunnel liner segments, gutter covers, acid waste storage tanks, septic tanks, panels, piles, etc. (Fowler, 1999; Reis, 2011).

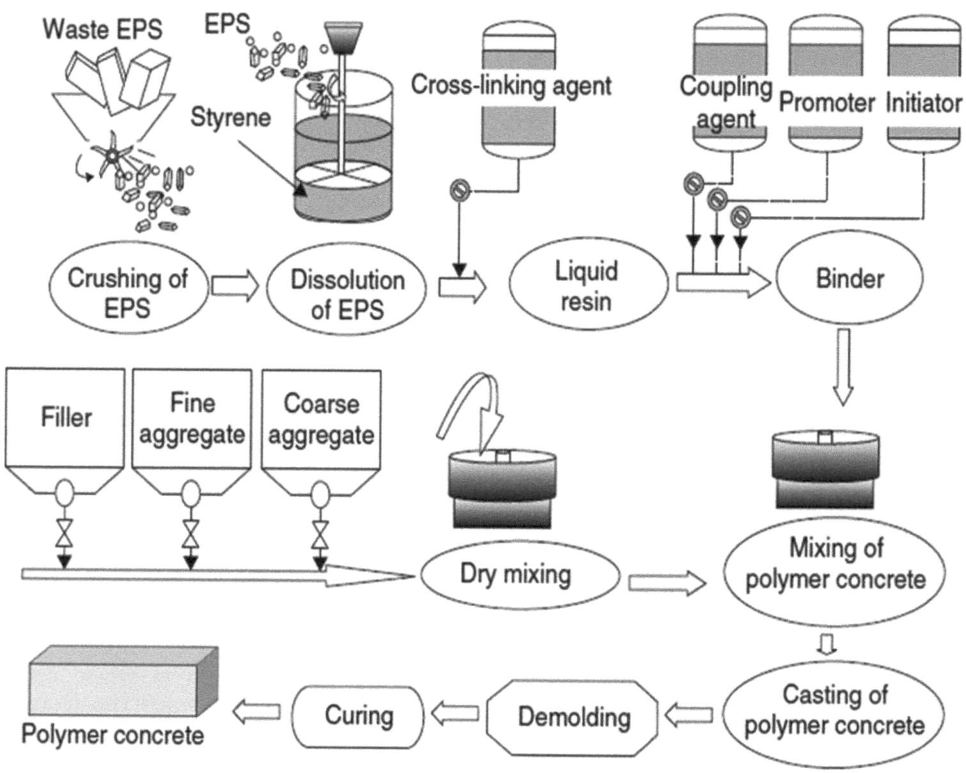

FIG. 12.1 Manufacturing process for precast polymer concrete products (Mindess, 2008).

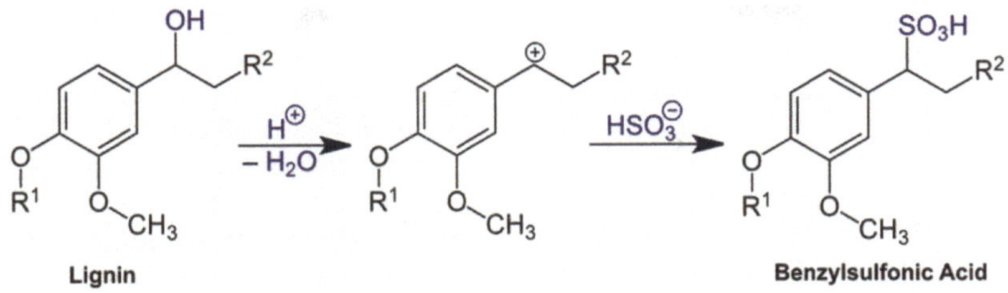

FIG. 12.2 Synthesis process of Lignosulfonate.

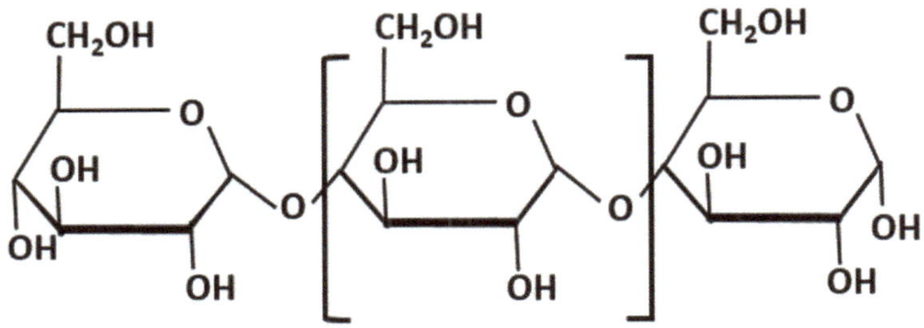

FIG. 12.3 Chemical structure of Amylose.

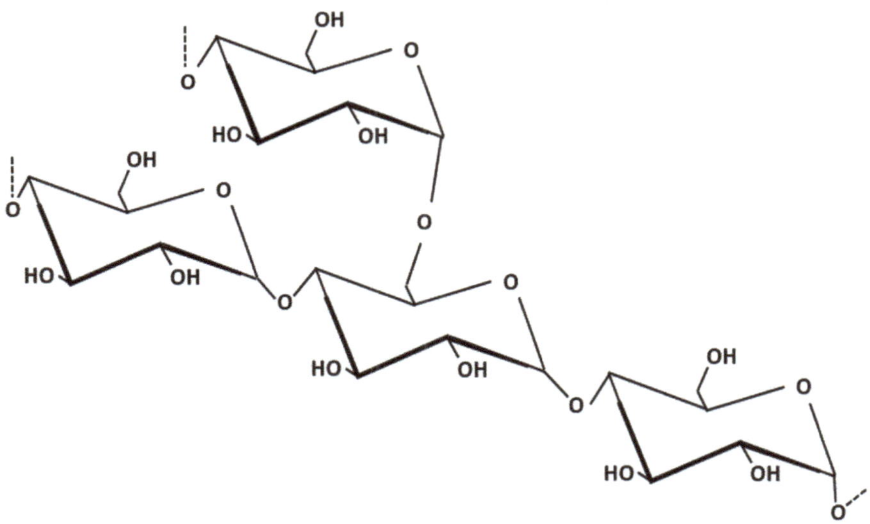

FIG. 12.4 Chemical structure of Amylopectin.

3 LIGNIN-BASED BIOPOLYMER

Lignin is a kind of organic polymer that makes cell walls of plants at 20%–30%. Pure lignin is water-insoluble. By processing lignin chemically, one can obtain lignosulfonate, which is widely used in construction. They act as plasticizers which can enhance processability and workability via increasing the mobility of concrete. It is the highly used admixture in the geotechnical construction field by volume. The main process for obtaining lignosulfonates is the sulfate pulping method (Fig. 12.2). In this method, a hot sulfite solution is used for the degradation of dissolved lignin and hemicellulose. For preparing raw liquor, sulfur dioxide is generally mixed with a base solution where it forms sulfurous acid that helps in degradation. It replaces the hydroxyl group by the sulfonate group. The spent sulfite liquor carries sugars, organic acids, and lignosulfonate. For different uses, different methods of purification are performed. Lignosulfonates are used primarily serving two purposes, such as improving processability and strength of the concrete. Primary sugars are removed for permitting lignosulfonate to work as water-reducing concrete additive. This helps in increasing the compressive strength as well as enhancing the mechanical properties of concrete. Reduction of retardation is needed for application as plasticizers that can be achieved by ethoxylation of hydroxyl groups. In this case, they mitigate the detachment of a concrete mix and increase the density of the mixture. It also helps in reducing cement consumption by 8% −10% (file 500).

Lignosulfonates are adhesive, surface active, and have the binding quality that makes them suitable for use in the production of chipboard, panels, and mineral wood fiber. Normally, reinforcing additives are rare and costly. Sometimes they can be toxic. By using it as a binder, at least half toxicity can be reduced, and it also makes a more viable and water-resistant product. They are being used in oil-extracting and oil-refining industries as a reagent in drilling oil and gas wells that maintain basic parameters of drilling fluids. For strengthening purpose, technical lignosulfonates are availed. In mining industries, they are used as flotation reagents. As they are cheap flotation reagent, their demand is increasing day by day in the mining industries, where they are also used in raising drill wells. They are also demanding in chemical industries, where they are utilized as a dispersant and suspension stabilizer in the creation of pesticides and other related chemicals.

With the increasing concern of environmental sustainability and eco-efficiency, application of lignosulfonate, as a replacement of chemical soil stabilizers, is increasing day by day (Çoban et al., 2018; Li et al., 2019; Ta'negonbadi and Noorzad, 2017, 2018). Chemical stabilizers such as gypsum, lime, alum, fly ash, etc., are being used as soil stabilizers during geotechnical construction. These chemical stabilizers showed excellent compressive strength enhancement but caused brittleness of the soil and environmental pollution via deteriorating groundwater. The brittleness of soil can cause seismic instability of geotechnical constructions. Lignosulfonate showed comparable properties with chemical stabilizers with low to no toxicity to the environment. In addition, as lignosulfonates are by-products, these are cheap compared with traditional chemical stabilizers. Researchers have been trying to investigate their behaviors and found that lignosulfonate-stabilized soil exhibits lower double layer thickness via surface neutralization and formation of a stable grain cluster. Lignosulfonates mostly used to harden silty clay, sandy soil, low plasticized clay, etc., are reported in the literature. B. Ta'negonbadi and R. Noorzad (2017) investigated lignosulfonate treated with high plasticized clay with various concentrations of lignosulfonates (LS) and observed that the synergistic effect of LS and clay improved the strength of the soil that resulted in enhanced long-term life cycle. This can be because of the improved cohesion between soil particles. But Zhang et al. reported that lignin-modified silty clay exhibited some disadvantages and limitations to wide-range applications. They found that the optimum moisture content of lignin should be 12% to enhance the properties of silty clay. Besides, it takes 4 weeks to harden and may modify the physicochemical behavior of sand minerals (Zhang et al., 2015). Those drawbacks trigger the necessity to use alternative biopolymer additives as soil stabilizers and/or building materials.

Though lignosulfonate has tremendous potential and is used largely in the construction field, it possesses some drawbacks. Plasticization and water reduction ability of lignosulfonate are limited in some cases such as highly plasticized concrete and more intense water reduction. To overcome this problem, researchers developed synthetic superplasticizers that are mainly polycondensate- or polycarboxylate-based compounds. For example, polycondensates include polymelamine formaldehyde sulfite or β-naphthalene sulfonic acid formaldehyde condensates with molecular weights less than 10,000 Da (Spiratos et al., 2003). Polycarboxylate-based superplasticizer includes methacrylic acid-methacrylate ω-methoxy polyethylene glycol ester copolymer (Shonaka et al., 1997). To get the highest properties, a blend of lignosulfonates and superplasticizers or pristine superplasticizers is recommended.

4 STARCH-BASED POLYMER

Starch is a carbohydrate that contains two polysaccharide macromolecules, amylose (soluble) and amylopectin (insoluble), and is produced from all green plants. Amylose is composed of linear α1−4 linked polymer chain of glycosidic bonds, whereas amylopectin is made of larger, branched polymer chains of α-glucose units with α1−4 and α1−6 bonds shown in Figs. 12.3 and 12.4 (da Cruz, RMS. 2019; Rangari et al., 2019).

Starch has been getting much attention for developing sustainable materials because of its fully biodegradable and renewable nature with easy availability and low cost. Starch derivatives offer a potential environment for fresh concrete to gain stable conditions and flow properties (Zhang et al., 2007). Corn starch showed excellent properties as a potential candidate for replacing aggregates. It produced lightweight concrete with comparable shrinkage properties. In construction, starch-based biopolymers are used extensively as specially pregelatinized starches in starch admixtures fabrication. Starch admixtures exhibited high durability, comparable shrinkage, and creep properties (Akindahunsi et al., 2013). Adhesion and adsorption of starch on the surface of the concrete play the role to harden and solidify the fresh concrete. Several process parameters may tune the properties of the starch-based concrete such as temperature, particle size, morphology, fineness of the grains, adhesion, etc. (Akindahunsi and Uzoegbo, 2015). Pregelatinized starches are prepared by proper modification called pregelanitization by which starch grows an ability to form a cold water paste. They can hold nearly all functional properties and viscosity of the base material, which is important in construction. They are also used as a cheap and low-temperature fluid loss intermediary for water-based drilling fluids. It usually performs at low temperatures, especially below 100°C. For getting API fluid loss of 2–5 ML at room temperature and 7 bar differential pressure, 1%–2% of pregelatinized starch is used. It is also useful for manufacturing plasterboard. It helps to connect the cardboard to the surface made of gypsum. By increasing the temperature, pregelatinized starch is converted to carboxymethyl starch. It contains a negatively charged functional group (CH_2COO^-). The ordered structure of starch is disturbed by the incorporation of carboxymethyl starch that produced starch with decreased gelatinization temperature, increased solubility, and improved storage stability. So, it is used in oil well drilling as additives as it is an alkaline and can replace carboxymethyl cellulose. It has temperature stability up to 120°C, which is better than pregelatinized starch.

Traditional concrete experiences strain failure at different stages of stress. This happened due to accumulation of stresses into a cement matrix. Besides, there are also microcracks near to the interfacial boundaries between the matrix and aggregates. These microcracks play the role to originate creep failure as it grows with applied stress. Continuous applied stress, as obtained strain can exceed the elastic strain limit of the concrete, results in time-dependent deformation with shrinkage and creep failure. Starch-based admixtures exhibited excellent performance against creep strain. A. A. Akindahunsi et al. (2019) investigated the creep behavior of cassava and maize starch-infused admixtures at different percentages. They fabricated concrete prisms with varying amounts of starch such as 0%–2% by weight and found that the starch-based concrete showed higher strength and minimum deformation compared with the neat system. They also observed that there is no adverse effect of starch-based concrete for an extended period of time.

5 PROTEIN-BASED BIOPOLYMER

Proteins consist of long chains of amino acid made by the polymerization reaction. Protein is obtained from many foods, like meat, fish, poultry, eggs, legumes, and dairy products. Protein is also found in oilseeds (Wool and Sun, 2005, p. 2), containing keratin, casein, collagen, and enzymes. They all are used in industrial applications. Casein is a Latin word meaning cheese and is made up to 80% protein of cow's milk (Burris, 2004). It is negatively charged and has a pH value of 6.6. When milk tastes sour, curd and whey can be obtained from it, but on further water removal, casein forms. Another way to get casein from milk is to lower the pH by adding some mineral acids into milk (Dahlberg, 1918). The quality of casein depends on fat content in milk and the amount of acid during processing. A low percentage of fat and acid can provide high-quality casein (*Casein Glues*, 1950). In the final step, casein should be washed with water to remove residual acids. Cheese like casein is insoluble in water but the salt of casein (caseinates) is soluble and forms paste-like solution shown in Fig. 12.5 (Fatehi et al., 2018).

Fatehi et al. (2018) investigated the effect of casein and sodium caseinate on sand geotechnical and structural properties. They investigated different process parameters such as curing time, temperature, fat content, and concentration of additives and found that higher biopolymer concentration and curing time gave high compressive strength sand. The temperature has also a positive effect on strength enhancement and optimum temperature was found as 60°C for best quality biopolymer-modified sand. At higher temperatures above 120°C, protein-based biopolymer molecules decomposed and caused the reduction of the strength of the composite sand. The optimum curing time was 7–14 days for casein and sodium caseinate, respectively. The sticky behavior of sodium caseinate showed higher compressive strength compared with cheesy casein due to the uniform distribution of the biopolymers. Schematic interaction model between sand and biopolymers is shown in Fig. 12.6 and attributed that van der Waals and electrostatic charges of casein and sodium caseinate help to adhere and attach with soil molecules, resulting in high-strength mixtures (Fatehi et al., 2018).

Casein is also used in paint industries as a form of caseinates that are prepared from casein by making them neutral with the help of alkali hydroxides. Caseinates are used there as they have two properties together, which is desired for paints. One is the viscous nature of the additive and another one is imparting self-leveling properties to gain a leveled layer of paint, but they

FIG. 12.5 Casein and Sodium Caseinate salt biopolymer from milk (Fatehi et al., 2018).

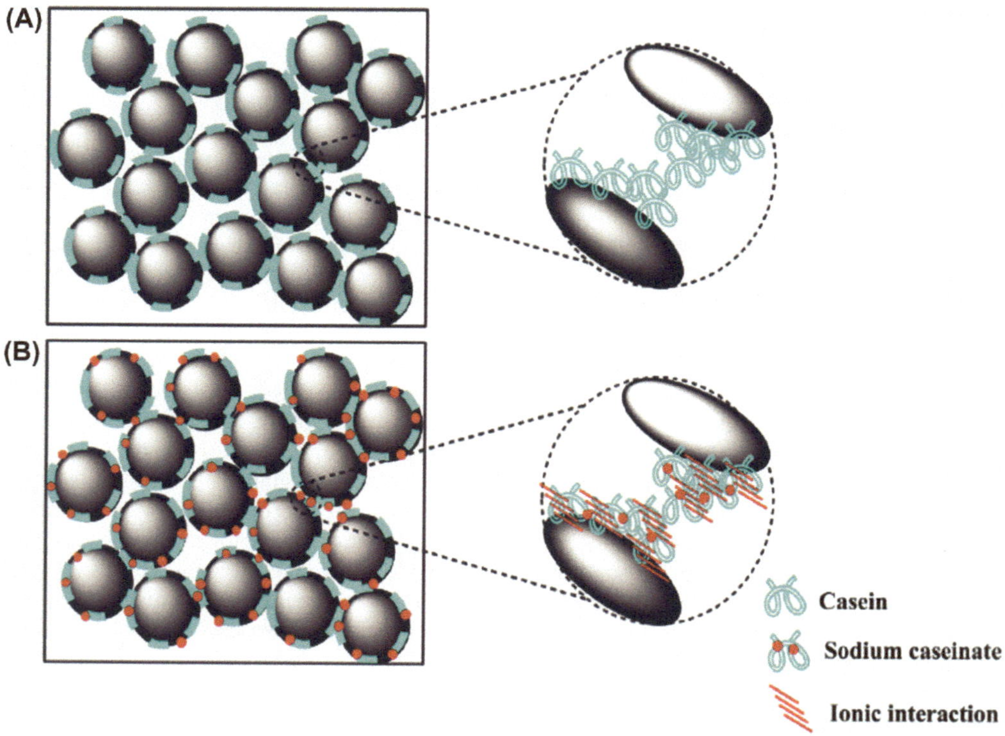

FIG. 12.6 Schematic of dune sand and **(A)** casein and **(B)** sodium caseinate interaction model (Fatehi et al., 2018).

should use it in a short time after completing the preparation. Otherwise, it may create an obnoxious smell and also can be degraded. Though there are some nondegradable synthetic paints available, casein-based paints are biodegradable, nontoxic, and cheaper.

Protein hydrolyzates are produced by baking of animal hair, hides or hoofs, blood etc., especially from cattle in the presence of sulfuric acid or enzymes. It helps in decreasing the surface tension of water remarkably, which makes it suitable for manufacturing foam concrete. For this purpose, foam and concrete are mixed, where foam is with very low density (0.5 kg dm^{-3}). These are mainly used in road construction as a basement via filling ditches and prefabricated dividers during building construction. Protein hydrolyzates produced spherical bubbles that offer 20%–30% higher strength compared with hexagonal bubbles generated from synthetic foamers. Therefore, the construction industry prefers protein-based foamers over synthetic foamers, where it required lightweight concrete with high compressive strength (Plank and Winter 2003). Though it has some disadvantages, like its application process is complex, it is more likely used because of its excellent quality.

6 BIOPOLYMER FROM SOIL

Many biopolymers are extracted from earth crust or soil, which are used in construction such as lignite, which is a kind of coal. It is basically the sediment coal formed in a long time through coalification in the vicinity of the earth's surface (Teichmüller, 1989; Thomas, 2013). Swamp forest plants and residues did not undergo aerobic degradation when flooded and initiated lignite formation via coalification. With long time accumulation and partial degradation via geological processes, plant sediments turned into lignite with the release of methane and/or carbon dioxide gas (Hatcher and Clifford, 1997). Four different types of coal such as lignite, subbituminous coal, bituminous coal, and anthracite are form based on process conditions (Ilg and Plank, 2017). Lignite is dark brown and has the lowest amount of carbon and the highest amount of water in all forms of coal. Chemically, it is a heterogeneous complex mixture with a wide range of organics such as bitumen, charred lignin, cellulose residues, humic substances, and inorganic components. Lignite is therewithal degradation of humic acid. Humic acid is a constituent of graft polymer using in oil well cementing. A process of grafting 2-acrylamido-2-methylpropane sulfonic acid, *N*, *N*-dimethylacrylamide, and acrylonitrile monomers on a humic acid backbone was patented by Giddings and Williamson (1987). By this method, excellent filtration-controlled property of graft polymer was seen in well cementing. M. Ilg. et al. developed a superplasticizer on lignite with acrylic acid and 2-acrylamido-2-tert butyl sulfonic acid graft copolymer shown in Figs. 12.7 and 12.8 (Ilg and Plank, 2017). They used the graft copolymerization method to make the specific modification of lignite, which helps lignite backbone to attach the monomers successfully. The copolymer exhibited excellent dispersing force, high slump retention time, and high sulfate resistance properties. This approach shows a road map to apply less expensive, abundant, and diverse lignite biopolymer as raw materials of concrete admixtures. The conclusion stated that the synthesis of lignite biopolymer-based superplasticizer could be a pathway to fabricate low-cost concrete admixtures for geotechnical constructions.

Lignin is mostly used in oil well drilling as a fluid additive. Previously, they are only considered as a diluent for water-based muds, but now they are used for oil emulsification, filtrate remediation, and balance of mud properties against the high-temperature environment. When it is mixed with lignosulfonate, it creates a very good quality water-based boring liquid system named CL-CLS mud, which is extensively used in the United States, which contains a high amount of bentonite (3%–10%) in which both lignite and lignosulfonate dispersed the solids. At temperatures up to 225°C, the chromium salts gave optimum stability (Darley and Gray, 1988). From the 1980s, CL-CLS muds have replaced cellulose and synthetic-based polymers inchmeal. But chromium has a bad effect on nature as it is toxic, so it became a problem. However, this difficulty was overcome by using zirconium salts or synthetic polymers, which are not as toxic as chromium salts for high temperatures. In case of low temperature, they are still used by vitrified lignite fly ash for the construction of a plastron plasma reactor. Other than those, Huddleston and Williamson developed water-based boring liquid and oil-based cement slurries based on vinyl lignite (Huddleston and Williamson, 1990). The fabricated composite significantly decreases the loss of liquid during drilling fluid aged at 350°F. Recently, the application of lignite-based biopolymers in oil drilling has declined considerably, but still, approximately 50,000 tons are used every year by industry.

7 XANTHAN GUM

Sticky xanthan gum is one of the microbial biopolymers that exhibits excellent potential over traditional binders in geotechnical constructions (Cabalar et al., 2017; Chang et al., 2016, 2015; Lee et al., 2017). The microorganism *Xanthomonas campestris* produces xanthum gum, which is an extracellular heteropolysaccharide comprising of (1,4)-β-D-glucopyranose with a trisaccharide side chain on every other glucose residue linked via C3 position (shown in Fig. 12.8 (Pacheco-Torgal et al., 2016). Xanthan gum is the first microbiological polymer used as a building material.

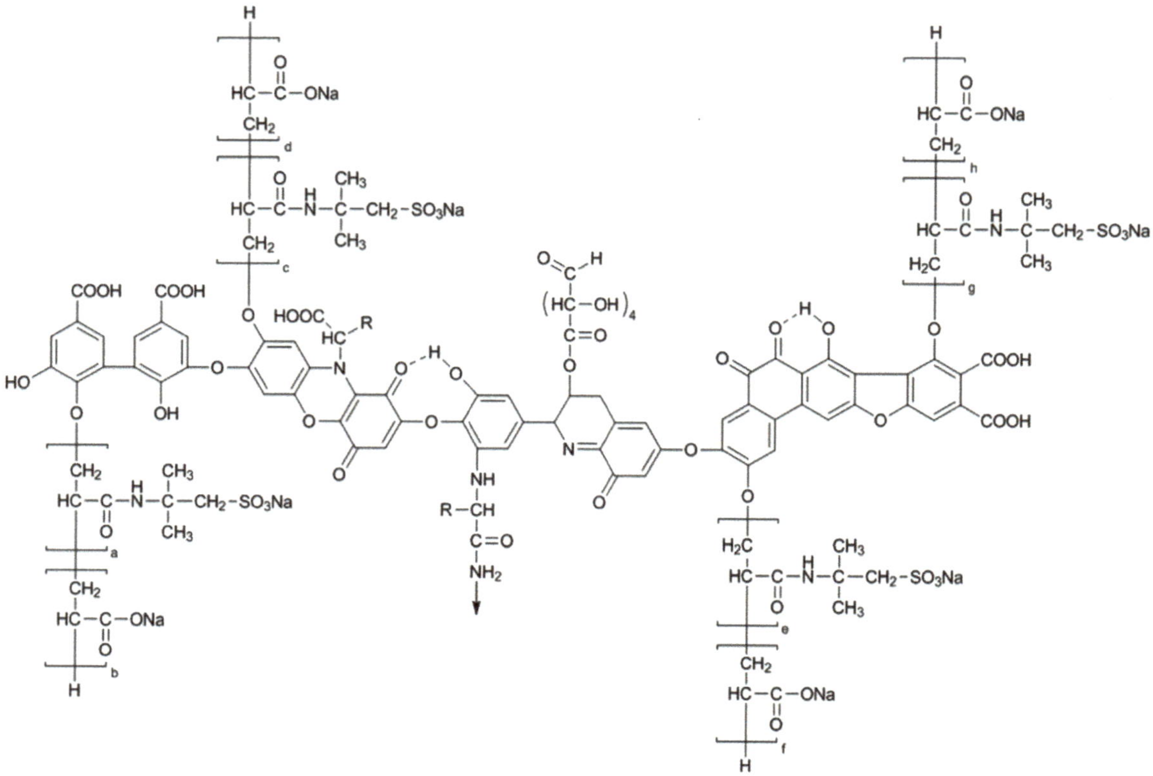

FIG. 12.7 Proposed Chemical structure of lignite {ATBS-co-AA} graft copolymer (Ilg and Plank, 2017).

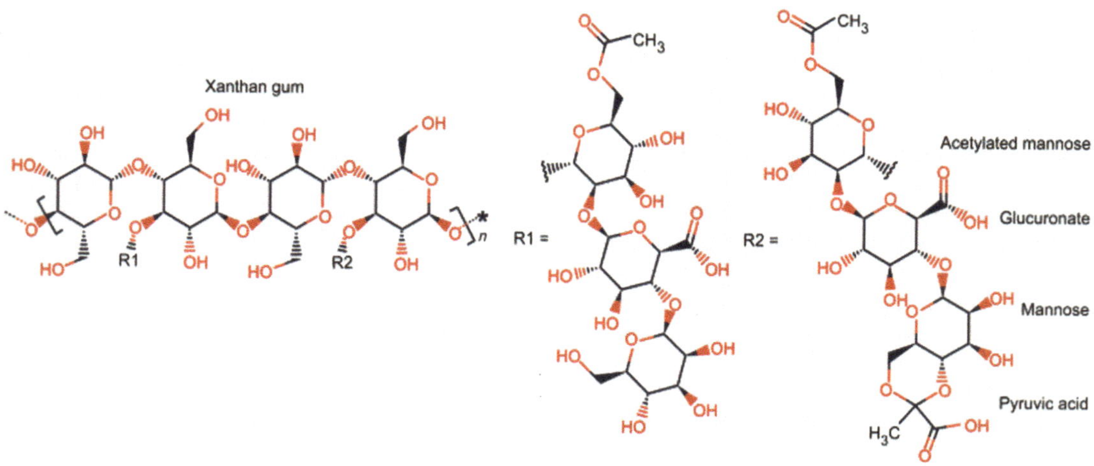

FIG. 12.8 Chemical structure of Xanthan Gum (Pacheco-Torgal et al., 2016).

For construction engineering, low plastic viscosity and high yield point are needed. If cellulose or starch-based biopolymers are dissolved in water, they will enhance the plastic viscosity, whereas the aqueous solution of xanthan gum decreases plastic viscosity and increases yield point. In another case, it can give a different type of viscosity for different cases. At the time of pumping through hole, the speed of the viscosity should be low for drilling fluids, and for closed circulation, high viscosity is needed. Xanthan gum

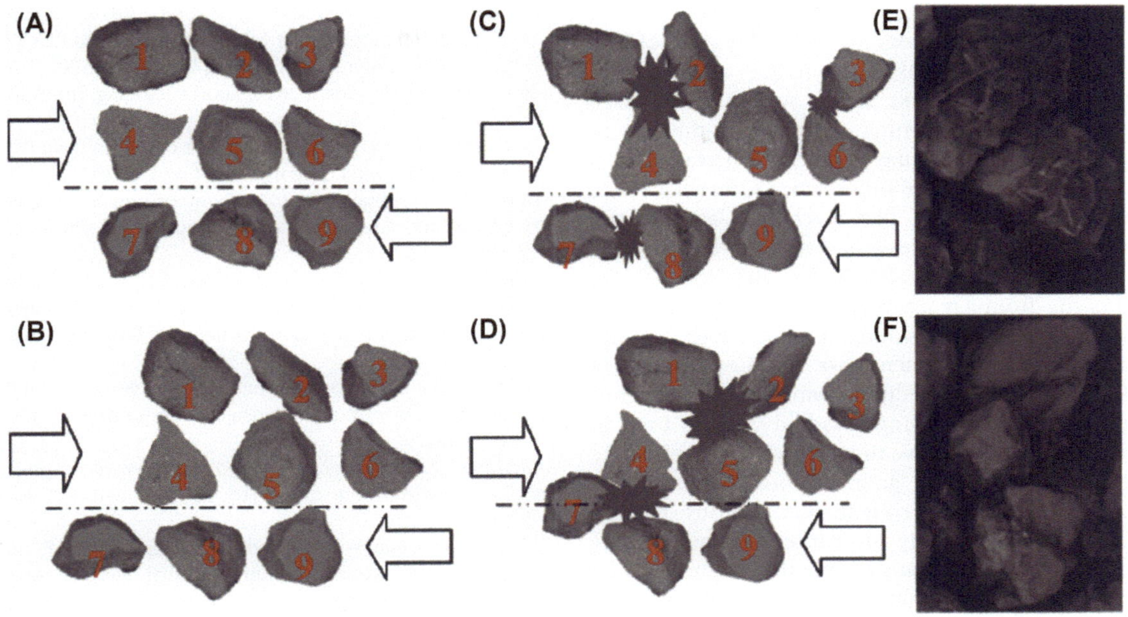

FIG. 12.9 Schematic of sand motion during shearing: **(A)** the pristine condition of clean sand without shearing, **(B)** the movement of clean sand during shearing, **(C)** the pristine condition of biopolymer-modified sand without shearing, **(D)** the movement of biopolymer-modified sand during shearing, and **(E, F)** different connections of biopolymer-modified sand (Chen et al., 2019).

shows this property accurately. If cellulose or starch-based biopolymers are dissolved in water, they will enhance the plastic viscosity, whereas aqueous solution as well as coarse particles may settle down at the bottom part of the screed, resulting in inhomogeneous products with poor surface smoothness and low compressive strength. Xanthan gum (0.03%–0.1% weight of binder) is probably infused to the dry formulation to avoid the bleeding problem and get homogeneous floor screed. Xanthan gum also enhances plaster adhesion, so it is also used as a wall plaster. Curing time also plays a vital role during biopolymers' modified soil, as discussed earlier, where cohesion dominates. C. Chen et al. (2019) studied the strength of xanthan adapted soil at varying drying conditions. They investigated the interaction between xanthan gum and soil molecules at different water content. Results showed that xanthan-modified soil with less than 33% of water content exhibited higher strength. This can be attributed due to strong interaction and bonding between xanthan gum and soil molecules. Complete water evaporation with increasing temperature, xanthan gum exhibited the highest strength but became brittle, which reduced cohesion with soil.

The hypothesized motion and distribution of clean and biopolymer-modified sand is depicted in Fig. 12.9 with SEM images. It showed that the neat sand had no cohesion. Due to the strong interaction between biopolymer-modified sand particles they exhibited higher shear strength after biopolymer treatment of sand. They also found the small sand particles made aggregates due to attraction and helped to improve sand strength as shown in Fig. 12.9E and F. The illustration also suggested that in dry condition, the biopolymer-modified sand particles show different ways of movement and rotation under shearing, resulting in the variation of shear strength of modified.

8 CONCLUSIONS

Polymer concrete or bioadmixtures have achieved tremendous attention in geotechnical construction and building industries. Biofeedstock origin admixtures exhibited significant advantages over synthetic-based admixtures. In some cases, inhabitants encountered sick house syndrome due to toxic emission and leaching of synthetic products from synthetic-based admixtures. If high strength and durability are the main issues, polymer concrete is suitable to use there. Methods for the successful production of polymer concrete, as well as their various applications in building materials, are described in this chapter. Lignosulfonate is also used as a building material. But, use of it will be decreased in the upcoming years because of the shortage of raw material (sulfite liquor) of lignosulfonate. On the other hand, starch- and cellulose-based biopolymers are becoming attractive as the use of highly developed building products is increasing, and they require this kind of biopolymer. Other than that, advanced technologies like machine plastering use these biopolymers instead of conventional constructional materials nowadays, which also changes the approach from manpower to more automated machine-based production. Casein and sodium caseinate salt biopolymers have also been used as a building material to reduce environmental concerns. Xanthan gum is satisfying as additives. It is used in soil treatment to improve the quality of the soil before use in construction. As "go green" becomes an important issue, the future construction industry will be occupied by high-quality admixtures for which people should not sacrifice nature, and the output will be completely nontoxic and safe. So, biopolymers will contribute to the construction industry as a principal material.

REFERENCES

Akindahunsi, A.A., Uzoegbo, H.C., 2015. Strength and durability properties of concrete with starch admixture. International Journal of Concrete Structures and Materials 9 (3), 323–335.

Akindahunsi, A.A., 2019. Investigation into the use of extracted starch from cassava and maize as admixture on the creep of concrete. Construction and Building Materials 214, 659–667.

Bulut, H.A., Şahin, R., 2017. A study on mechanical properties of polymer concrete containing electronic plastic waste. Composite Structures 178, 50–62.

Burris, K.P., 2004. Antimicrobial Activity of Trypsin and Pepsin Hydrolysates Derived from Acid- Precipitated Bovine Casein.

Cabalar, A.F., Wiszniewski, M., Skutnik, Z., 2017. Effects of xanthan gum biopolymer on the permeability, odometer, unconfined compressive and triaxial shear behavior of a sand. Soil Mechanics and Foundation Engineering 54 (5), 356–361.

Chang, I., Im, J., Cho, G.-C., 2016. Geotechnical engineering behaviors of gellan gum biopolymer treated sand. Canadian Geotechnical Journal 53 (10), 1658–1670.

Chang, I., Im, J., Prasidhi, A.K., Cho, G.-C., 2015. Effects of Xanthan gum biopolymer on soil strengthening. Construction and Building Materials 74, 65–72.

Chen, C., Wu, L., Perdjon, M., Huang, X., Peng, Y., 2019. The drying effect on xanthan gum biopolymer treated sandy soil shear strength. Construction and Building Materials 197, 271–279.

Çoban, O., Bora, M.Ö., Kutluk, T., Özkoç, G., 2018. Mechanical and thermal properties of volcanic particle filled PLA/PBAT composites. Polymer Composites 39 (S3), E1500–E1511.

da Cruz, Rui, M.S. (Eds.), 2019. Food Packaging: Innovations and Shelf-Life. CRC Press.

Dahlberg, A.O., 1918. The Manufacture of Casein from Buttermilk or Skim Milk. Department of Agriculture, US.

Darley, H.C., George, R.G., 1988. Composition and properties of drilling and completion fluids. Gulf Professional Publishing.

Fatehi, H., Abtahi, S.M., Hashemolhosseini, H., Hejazi, S.M., 2018. A novel study on using protein based biopolymers in soil strengthening. Construction and Building Materials 167, 813–821.

Fowler, D.W., 1999. Polymers in concrete: a vision for the 21st century. Cement and Concrete Composites 21 (5–6), 449–452.

Giddings, D.M., Charles, D.W., 1987. Terpolymer composition for aqueous drilling fluids. U.S. Patent 4, 678, 591, issued July 7.

Global Status Report 2017. n.d. Retrieved August 19, 2019, from World Green Building Council website: https://www.worldgbc.org/.

GSP, U. n.d. United Nations Secretary-General's High-Level Panel on Global Sustainability (2012): Resilient People, Resilient Planet. A Future Worth Choosing. New York: UN GSP.

Hatcher, P.G., Clifford, D.J., 1997. The organic geochemistry of coal: from plant materials to coal. Organic Geochemistry 27 (5–6), 251–274.

Herczeg, M., McKinnon, D., Milios, L., Bakas, I., Klaassens, E., Svatikova, K., Widerberg, O., 2014. Resource Efficiency in the Building Sector: Final Report to DG Environment. European Commission, Brussels, Belgium. See. http://ec.europa.eu.

Hoque, M.E., A, W.M., G, D.J.M., L, C.Y., 2019. Polycaprolactone (PCL) based synthetic biopolymers for modern scaffold-based tissue engineering. Int. J. Appl. Sci. - Res. Rev. https://doi.org/10.21767/2394-9988-C2-006.

Huddleston, D.A., Charles, D.W., 1990. Vinyl grafted lignite fluid loss additives. U.S. Patent 4, 938, 803, issued July 3.

Ilg, M., Plank, J., 2017. A novel kind of concrete superplasticizer based on lignite graft copolymers. Cement and Concrete Research 79, 123–130. https://doi.org/10.1016/j.cemconres.2015.09.004.

Lee, S., Chang, I., Chung, M.-K., Kim, Y., Kee, J., 2017. Geotechnical shear behavior of xanthan gum biopolymer treated sand from direct shear testing. Geomechanics and Engineering 12 (5), 831–847.

Li, Y., Zhang, Y., Ceylan, H., Kim, S., 2019. Laboratory evaluation of silty soils stabilized with lignosulfonate. In: Airfield and Highway Pavements 2019: Testing and Characterization of Pavement Materials. American Society of Civil Engineers, Reston, VA, pp. 531–540.

Mindess, S., 2008. Developments in the Formulation and Reinforcement of Concrete, vol. 263.

Navarrete, R.C., Seheult, J.M., Coffey, M.D., 2001. New biopolymers for drilling, drill-in, completions, spacer, and coil-tubing fluids, part II. In: SPE International Symposium on Oilfield Chemistry. Society of Petroleum Engineers.

Pacheco-Torgal, F., Ivanov, V., Karak, N., Jonkers, H., 2016. Biopolymers and Biotech Admixtures for Eco-Efficient Construction Materials. Woodhead Publishing.

Plank, J., Winter, C., 2003. Adsorption von Fließmitteln an Zement in Gegenwart von Verzögerern. GDCh Monogr 27, 55–64.

Plank, J., 2004. Applications of biopolymers and other biotechnological products in building materials. Applied Microbiology and Biotechnology 66 (1), 1–9.

Rangari, V.K., Manik, C.B., Boniface, J.T., 2019. Biodegradable polymer blends for food packaging applications. In: Food Packaging: Innovations and Shelf-Life, p. 151.

Reis, J.M.L., 2011. Effect of aging on the fracture mechanics of unsaturated polyester based on recycled PET polymer concrete. Materials Science and Engineering: A 528 (6), 3007–3009.

Shonaka, M., Kitagawa, K., Satoh, H., 1997. Chemical structures and performance of new high-range water-reducing and air-entraining agents. Special Publication 173, 599–614.

Spiratos, N., Pagé, M., Mailvaganam, N.P., Malhotra, V.M., Jolicoeur, C., 2003. Superplasticizers for Concrete: Fundamentals, Technology, and Practice.

Ta'negonbadi, B., Noorzad, R., 2017. Stabilization of clayey soil using lignosulfonate. Transportation Geotechnics 12, 45–55.

Ta'negonbadi, B., Noorzad, R., 2018. Physical and geotechnical long-term properties of lignosulfonate-stabilized clay: an experimental investigation. Transportation Geotechnics 17, 41–50.

Teichmüller, M., 1989. The genesis of coal from the viewpoint of coal petrology. International Journal of Coal Geology 12 (1–4), 1–87.

Thomas, L., 2013. Coal Geology. Wiley Online Library.

Vakili, A.H., Ghasemi, J., bin Selamat, M.R., Salimi, M., Farhadi, M.S., 2018. Internal erosional behaviour of dispersive clay stabilized with lignosulfonate and reinforced with polypropylene fiber. Construction and Building Materials 193, 405–415.

Wahab, A., Islam, N., Hoque, M.E., Young, D.J., 2018. Recent advances in Silver Nanoparticle containing Biopolymer Nanocomposites for infectious disease control – a mini review. Current Analysis Chemistry 14, 198–202.

Wool, R.P., and Sun X.S. Bio-based polymers and composites. 2005. New York: Elsevier. Bledzki, AK, and J. Gassan, Progress in Polymer Science 24 (1999): 221–274

Zhang, D.F., Ju, B.Z., Zhang, S.F., He, L., Yang, J.Z., 2007. The study on the dispersing mechanism of starch sulfonate as a water-reducing agent for cement. Carbohydrate Polymers 70 (4), 363–368.

Zhang, T., Songyu, L., Guojun, C., Anand, J.P., 2015. Experimental investigation of thermal and mechanical properties of lignin treated silt. Engineering Geology 196, 1–11.

CHAPTER 13

Biopolymers for Sustainable Packaging in Food, Cosmetics, and Pharmaceuticals

TILOTTOMA SAHA • MD ENAMUL HOQUE • TARIQ MAHBUB

1 INTRODUCTION

1.1 Biodegradable Plastics

Plastic materials that under certain environmental conditions and with the help of living organisms degrade to become completely natural elements like carbon dioxide, water, and compost are called biodegradable plastics. Biodegradable plastics can be of a natural origin, bio-based, or produced from fossil resources (Chauhan and Chauhan, 2015). Different biodegradable plastics used in food, cosmetics, and pharmaceutical industry are shown in Fig. 13.1.

1.2 Bioplastic

It is a plastic that is prepared from natural raw materials (such as sugar, starch, cellulose, potatoes, cereals, molasses, soybean oil, corn, etc.). It is prepared using a sustainable process, and it may or may not be biodegradable (Byun and Kim, 2013).

1.3 Compostable

A material is compostable when it biodegrades within a certain time under specific conditions that define composting. Composting is a process carried out by humans, advancing the natural processes of biological decomposition and giving products like carbon dioxide, water, and mineral fertilizer (compost). A compostable material should meet with European standard EN 13432 (US ASTM D6400-04), which states decomposition of 90% or greater of the initial quantity in less than 6 months (Cinelli et al., 2019).

1.4 Biopolymers

Biopolymers are polymers that are produced in living organisms. A large number of monomeric units are bonded covalently to form a large structure (Franceschi et al., 2014; Jipa et al., 2012; Paridah et al., 2016). The reason why biopolymers got high attention in food, cosmetic, and pharmaceutical sectors are the following:

(1) Biopolymers have vast diversity and a small modification in mechanical properties that can pave the way of packaging and preservation in those sectors.
(2) Biopolymers can be sustainable, carbon neutral, and renewable. Their sources are agricultural nonfood crops. So, environmentally, soundness makes them a reliable source in these sectors.
(3) Biopolymers are biodegradable and some are also compostable (90% breakdown within 6 months). So, if they are used, it can prevent synthetic polymer-causing environmental pollution.

2 BIOPOLYMERS IN FOOD PACKAGING

Biopolymers are directly derived from plant or animal biomass in the form of polysaccharides and protein. Polysaccharides are found during the growth cycle of all organisms, and protein is an important component of every cell in the body. Both of them are used to produce biodegradable materials (Chauhan and Chauhan, 2015; Kaplan, 1998; Larotonda et al., 2016). Biodegradable polymers can be divided into three different categories in the food packaging sector as presented in Table 13.1.

Packaging materials can be manufactured from biopolymers like proteins, lipids, polysaccharides, or sometimes a combination of them. It has high potential to replace presently available synthetic plastics (Balasubramanian et al., 2009; Lagarón et al., 2016; Lazic and Hromiš, 2018; Liu, 2006; Neo et al., 2013; Peelman et al., 2013; Pellicer et al., 2017; Tang et al., 2012;

Advanced Processing, Properties, and Applications of Starch and Other Bio-Based Polymers. https://doi.org/10.1016/B978-0-12-819661-8.00013-5

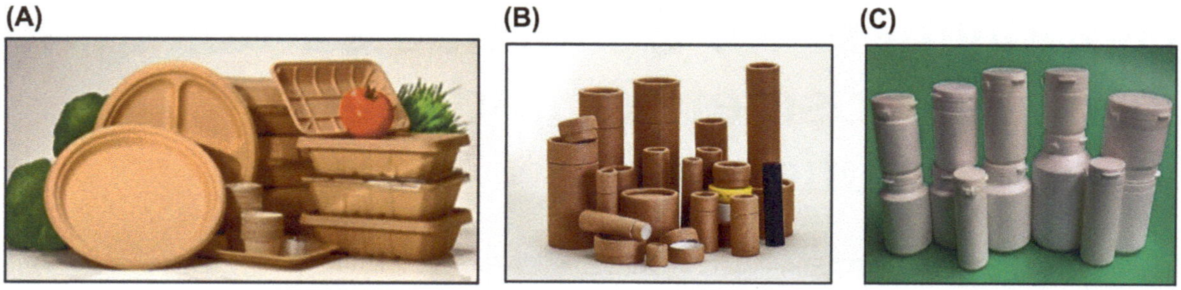

FIG. 13.1 Biodegradable packaging used in **(A)** food industry, **(B)** cosmetics and toiletries industry, and **(C)** pharmaceutical industry.

TABLE 13.1
Biopolymers for Food Packaging.

Polyxsaccharides	Protein	Aliphatic Polyester
• Alginate	• Gelatin	• Polylactic acid (PLA)
• Cellulose	• Collagen	• Polyhydroxybutyrate (PHB)
• Carrageenan	• Soy protein	
• Chitin/chitosan	• Whey protein	
• Gellan	• Zein	
• Pullulan		
• Pectin		
• Starch		
• Xanthan		

Tolstoguzov, 2000; Vartiainen et al., 2016). Food packaging made from a biopolymer can protect food items from any kind of influence through the process of increasing time for food decaying, controlling, and extending the shelf life of foods (Gabor Daniela, 2012).

2.1 Solid/Dry Food Packaging

Solid or dry food products are especially sensitive to moisture and must be protected against remoistening to avoid caking, loss of crispiness, or stickiness. Due to their high sensitivity to water, monolayered or unmodified biopolymers are not suitable for dry food packaging. Biopolymers such as polylactic acid (PLA), polyhydroxyalkanoates (PHAs), and their derivatives display the lowest water vapor permeability compared with expensive agro-based polymers such as wheat gluten, starch, and their derivatives. A good alternative can be to combine biopolymers and low-cost agropolymers (i.e., fiber) in order to obtain bilayer or composite films with lower sensitivity to water and reduce cost. Biopolymers are already widely used for packaging dry food goods with a midterm or long-term shelf life (e.g., flour, coffee grain, bakery, and pastry products). The biopolymers currently used for such applications are mainly based on PLA, paper, and other cellulose-based fibers. More development of biodegradable packaging is needed to protect dry or oxidation-sensitive products against oxygen transfer from the surrounding atmosphere.

2.2 Liquid Food Packaging

For the liquid food packaging, the biopolymer should have similar resistance, barrier, inertia, long-term stability, and transparency as like conventional plastic such as polyethylene terephthalate (PET). Until now, only the PLA and its composites are able to meet these specifications for liquid food packaging. PLA made bottles are used for water, soda water, coke, organic milk, etc., containment. There is also a great interest in the eco-friendly drink bottle from the consumer for reducing landfilling. Unfortunately, that type of product has some limitations. PLA needs to be degraded for industrial composting. Moreover, the heat resistance of the PLA bottle is not at a satisfactory level. For these

TABLE 13.2
Commercial Biopolymer Packaging for Solid and Liquid Foods (Ramos et al., 2018).

Materials	Applications	Food Examples
PLA	Thermoformed films and trays (transparent)	Fruits and vegetables (strawberries, lettuce, peppers, etc.)
	Bags	Bread, tea bags
	Cups (PLA-based foam)	Hot drinks (coffee and tea from vending machines)
	Cups and bottles	Dairy products (e.g., milk, custard, cheese, yogurt, etc.)
	Trays	Ice cream
PHA	Trays and films	Fresh meat and frozen foods, etc.
Starch	Semitransparent film	Fruits and vegetables (e.g., apple, carrots)
	Covering films with barrier function	Meat, fish, cheese
	Containers	Confectionary (plastic)
Cellulose	Cellophane	Fruits and vegetables (tomatoes, bell peppers, etc.)
	Covering films with barrier function (cellophane)	Meat, fish, cheese
	Packaging films	Coffee bean, bread, butter, etc.

reasons, plant-based PET or its mixture derivatives may have a better potential for such packaging. Various commercial biopolymer packaging materials along with their respective applications for solid and liquid foods are presented in Table 13.2.

2.3 Polysaccharides for Food Packaging

2.3.1 Alginate

Alginate is an indigestible polysaccharide that is produced naturally and generally collected from brown algae. The most commonly used algae are *Laminaria hyperborean*, *Macrocystic pyrifera*, *Ascophyllum nodosum*, etc. Apart from them, *Azotobacter vinelandii* of *Pseudomonas aeruginosa* is also able to produce alginate in the form of exopolysaccharide. Alginate molecular structure is unbranched and made of linear binary copolymers of β-D-mannuronic acid (M) and α-L-glucuronic acid (G) residues bonded by 1,4-glycosidic chain as shown in Fig. 13.2. An algal alginate structure has three uronic acid blocks. These are the homopolymeric region of M and G blocks with alternating MG blocks containing both polyuronic acids. Generally, O-acetyl groups are found in bacterial alginates, which are absent in the algal alginates. Moreover, bacterial alginates have higher molecular weights than algal polymers.

Due to improved quality and extended shelf life of fruits, vegetables, meat, poultry, cheese, and seafoods, alginate-based edible films are used extensively. These types of films gained attention due to its reduced dehydration rate, controlled respiration rate, enhanced product appearance, and improved mechanical properties (Lagarón et al., 2016; Tolstoguzov, 2003).

2.3.2 Carrageenan

Carrageenan is another excellent natural edible biopolymer candidate that has good film-forming abilities. This family of natural polymers usually extracted from red seaweeds and sometimes is cultivated especially in tropical countries. Carrageenan is frequently used in the food industry for the property of forming emulsions and gels to stabilize fat. Alternating 3-linked-β-D-galactopyranose and 4-linked-α-D-galactopyranose units form the molecular structure of carrageenan (Fig. 13.3).

There are three types of carrageenans such as κ-carrageenan, ι-carrageenan, and λ-carrageenan. *Kappaphycus alvarezii* algae are the source of κ-carrageenan, *Eucheuma denticulatum* is the source of ι-carrageenan, and finally, *Gigartina pistillata* or *Chondrus crispus* algae are the sources of λ-carrageenan. However, "hybrid" carrageenans are extracted from various types of algae that are quite uncommon and which can be considered as composite materials because their molecular structure is based on different types of carrageenans (ι- and κ-carrageenans) and structured in blocks (Larotonda et al., 2016). Pure κ- and ι-carrageenans have gelling ability to form high-quality films. Carrageenan-derived coatings are used for producing packaging for different types of foods. As a packing material, carrageenans produce antimicrobial agents for reducing the dehydration, oxidation, and degradation of food (Larotonda et al., 2016).

Hybrid carrageenans have different gel-forming capabilities than pure carrageenan (between pure ι- and pure κ-carrageenan). The film-forming solution

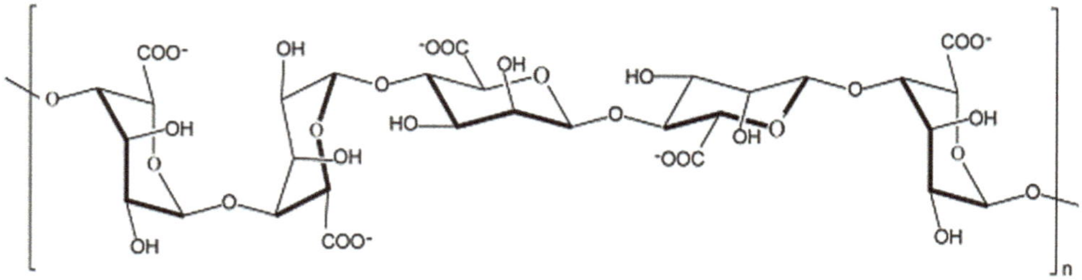

FIG. 13.2 Chemical structure of alginate.

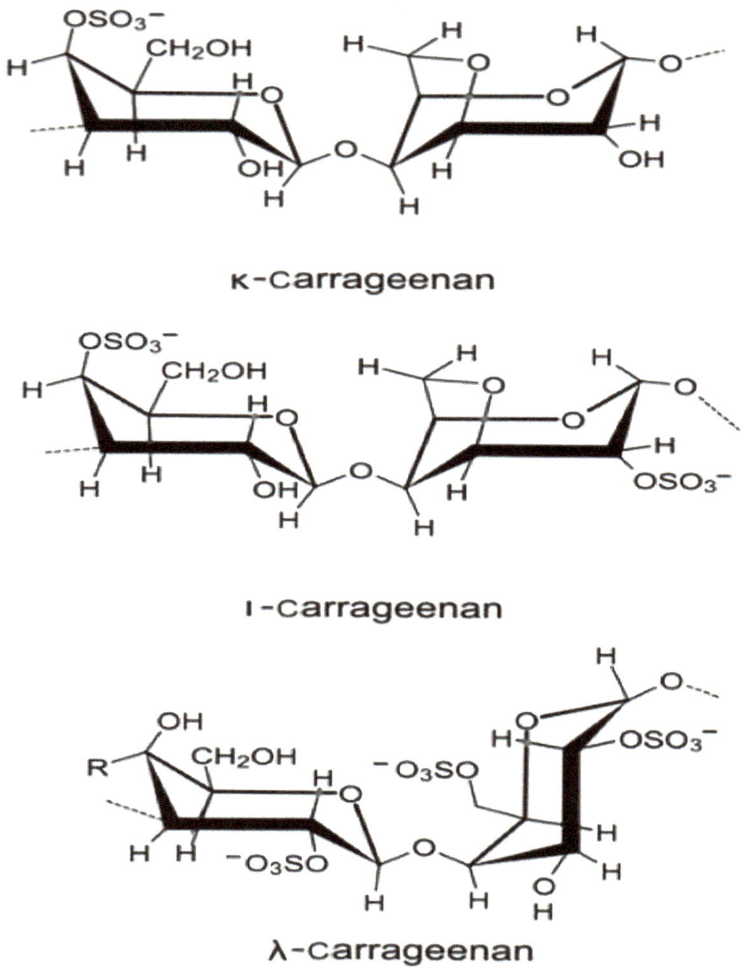

FIG. 13.3 Chemical structure of carrageenan.

produced from hybrid carrageenan can show rheological properties that are absent for only κ carrageenan-derived solutions (Tolstoguzov, 2003). Films of hybrid carrageenan have the following properties:

- Homogeneous, flexible, thin, and transparent smooth surfaces
- Enhanced UV barrier (compared with commercial carrageenan), higher water vapor permeability (hydrophobic)

These characteristics make hybrid carrageenans a unique material for bio-based coatings and biodegradable films.

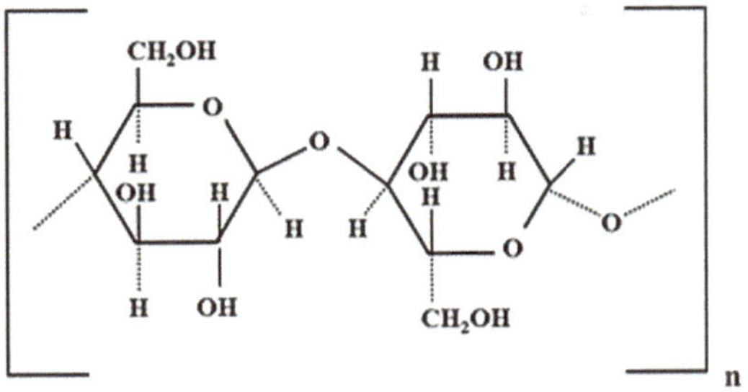

FIG. 13.4 Chemical structure of cellulose.

2.3.3 Cellulose

Cellulose is a member of the polysaccharides family and made of millions of β(1 → 4) linked D-glucose units as shown in Fig. 13.4. It is one of the most abundant organic polymers on earth and is a vital component of green plants, algae, and the oomycetes primary cell wall. There are some of the bacteria that can secrete it to form biofilms.

Materials based on biopolymers are rapidly gaining interest in recent years for the role of plastic alternatives and the development of sustainable materials for packaging. Basically, materials that are derived from plants such as cellulose are recently getting attention for its active biodegradability. The cellulose can be derived sustainably from a range of sources including kenaf fiber, wood pulp, and many more. It can possess up to 99% of bio-based material (Ketabchi et al., 2016). Cellulose films have widespread applications in the food industry as edible packaging materials. It has also gained its popularity due to its easy availability and abundance in nature. The main thing to maximize the shelf-life of cellulose packaging films is to avoid exposure to humidity and excessive heat, moisture, and light. These biofilms are also very stable when they are splashed with water. These films are very much suitable for dried fruits, biscuits, rice, dried beans, chocolates, bars of soap, wrapping prints or flowers, and so on. But they are not suitable for cold storage and not suitable for containing wet food products because it is not waterproof (Cinelli et al., 2019; Gabor Daniela, 2012).

2.3.4 Chitin/chitosan

Chitin is known as the second most abundant polysaccharides in nature just after the cellulose. It is safe, biocompatible, environment-friendly, biofunctional, and possesses antimicrobic attributes. Chitosan can also incorporate functional substances like minerals or vitamins and can show antibacterial property that makes it preferred over other bio-based packing materials. This is why chitosan is used widely for packaging material owing to the quality presentation of a variety of foods (Cristina et al., 2015; Shipra et al., 2008; Tripathi et al., 2008). Chitosan is mainly derived from the exoskeleton of crustacean shells (found in food waste). It has a linear polysaccharide chain consisting of (1,4)-linked 2-amino-deoxy-β-D glucan and deacetylated derivative of chitin as shown in Fig. 13.5.

It can possibly be modified into gels, films, fibers, sponges, beads, or nanoparticles (NPs) (Franceschi et al., 2014; Kaplan, 1998; Marsh and Bugusu, 2007; Shipra et al., 2008). When the covering material possesses antimicrobial activity, it can limit or stop the microbial growth period extending the lag period or reducing the growth rate of microorganisms. The antimicrobial packaging system has three primary purposes. These are (A) assuring safety, (B) maintaining quality, and (C) extending shelf life. Antimicrobial chitosan packaging can extend the food shelf life and thus improve the quality of the foods. Additionally, by combining chitosan films with other film-forming materials, its functional properties can be improved greatly

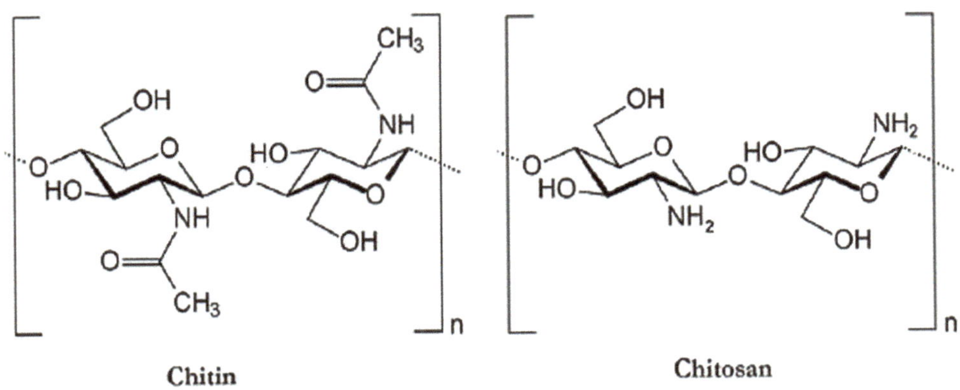

FIG. 13.5 Chemical structure of chitin and chitosan.

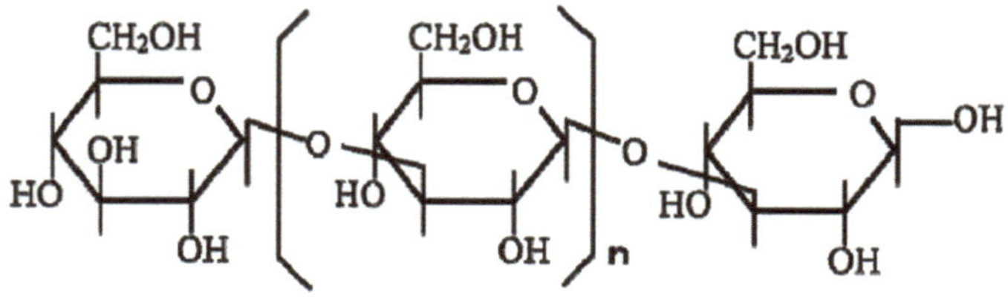

FIG. 13.6 Chemical structure of curdlan.

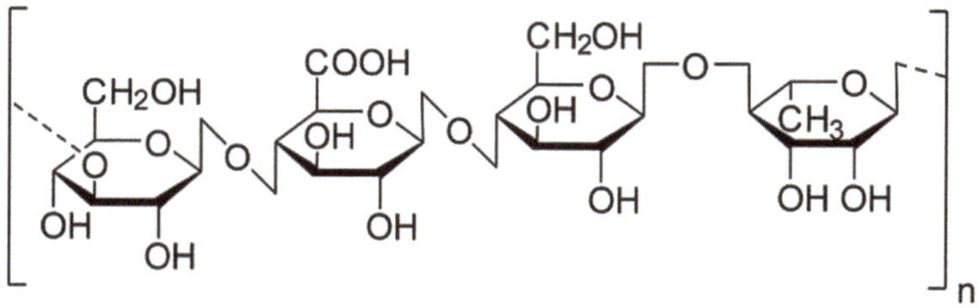

FIG. 13.7 Chemical structure of gellan gum.

(Berekaa, 2015; Chauhan and Chauhan, 2015; Reddy et al., 2006; Shipra et al., 2008).

2.3.5 Curdlan

Curdlan is used as a thickener, texturizer, and stabilizer in the food industry. It is a water-insoluble linear polysaccharide considered as a nutritional fiber with gelatin properties. Curdlan consists of β-(1 → 3)-linked glucose (Fig. 13.6) residues and can form elastic gels upon heating in aqueous solution. It is mainly produced from *Alcaligenes faecalis* var. myxogenes and nonpathogenic *Agrobacterium* species.

Some modifications of curdlan film (curdlan composite film) can show higher tensile strength, elongation, enhanced viscosity, and good moisture barrier. This film can be used to keep fruits fresh. For the rigidity and stability of the structure, the curdlan gels are used in the food industry for improving the food texture (Gabor Daniela, 2012; Mangolim et al., 2017).

2.3.6 Gellan

Biopolymer has a great role in reducing oil absorption for fried food. Gellan-made edible food films used in food preparation showed a decrease in oil absorption. The reason for its thermal gelling property is that gellan reduces oil absorption. Reducing dehydration, gellan offers less space during frying and provides different food texture at the time of frying. There are some important properties of gellan, which make it a polymer widely used in bioapplication including the food industry, such as biocompatibility, lack of toxicity, biodegradability, stability to an acidic environment and against enzymes, stability at a pH value between 2 and 10, temperature resistance, and film-forming ability that acts as a barrier to oil absorption (Gabor Daniela, 2012; Sapper et al., 2019).

This anionic polysaccharide (gellan) is derived from *Aeromonas (pseudomonas) elodea* bacterium, renamed *Sphingomonas paucimobilis.* Now it is produced commercially from the varieties of *S. paucimobilis.* Gellan gum is a linear tetrasaccharide and has high relative molecular mass (molecular weight), made of about 50,000 residues. Chemical structure of gellan is presented in Fig. 13.7. Alkali treatment is usually done for deesterification of these gums before using in food (Iurciuc et al., 2016).

2.3.7 Pectin

Pectin is a branched heteropolysaccharide, and the structural unit consists of the long chain of galacturonan fragments along with rhamnose, galactose, arabinose, xylose, and other neutral sugars. The matrix is made up of cellulose and hemicellulose, which has a contribution to the cell structure. Pectin has undefined molecular weight like other polysaccharides having several sugar units that are differently methyl esterified. Several distinct polysaccharides have been identified and characterized by the pectin group. The structure of pectin consists of an α-(1-4)-linked D-galacturonic acid polysaccharide backbone (Fig. 13.8). Source and extraction methodology control the preparation of substructural entities. Commercial extraction can cause extensive degradation and alter the neutral property of side chain made up of sugar (Süfer, 2018).

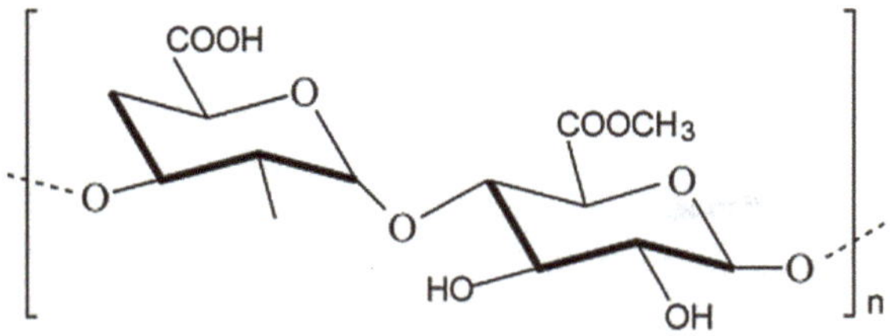

FIG. 13.8 Chemical structure of pectin.

Pectin has a homopolymeric structure consisting of partially 6-methylated and 2- and 3-acetylated poly-D-galacturonic acid residues ("smooth"). There are also substantial "hairy" nongelling areas of alternating α-L-rhamnosyl and α-D-galacturonosyl sections. These areas have branch points with side chains of mainly α-L-arabinofuranose and α-D-galactopyranose (rhamnogalacturonan I), which are mostly neutral. They are almost 1–20 residues in a single neutral side chain.

The sources of pectin are plentiful. Among them, commercial extraction of pectins occurs from citrus- and apple-based fruits. It can also be extracted from several other fruits as well as their by-products of mango peel, sugar beet pulp, sunflower head, soybean hull, *Akebia trifoliata* peel, passion fruit peel, peach pomace, banana peel, and many more (Liu et al., 2017). In food industries, pectin polymers are used for making digestible coatings and films. The films as a natural barrier control the exchange of gases, moisture, liquids, and volatiles between the environment and food. Another benefit of it is to prevent any kind of microbial contamination to the fruits and vegetables. Pectin is a significant renewable natural polymer found as the main component of all the biomass and is plentiful in nature. Pectin derivatives along with pectin alone are used in many bioperishable packaging materials due to their flexibility. Pectin provides moisture, oil, and odor barrier and reduces respiration and oxidation rate for the product. An emulsifier (food grade) is mixed during preparation of edible films. Another application of it is to use for fresh fruits, vegetables, foods, and other food products.

2.3.8 Pullulan

Pullulan is industrially produced by the fermentation of *Aureobasidium pullulans* fungus, which is found commonly from the soil, water, or plant materials. This chemical formula of this polysaccharide is $C_6H_{10}O_5$. It is a maltotriose timer of α-(1 → 6)-linked (1 → 4)-α-D-triglucosides as shown in Fig. 13.9. The intermediate behavior between the amylose and dextran of this compound is due to the coincidence of both α-(1 → 6) and α-(1 → 4) simultaneously in a compound. This special structure gives pullulan its flexibility. Pullulan has a high solubility in water but insolubility in a biological solvent (Farris et al., 2014; Sapper et al., 2019).

Pullulan is nontoxic, nonmutagenic, biodegradable, noncarcinogenic, edible, and nonhygroscopic, decomposes at high temperature(250-280⊠C), has high film-forming ability, is comparatively mechanically strong, and possesses the ability to form thin films, NPs, and nanofibers, and is also used for flexible coating. Sometimes edible pullulan films are coated with antimicrobial compounds for prolonging the shelf life along with for improving the protection of food products. Moreover, pullulan composite films show improved mechanical properties, control the moisture, oxygen, and other gas exchanging rates, resist heat, and provide transparency, which is thereby desirable in the applications for the food packaging systems. The packaging films have antimicrobials properties that are advantageous to meat and poultry products. These antimicrobials show prolonged inhibitory activity and thus allow the limited transfer of the antimicrobial compound from the packaging to the products (Trinetta and Cutter, 2016).

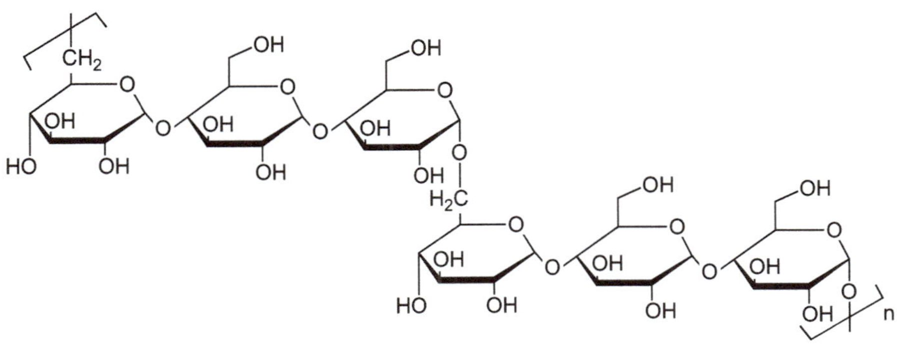

FIG. 13.9 Chemical structure of pullulan.

2.3.9 Starch

Starch is extracted from many agricultural plants such as bean, corn, potatoes, rice, sago, wheat, etc (Hoque et al., 2013). The main constituent of it is more than 60% cereal kernels. Starch has the main benefit that it can be easily separated from other components, and depending on its various botanical source, the granules have different shapes, sizes, structures, and chemical compositions. Starch is made of two essential polysaccharides, namely amylose and amylopectin. They also have a small number of other elements like lipids and proteins (Arcan et al., 2017; Franceschi et al., 2014; Pellicer et al., 2017). The amylose is responsible for the film-forming properties of starch. It is a linear chain polymer of α-(1-4)-anhydroglucose units and its molecular size varies between 200 kg/mol and 800 kg/mol. Its chemical structure is shown in Fig. 13.10. It accounts for about 20%–25% of most granular starches. Amylopectin is a highly branched polymer of short α-(1-4) chain linked by α-(1-6)-glycosidic branch containing 20–30 glucose moieties. It has very high molecular weight (5,000–30,000 kg/mol).

When the gelatinous starch melt is cooled from its hot state, the dispersed amylose molecules are reassociated by the retrogradation process to form flexible gels. These gels have a three-dimensional matrix of the continuous phase of amylose molecules containing the homogenously dispersed filler. The heterogeneous nature of starch is beneficial to industrial use including food packaging sectors. The starch molecule is insoluble in water due to the hydrogen bond. For a good and uniform film-forming solution, starch needs to be gelatinized in excess water (Balasubramanian et al., 2009; Lagarón et al., 2016). The main advantages of the starch-based film in food sectors are as follow:

- Comes from renewable plant resources.
- Does not contain toxins or other harmful compounds.
- Production of this plastic produces much less greenhouse gas emission compared with petroleum-based plastic (68%).
- This starch-based plastic can easily be facilitated for industrial composting.

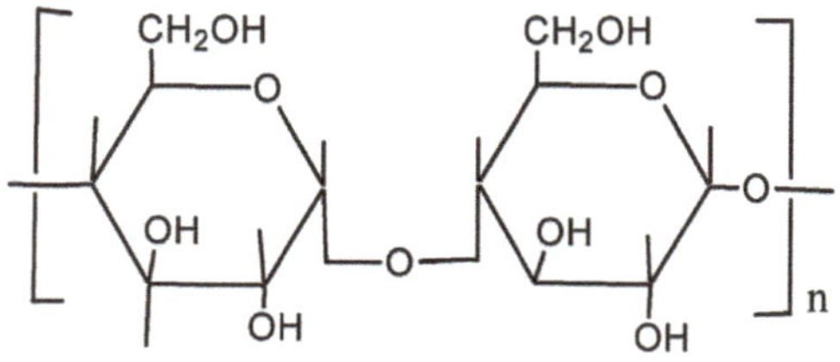

FIG. 13.10 Chemical structure of starch.

Starch-based biofilms possesses good oxygen barrier property that can be exploited to be used as an edible coating for foods (i.e., fruits and vegetables) with a high respiration rate. The films can suppress respiration resulting in delayed oxidation. The poor moisture barrier properties due to hydrophobic nature have a good effect on the coating when washed off after use. Moreover, the other advantages of these inexpensive edible starch-based films are generally tasteless, colorless, and odorless (Chauhan and Chauhan, 2015).

2.3.10 Xanthan

Xanthan gum, an extracellular polysaccharide, has relatively high molecular mass and is commercially extracted from the bacterium *Xanthomonas campestris*. It is known as an important commercial microbial hydrocolloid used in the food industry as a thickener, stabilizer, etc. The chemical structure of xanthan is composed of a linear β-1,4-linked D-glucose chain substituted on every second glucose unit by a charged trisaccharide side chain with gluconic acid residue between two mannose units (Fig. 13.11).

Xanthan can influence the mechanical properties and moisture absorption of cassava starch films. This gum can enhance film traction but can generate less deformable matrix (Ross-Murphy, 1995). The formation of gum-starch hydrogen bonds can interfere with amylose packing and retard the formation of polymer-water hydrogen bonds in the amorphous areas. In this way, it can reduce the water absorption capacity in the films (Sapper et al., 2019).

2.4 Proteins for Food Packaging

2.4.1 Collagen

Collagen is known to be the most commercially successful edible protein film. Collagen has been used in the meat industry for producing edible sausage casings due to its fill forming ability. Collagen is fibrous in nature, and the sources of this protein are animal tissue, skin, bones, and tendons, which comprise about 30% of the total mass of the body. Collagen is easily available, is nontoxic, and can provide an excellent basis for biomaterials. The basic amino acids in collagen are glycine, hydroxyproline, proline, and alanine. The ordered triple helical structure of collagen is stabilized by both intrachain hydrogen bonds and by structural water molecules. Collagen fibrils are produced by self-assembly of collagen molecules in the extracellular matrix, and that is the source of tensile strength to the animal tissue (Lalit Jajpura, 2015; Pellicer et al., 2017). Collagen is used to produce eatable films and coatings from animal proteins, and they can be

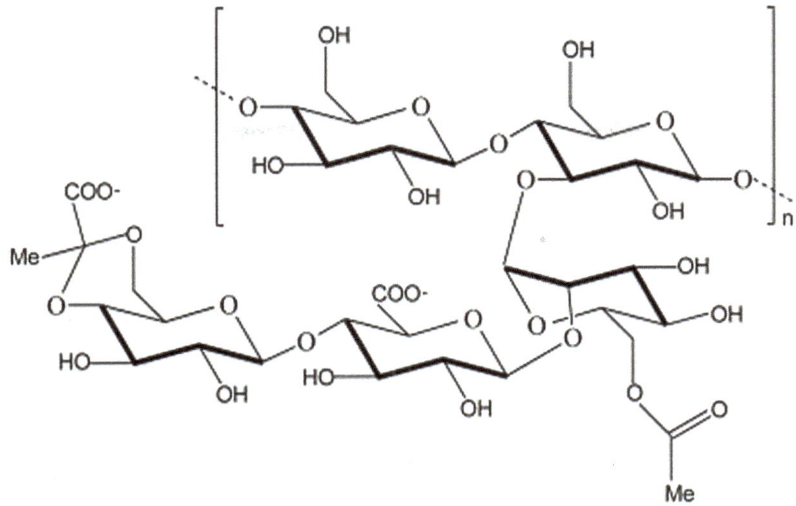

FIG. 13.11 Chemical structure of xanthan.

dissolved in dilute acid, alkali, or natural solutions. Gelatin is produced by the hydrolysis of collagen. Gelatin-based coatings can reduce the migration of oxygen, moisture, and oil. Collagen films at 0% relative humidity can show excellent oxygen barrier property but the permeability increases rapidly with increasing relative humidity. Various types of cross-linking agents have been used for improving the thermal stability of the films. Carbodiimide, microbial transglutaminase, and glutaraldehydrate are usually used as cross-linking agents (Alizadeh and Behfar, 2013).

2.4.2 Gelatin

Besides many biomedical and pharmaceutical applications (Nuge et al., 2017; Nuge et al., 2020), gelatin is utilized for making edible coatings and films. It provides biodegradability as well as good gelling, film-forming properties, and desirable packaging properties. Moreover, various film-forming gelatin sources have different properties. These properties depend on amino acid structure, molecular weight, extraction method, and pretreatment. Usually, gelatins are extracted commercially from different natural sources such as porcine and bovine. The sources of gelatin raw materials are porcine skin (80%), bovine hide (15%), and the rest of gelatin comes from porcine bone, cattle bone, and fish (5%). Nowadays, much attention is paid to gelatin due to its perishable nature, excellent gel-forming ability, and superior film formability. The coating and films of gelatin provide a barrier to gas, water, oil, aroma, and flavor, which are exploited to protect foods from their surrounding environment (Ramos et al., 2016). Other biopolymers ingredients are added during the film and coating formation to improve the packaging film properties as well as provide additional functional properties that lead to versatile applications of gelatin films and coatings. When gelatin films are modified with some ingredients, these films can show high potential in active packaging systems for decreasing oxidative reaction rate, ensuring food safety, controlling microbial growth, and expanding product shelf-life. The other benefit of this intelligent type of packaging system able to provide a quantitative message about the quality of the packaged foods through changing its coloration indicates consumers without damaging the package (Schmitt and Turgeon, 2011).

2.4.3 Soy protein

As a packaging material, soy protein has high potential due to its inherent properties of renewability, biocompatibility, biodegradability, and film-forming capacity. Soy protein has been used since the 19th century in a variety of foods. Emulsification and texturing are the two main characteristics of soy proteins for which they is highly used in the food industry. Soy protein is found in three distinct forms such as soy flour (SF), which is relatively pure, soy protein concentrate, and soy protein isolate (SPI). Among them, SPI contains 70% proteins and 18% carbohydrate, and SF contains about 52% proteins and 32% carbohydrates. SF is used because it is the most inexpensive variety than the others (Lagarón et al., 2016; Schmitt and Turgeon, 2011). Soy protein can be used in the food manufacturing industry as plastics, adhesives, and packing materials and can be a good option of the petroleum-based polymer (Liu et al., 2017; Swain et al., 2004).

2.4.4 Whey protein

This globular protein can be unfolded and cross-linked to a new polymer under certain conditions that makes them a superior ingredient for food packaging films and coatings. Cross-linked protein films are considered more stable and have longer durability than polysaccharide-based counterparts of this protein. Whey protein with suitable functionalities as food packaging materials has gained attention due to its optical and barrier properties compared with existing biopolymers (Schmitt and Turgeon, 2011).

When proteins have different types of globular proteins: about 19% α-lactalbumin(α-La), 57% β-lactoglobulin (β-Lg), 7% bovine serum albumin, 13% several immunoglobulins, and 4% polypeptide proteose-peptone. Among all the protein, β-Lg is responsible for gelation and compaction agent of whey protein as it holds a major portion. Films and coatings of whey protein are very well-known as a replacement for synthetic polymers used for the packaging of food products (Schmid and Müller, 2018).

2.4.5 Zein

Zein is a prolamin group protein found in maize. This hydrophobic protein forms around 45%–50% of corn-based proteins. Other sources of zein are barley (hordein), wheat (gliadin), rye (decalin), and sorghums (kafirin). Four different types of zein are α-, β-, γ-, and δ-zein, in which the most abundant is α-zein in nature (around 80% of whole fractions). Zein is one of the major edibe film-forming biopolymers in the food industry, and in the near future, it is possibly convertible into one of the most extensively produced agroindustrial crops of the world. The most important characteristics of the zein-based biopolymers are

- It has a unique film-forming ability.
- It can be soluble in organic solutions i.e., ethanol.
- It has compatibility with most of the natural antioxidants and antimicrobials.
- It is suitable for gas permeability in the films and coatings and thus makes them useful in fresh fruits and vegetable packaging for modified atmosphere packaging as well as coating purpose (Arcan et al., 2017).

Apart from this, zein is easily employed as a food coat, by brushing or spraying film-forming solutions on the food surface or immersing food items into film-forming solutions, where the solidified zein turns into the coating. Dry cast zein films could also be used for covering foods or by placing these films above the food surfaces or inserting the food layers (Neo et al., 2013).

3 ALIPHATIC POLYESTERS FOR FOOD PACKAGING

3.1 Polylactic Acid

PLA is used for the packaging of foods because it is biodegradable, nonpoisonous, and compostable. It is extracted from starch and/or sugar. It is mechanically strong but plastic in nature, which leads to wider applications including biomedical, pharmaceutical, packaging, and so on (Hoque et al., 2005; Lagarón et al., 2016; Sharif et al., 2019). PLA has a specific structure of lactic acid (2-hydroxy propionic acid) and pendant a methyl group with the α-carbon. When PLA is mixed with natural antimicrobial agents such as hisin, silver zeolite, and lysozyme, it shows inhibiting effects against some defined microorganisms. Among these microorganisms *Escherichia coli, Listeria monocytogenes, Staphylococcus aureus*, and *Micrococcus lysodeikticus* are more significant. Again, native antimicrobial substances have been incorporated into coatings on the surface of the PLA to make them effective against spoilage and pathogenic microorganisms. PLA-based antimicrobial packaging with little or no preservatives show minimum microbial contamination. Carbon nanotubes, silica NPs, and nanoclays can widen the use of bio-based films like PLA films (Süfer, 2018).

3.2 Polyhydroxybutyrate

Polyhydroxybutyrate (PHB) is a natural thermoplastic polyester that possesses a good number of mechanical properties that are comparable to synthetic degradable polyester such as the poly L-lactides. It is a well-known biodegradable PHA and is a naturally occurring β-hydroxy acid (a linear polyester). Depending on the type of bacteria and the feed used to produce PHB, the general structure of the repeating units can vary. Typically, $-(CH_2)_n-CH_3-$ is the most naturally occurring PHAs (Bucci et al., 2005). Among many of the microorganisms that can accumulate PHB, *Ralstonia eutropha* is most widely studied, due to its large PHB accumulation ability. Other PHB accumulation microorganisms are *Halomonas boliviensis, Haloferax mediterranei, Bacillus megaterium*, and others (Chea et al., 2016). PHB-based polymers or composites are used extensively in the food industry for packaging purposes. Examples of bio-based food packaging used in industry today are shown in Fig. 13.12.

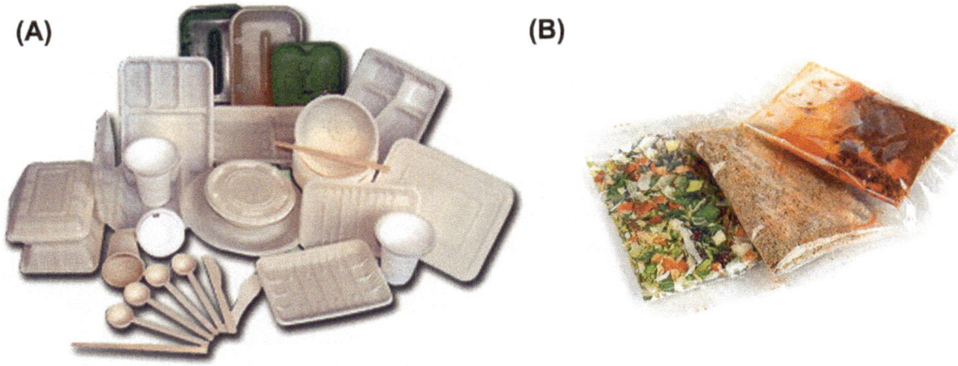

FIG. 13.12 Different types of bio-based food packaging: **(A)** tray, cups, spoons, etc., and **(B)** Transparent bags.

4 BIOPOLYMERS IN COSMETIC PACKAGING

Consumer awareness about the petrochemical-based plastic packaging, waste, and bioeconomy regulations is pushing us for the use of biodegradable and bio-based packaging materials. So, cosmetic packaging is also looking for an alternative solution other than petroleum-based polymer packaging. Modified bio-based and biodegradable polymer packaging is used for meeting the challenges of cosmetic preservation as well as sustainability and biodegradability. Some bio-based and bioperishable plastics used for cosmetic packaging are poly (lactic acid), PHAs, polysaccharides, etc. They are already used for both rigid and flexible packaging. Though we use bioplastics for creams, lipsticks, packaging, etc., we can use PLA, bio-PE (polyethylene), and bio-PET as renewable materials for packaging. Some cosmetics require several properties like biodegradability or recyclability as well as improved functionalities of the packaging material targeting special applications. Specific requirements are obvious for cosmetic packaging due to the intrinsic properties of cosmetic products. So, packaging must have the ability to contain the inherent properties of cosmetics products stopping deterioration prior to the expiration date and preventing degradation of cosmetic products from any kind of external contamination. Additives are added to increase protection from UV-light; they are specially added when the packaging needs to be transparent to prevent product conversion by photoactivated processes. Moreover, pigments and dyes are added to improve the esthetic properties of packaging. Moreover, other additives used in processing operations (processing catalyst) or as antioxidants are usually found in plastic materials (Gabor Daniela, 2012; Kumar and Kumar Gupta, 2012).

Plasticizers added to plastic materials must be selected considering the case of oxygen and water vapor exchange between product and packaging. The migration of elements from the packaging to the product can change product chemistry along with effectiveness and quality and in the reverse process (exchange of the elements between product and container). This may also alter the container's characteristics decreasing its resistivity as well as endurance. Nowadays, bioplastic packaging has a high and increasing demand. It is hoped that bioperishability and biopolymer matrices can provide some extra benefits over the petroleum-based polymer packaging, which has no biorecyclability (compostability or biodegradability) (Cinelli et al., 2019; Gabor Daniela, 2012; Kumar and Kumar Gupta, 2012).

Different types of biopolymers involved in cosmetic packaging:

4.1 PolyLactic Acid

PLA is a type of thermoplastic materials, and its manufacturing process involves injection molding, blow molding, extrusion, and thermoforming. PLA shows semicrystallinity structure with an approximate tensile modulus of 3 GPa, the tensile strength of 50–70 MPa, flexural Strength of 100 MPa, and elongation at breaking load of about 4%. Other properties of PLA are also quite significant. Like, it has glass transition temperature of 50–60°C, the crystalline temperature of 90–140°C, and MP of 150–180°C. The property that made PLA important in the packaging industry is time-dependent compost and its degradability. Other parameters like temperature, MW, crystallinity, impurities, and catalyst concentration can also change the overall properties of PLA. PLA film has good ultraviolet light barrier properties too.

PLA-based packaging basically gains attention recently due to its good rigidity and mechanical resistance. Sometimes polybutylene adipate-co-terephthalate like polyesters are used to enhance the toughness of PLA. For hot liquid filing, PLA-based polymeric materials show improved mechanical strength above the glass transition temperature of PLA. According to this consideration, it is one of the most important materials. When PLA is blended with cellulose acetate, it can provide very promising rigidity and full biodegradability depending on the quantity of acetylation degree of cellulose acetate and of its blending content. PLA reinforced plasticized cellulose acetate composite shows high toughness and high Young's modulus compared with raw PLA (Ullah et al., 2016). All of these properties are important in the packaging industry.

For flexible packaging, ductility and toughness of PLA can be increased using plasticizer. When plasticizers are added, glass transition temperature (T_g) of PLA, decreased yield stress, and higher elongation at break are observed at room temperature. But both properties are important for the improved flexibility of sheets and films. Some effective plasticizers are acetyl tributyl citrate, triacetin, and oligoethers, oligo lactic esters, oligo adipic esters for PLA, and its blends. Good barrier properties of PLA need to be particularly high for both rigid and flexible packaging that is generally used for liquid and pasty cosmetics. This property can be modified by different types of inorganic additives and clays in the PLA cosmetic bottles. The appropriate thickness for the PLA-based tubes is also important for cosmetic packaging.

PLA-based packaging durability for cosmetic products depends on oil and water. This packaging tends to hydrolyze easily. The shelf life of the packed products is influenced by this hydrolysis mechanism. PLA-based composites can be suitable for suitable cosmetic packaging. Exchange of substance from the packaging to the product can somewhat change its content.

4.2 Polyhydroxyalkanoates

PHAs are important as a packaging material for its high perishability in various situations. PHAs can be processed and formulated for use in many versatile applications such as packaging, paper coating, nonwoven fabrics, molded goods, adhesives, performance additives, and films. Bacteria and other cultivated biomass are the main sources of PHAs polymers. A variety of useful products are processed from them because their biodegradability and naturalness are useful for cosmetic packaging applications. One of the best-characterized and most widespread members of the PHAs family is poly(3-hydroxybutyrate) (PHB), 3-hydroxybutyrate homopolymer (Barletta et al., 2016). PHA monomeric unit is shown in Fig. 13.13.

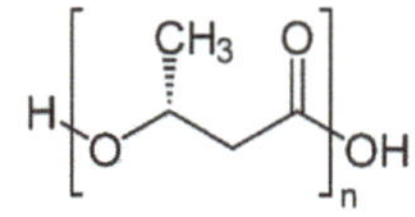

FIG. 13.13 Polyhydroxyalkanoate monomeric unit.

This polymer family has interesting properties for packaging compared with low-density polyethylene. PHB has low water vapor permeability, and it is generally used for mass shrinkage packaging and malleable intermediate containers for packaging of food (Ortega-Toro et al., 2016). Copolymers of hydroxybutyrate and hydroxy valerate incorporating poly(β-hydroxybutyrate-co-β-hydroxy valerate) (PHBγ), which is commercially marketed, have thermoplastic properties like polypropylene as well as good mechanical properties. Thus, the presence of 3Hγ or 4Hβ comonomers in the chains of PHA causes an extensive difference in mechanical behavior. By adding comonomer, increased toughness and decreased stiffness and tensile strength can be obtained. Today, PHAs are used as an effective alternative for some petrochemical-based polymers for similar physical and chemical characteristics.

Due to the lower production rate of PHA than PLA, it is comparatively expensive than PLAs, but it has several properties that make PHAs more useful. It has no harmful effect when it is applied in contact with skin, emits lower greenhouse gas, and has very high biodegradability as well as excellent biocompatibility in a different environment (Vartiainen et al., 2016). It shows biodegradability in the presence of many aerobic and anaerobic microorganisms (cyanobacteria, bacteria, and fungi). They can degrade PHAs in several environments like soil, industrial/domestic humus, freshwater, and several coastal ecosystems as raw material and polymeric matrix in biocomposites. So, PHA-based packaging is very encouraging and being tested in many sectors where biocompatibility is a major concern. Moreover, useful modification can make them properly durable and give them possible shelf life when used as packaging for cosmetic products (Cinelli et al., 2019).

4.3 Polysaccharides

Starch and cellulose by-products are mostly used polysaccharides in the field of cosmetic packaging. Along with that, recently chitosan and chitin are also exploited as active packaging due to their antimicrobial property. Raw cellulose has difficulty in use as a packaging material due to its highly hydrophobic nature and is consequently unsuitable as the packaging material as a moisture barrier is one of the basic characteristics required for packaging material. Along with that, high crystallinity makes them brittle and provides relatively low thermal degradability. Mostly recycled cellulose by-products are polysaccharides, made of linear chains of β-(1 → 4)-glycosidic moieties with hydroxypropyl, methyl, or carboxyl substituents (hydroxypropyl cellulose, hydroxypropyl methylcellulose, carboxymethyl cellulose, or methylcellulose).

It can be used as composite packaging material or in multilayer packaging materials even though materials have a low moisture barrier property (Kumar and Kumar Gupta, 2012; Liu et al., 2017). To make starch useful, it needs to be adjusted either by plasticization or by mixing it with other polymeric materials. Sometimes chemical and physical modification or linking different approaches can be useful for making starch processable. Starch-based materials consist of a blend of starch and other polymers including poly(vinyl alcohol), polycaprolactone, or poly (ethylene-co-vinyl alcohol). Starch derived thermoplastics have a wider range of industrial applications from extrusion application, blow molding, injection molding, film blowing, and forming. Though they are modified, they show limited resistance to water to water vapor because water swells. The starch portion of the wheat blends are used for film preparation and these packaging or films are good for only hydrophilic dry products (Jha et al., 2017; Sapper et al., 2019).

In general, chitosan is well-known for its special characteristics like lack of toxicity, oxygen barrier property, film formability, biodegradability, antifungal, and antioxidant characteristics. For this reason, chitosan is largely used in many applications. The antifungal properties of chitosan alone or combined with other antioxidants and essential oils like antifungal substances are largely investigated. Due to these properties, chitosan is generally used as a coating material on a PLA film for producing flexible packaging with bioperishability and antioxidants properties of different beauty products. Other benefits are the existence of antimicrobial substances of biopolymers to modify cell vitality and their generation. It increases the possibility of transforming some general hygienic elements promoting the health and beauty of the skin. Chitosan- and chitin-based packaging increases the antimicrobial and skin-regenerative properties of cosmetics as well as increases the product life. When active molecules are incorporated in chitin, nanofibrils along with their derivatives are used for preparing bioplastic active materials and surfaces that have potential applications in the cosmetic industry (Cinelli et al., 2019; Jipa et al., 2012; Pellicer et al., 2017; Schmitt and Turgeon, 2011).

Cosmetic packaging is not usually used more than once or reprocessed, compostable, or environmentally bioperishable materials for cosmetic packaging can be a sustainable plan. It can be a balance of challenges and opportunities for sustainability. Inventive bio-based and compostable materials are already advanced and suitable to produce cosmetic packaging while some have remained under development with very promising characteristics and perspective. Awareness and support from the consumers are important for the development along with sustainability in cosmetic packaging. Moreover, policymakers should take steps for saving our environment. Examples of packaging made of biopolymer used in cosmetic industry are demonstrated in Fig. 13.14.

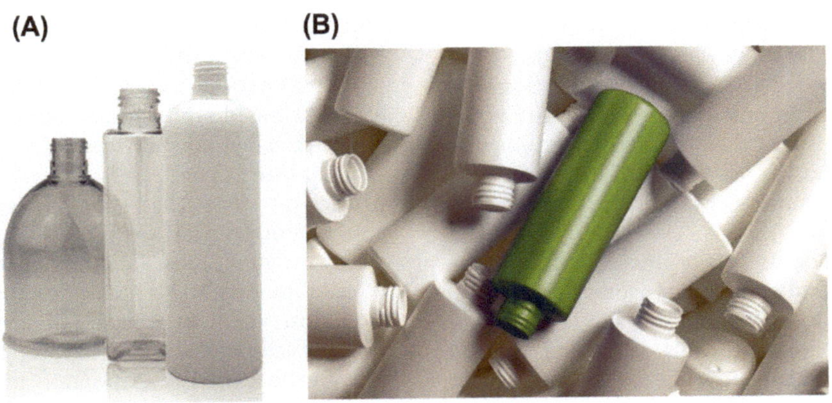

FIG. 13.14 Different types of bio-based cosmetic packaging: **(A)** bottles and **(B)** tubes.

5 BIOPOLYMERS IN PHARMACEUTICAL PACKAGING

Packaging requirements from pharmaceutical products are complex than those of other products. Pharmaceutical packaging is a combination of art and science of preserving and protecting the pharmaceutical product from damage by enclosing them (Kumar and Kumar Gupta, 2012). According to the "Good Manufacturing Practice" for the drug industry, the high-quality packaging method and packaging materials must be used, which have the following characteristics:

- Should not allow the penetration of any unusual substances.
- Must not react with the medicine ingredients.
- Must be strong enough to protect the inside environment from the influences of external environmental factors such as dampness, heat, moisture, light, oxygen, biological and mechanical contamination, etc.

6 BIODEGRADABLE PHARMACEUTICAL PACKAGING MATERIALS

Paper is a very well-known packaging material which is biodegradable and recycled easily. So, they are suitable to produce eco-friendly pharmaceutical packaging material. Lightweight, the permeability of gases and moisture, easy tear ability, etc., are the main advantages of paper packaging. There are different types of paperboard and some of them are as follows (Ortega-Toro et al., 2016):

(a) **White paperboard**: Chemically bleached pulp is usually used to manufacture white paperboards. To adjust some basic properties like heat sealability, lamination is applied on this board. Usually, PE or waxes are used for lamination. The laminated white paperboards are used for primary packaging. Sometimes, the internal coating of the carton can also be done by using white paperboards.

(b) **Chip paperboards**: Chip paperboard is usually used to make an outer covering of a carton. This paperboard is made of recycled paper. By using white paperboards, the lining of the chipboard is done, which improves its mechanical strength and appearance.

(c) **Solid paperboards**: It is composed of multiple layers of boards. Bleached sulfate boards are commonly used to manufacture solid paperboards. These boards are generally laminated with PE. Edible films are used to make the covering water or another liquid resistant.

(d) **Fiber paperboards**: There are two types of fiber paperboards such as corrugated and solid fiber paperboards. Corrugated paperboard is composed of two layers of craft paper, which has corrugated material between them. The fiber paperboard possesses high resistance against impact. It is commonly used for packaging materials that are transported by ships. The other one is solid paperboard that is composed of two layers, an inner and outer layer. White paperboards constitute the inner layer and craft paperboards fabricate outer layer. Lamination of solid paperboard can be done by bioplastics for keeping the packed product in dry form.

(e) **Cork**: Cork is extracted from the bark of an oak tree. The chemical inertness of cork makes it suitable for pharmaceutical packaging. It also does not impart flavor or odor to the packed product.

There are many other sources for raw materials of biodegradable pharmaceutical packaging. Some of them are discussed below:

Starch: Starch is used to prepare biodegradable pharmaceutical packaging material with a controllable lifetime. Basically, starch is formed into sheets and used for protecting the glass dishes, trays, etc. Starch-based bioplastics are mainly bags and sacs that are used as flexible and rigid packaging. In preparing starch packaging material, plasticizers are used to make starch material less brittle. However, dry starch is not suitable for packaging as it is not thermoplastic. Along with this, its granular form is also not suitable for the plastic industry as it creates process difficulties during injection molding and extrusion. Biodegradable plasticizers such as glycol, polyether, urea, and polyhydroxy components are used to minimize the brittleness. The other benefit of the plasticizer is to inhibit the growth of microbes. Four types of starch-based polymers are

- Starch-based thermoplastic
- Starch-based polyvinyl alcohol
- Starch-based synthetic aliphatic polyester
- Starch-based polybutylene succinate

Cellulose: Cellulose or its derivative-based bioplastics are very attractive for packaging in biomedical and pharmaceutical applications. It is found abundantly in nature. There are various commercial products of cellulose derivatives including ethylcellulose, methylcellulose, cellulose acetate, carboxyl methylcellulose, hydroxy cellulose, and hydroxypropyl cellulose. Cellulose acetate is used as a laboratory and pharmaceutical packaging material.

Xylan: This type of carbohydrate can be found in the different types of plant cell walls and algae. This biodegradable xylan can be exploited for preparing eco-friendly pharmaceutical packaging material.

Chitin: Invertebrates, insects, and yeast are the main sources of chitin. Chitin possesses antimicrobial characteristics that make its use in packaging material for the pharmaceutical sector widely. Heavy metal ion impurities are absorbed by chitin spontaneously, which makes it perfect to prepare as edible packaging material. Chitosan is another derivative of chitin that can also be used in pharmaceutical packaging. Usually, biodegradable laminations are prepared by chitosan cellulose. The chitin is a hard, white, nitrogenous, inelastic polysaccharide. Biocompatible chitin is used in various biomedical applications. Chitosan as a nature-friendly material can be used in wastewater purification, paper finishing as a proton-conducting polymer in batteries, cosmetics for the preservation fruits, and to prepare packaging films. Polyethylene glycol (0.25%–0.5%) is added in chitosan-modified films to reduce water vapor transmission from the film.

Protein: Packaging materials coming from protein are processed by modifying their side chains in structure. For making a biodegradable packaging film, cross-linking is done between natural protein and the synthetic monomer unit. Agricultural feedstocks are commercially used for making protein-based packaging materials. Renewable plants and animals are two easily available sources of protein. Proteins obtained from the plant include gluten, zein, soy, etc., whereas animal-based protein includes whey, casein, gelatin, collagen, keratin, etc. The followings are some proteins that are commonly used to prepare pharmaceutical packaging material.

Gluten: Corn and wheat are the main sources of gluten. This type of protein possesses plastic-like characteristics (i.e., resistance to water). It is cheap in cost and can be made easily useable in the manufacturing of edible packaging films.

Zein: Zein is an alcohol-soluble protein. It is used for the coating purpose of pharmaceutical products and in the manufacturing of biodegradable packaging materials.

Soy: There are three types of commercial soy. These are soy isolate, soy concentrates, and SF. It is exploited for manufacturing biodegradable plastics and films. This is also used as a coating substance for pharmaceutical products.

Casein: It is a type of animal-oriented protein. Milk is the main source for casein, which can be easily processed out. Casein is used to manufacture a thermoset type of plastic. Due to its good adhesive ability, casein is used for labeling purposes.

Whey: Whey is a coproduct of the cheese industry. Packaging films and edible coating are made from whey on pharmaceutical products.

Gelatin: Bones and skin of animals are the two main sources of gelatin. Gelatins are used to prepare microspheres and to microencapsulate vitamins. A film of gelatin is usually used to prepare to encapsulate shells and in the manufacture of tablets.

Collagen: Collagen is a fibroid protein; animal tendons, skin, and bones are the main sources of it. Packaging materials with good properties are made from collagen.

Keratin: Animal hair and nails are the main sources of keratin. Keratin is the cheapest protein among animal-based proteins. Generally, bioplastics are prepared from keratin.

PLA: PLA-based biodegradable plastics can also be used in the pharmaceutical industry.

Pullulan: Pullulan is a viscous polysaccharide that consists of maltotriose units. Pullulan-made films are colorless, transparent, and oil resistant. It can also be used as packaging films for pharmaceutical industries.

Miscellaneous biodegradable polymers: PHA, the thermoplastic polyester, can be produced from a simple fermentation process. Its derivative PHB can also be suitable as an alternative to synthetic plastic in the pharmaceutical industries (Lyashenko et al., 2018). Biodegradable polymers are useful for shelf life extension of pharmaceutical products. Nowadays, multicomponent coatings and films have gained attention due to their joint functional attribute. Eco-friendly packaging materials should follow drug packaging legislation, which is a major important requirement for them (Cristina et al., 2015). Specifically, these should not be incompatibility between drug products and bioplastic packaging material. Together these desired types of features are strictly maintained in packaging material quality and eco-safety. Some applications of biopolymers used in pharmaceutical industry are presented in Fig. 13.15.

7 CONCLUSIONS

Over the last few decades, polymeric materials have had hype in the global industry due to its adaptability, durability, and price. Nowadays, we can hardly imagine a product or service without polymeric constituent. With the progress of time, many synthetic polymers with excellent properties have been developed form fossil fuels. Unfortunately, environmental incompatibility hindered their widespread use since the natural

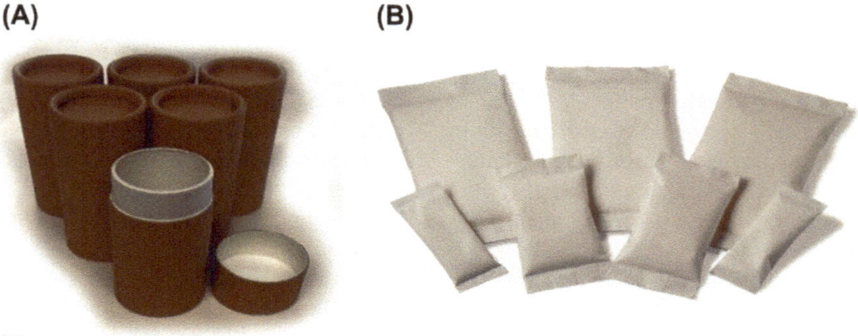

FIG. 13.15 Different types of bio-based pharmaceutical packaging: **(A)** cylindrical bottles with cap and **(B)** closed bags.

recycling system is unusual for the synthetic polymer. Global awareness regarding the negative impact of synthetic polymers on the environment leads the path for developing biopolymers from natural sources. Eco-friendly biopolymer packaging materials are expensive in some cases, but they can be a good alternative to synthetic plastic packaging by offering potential ways to use waste natural products and zero negative impact on the environment. Moreover, composite materials with two or more biodegradable agents offer superior chemical, physical and mechanical properties compared to noncomposite biomaterial property and even to synthetic polymers. So, though already used commercially, there is much room for further research for sustainable and low-cost packaging solutions for food, cosmetic, and pharmaceutical industries. We hope if the current interest in bio-based polymer continues to grow and different agencies fund the research eventually, the world will be free from the cursive effect of synthetic polymers.

REFERENCES

Alizadeh, A., Behfar, S., 2013. Properties of collagen based edible films in food packaging: a review. Annals of Biological Research 4, 253–256.

Arcan, İ., Boyacı, D., Yemenicioğlu, A., 2017. The use of zein and its edible films for the development of food packaging materials. Reference Module in Food Science 1–11. https://doi.org/10.1016/b978-0-08-100596-5.21126-8.

Balasubramanian, A., Rosenberg, L., Yam, K., Chikindas, M., 2009. Antimicrobial packaging: potential vs. reality—a review. Journal of Applied Packaging Research 3, 193–221.

Barletta, M., Puopolo, M., Tagliaferri, V., Vesco, S., 2016. Graphene-modified poly(lactic acid) for packaging: material formulation, processing and performance. Journal of Applied Polymer Science 133, 1–11. https://doi.org/10.1002/app.42252.

Berekaa, M.M., 2015. Nanotechnology in food industry; advances in food processing, packaging and food safety, International Journal of Current Microbiology and Applied Sciences 4 (5), 345–357.

Bucci, D.Z., Tavares, L.B.B., Sell, I., 2005. PHB packaging for the storage of food products. Polymer Testing 24, 564–571. https://doi.org/10.1016/j.polymertesting.2005.02.008.

Byun, Y., Kim, Y.T., 2013. Utilization of Bioplastics for Food Packaging Industry, Innovations in Food Packaging, second ed. Elsevier Ltd. https://doi.org/10.1016/B978-0-12-394601-0.00015-1.

Chauhan, A., Chauhan, P., 2015. Natural fibers and biopolymer. Journal of Chemical Engineering & Process Technology s6, 1–4. https://doi.org/10.4172/2157-7048.s6-001.

Chea, V., Angellier-Coussy, H., Peyron, S., Kemmer, D., Gontard, N., 2016. Poly(3-hydroxybutyrate-co-3-hydroxyvalerate) films for food packaging: physical-chemical and structural stability under food contact conditions. Journal of Applied Polymer Science 133, 1–8. https://doi.org/10.1002/app.41850.

Cinelli, P., Coltelli, M.B., Signori, F., Morganti, P., Lazzeri, A., 2019. Cosmetic packaging to save the environment: future perspectives. Cosmetics 6, 26. https://doi.org/10.3390/cosmetics6020026.

Cristina, M., Arantzazu, V., Marina, R., Nuria, B., Carmen del, G.M., Alfonso, J., 2015. Active edible films: current state and future trends. Journal of Applied Polymer Science 133. https://doi.org/10.1002/app.42631.

Farris, S., Unalan, I.U., Introzzi, L., Fuentes-Alventosa, J.M., Cozzolino, C.A., 2014. Pullulan-based films and coatings for food packaging: present applications, emerging opportunities, and future challenges. Journal of Applied Polymer Science 131, 1–12. https://doi.org/10.1002/app.40539.

Franceschi, E., Cezaro de, A., Ferreira, S.R.S., Kunita, M.H., Muniz, E.C., Rubira, A.F., Oliveira, J.V., 2014. Biopolymers based on renewable resources: a review. The Open Chemical Engineering Journal 5, 11–20. https://doi.org/10.2174/1874123101004010011.

Gabor Daniela, T.O., 2012. Biopolymers used in food packaging: a review. Food Technology 16, 3–19.

Hoque, M.E., Hutmacher, D.W., Feng, W., Li, S., Huang, M.-H., Vert, M., Wong, Y.S., 2005. Fabrication using a rapid prototyping system and in vitro characterization of PEG-PCL-PLA scaffolds for tissue engineering. Journal of Biomaterials Science, Polymer Edition 16, 1595–1610.

Hoque, M.E., Ye, T.J., Yong, L.C., Mohd Dahlan, K., 2013. Sago starch-mixed low-density polyethylene biodegradable polymer: synthesis and characterization. Journal of Materials 1–7. https://doi.org/10.1155/2013/365380. Article ID 365380.

Iurciuc, C., Savin, A., Lungu, C., Martin, P., Popa, M., 2016. Gellan. Food applications. Cellulose Chemistry and Technology 50, 1–13.

Jha, S., Rohilla, P., Singh, K., 2017. Starch based packaging materials: a review. International Journal of Research and Analytical Reviews 4, 2348–2350.

Jipa, I.M., Stroescu, M., Stoica-Guzun, A., Dobre, T., Jinga, S., Zaharescu, T., 2012. Effect of gamma irradiation on biopolymer composite films of poly(vinyl alcohol) and bacterial cellulose. Nuclear Instruments and Methods in Physics Research Section B: Beam Interactions With Materials and Atoms 278, 82–87. https://doi.org/10.1016/j.nimb.2012.02.024.

Kaplan, D.L., 1998. Introduction to biopolymers from renewable resources. Biopolymers from Renewable Resourses 1–29. https://doi.org/10.1007/978-3-662-03680-8_1.

Ketabchi, M.R., Khalid, M., Ratnam, C.T., Manickam, S., Walvekar, R., Hoque, M.E., 2016. Sonosynthesis of cellulose nanoparticles (CNP) from kenaf fiber: effects of processing parameters. Fibers and Polymers 17, 1352–1358. https://doi.org/10.1007/s12221-016-5813-4.

Kumar, S., Kumar Gupta, S., 2012. Applications of biodegradable pharmaceutical packaging materials: a review. Middle-East Journal of Scientific Research 12, 699–706. https://doi.org/10.5829/idosi.mejsr.

Lagarón, J.M., López-Rubio, A., José Fabra, M., 2016. Bio-based packaging. Journal of Applied Polymer Science 133. https://doi.org/10.1002/app.42971.

Lalit Jajpura, A.R., 2015. The biopolymer sericin: extraction and applications. Journal of Textile Science & Engineering 05, 1–5. https://doi.org/10.4172/2165-8064.1000188.

Larotonda, F.D.S., Torres, M.D., Gonçalves, M.P., Sereno, A.M., Hilliou, L., 2016. Hybrid carrageenan-based formulations for edible film preparation: benchmarking with kappa carrageenan. Journal of Applied Polymer Science 133, 1–10. https://doi.org/10.1002/app.42263.

Lazic, V.L., Hromiš, N.M., 2018. For Food Shelf-Life Prolongation. https://doi.org/10.1016/B978-0-12-811449-0/00008-6.

Liu, L., 2006. Bioplastics in food Packaging: innovative technologies for biodegradable packaging. Environmental Protection 1–13.

Liu, X., Song, R., Zhang, W., Qi, C., Zhang, S., Li, J., 2017. Development of eco-friendly soy protein isolate films with high mechanical properties through HNTs, PVA, and PTGE synergism effect. Scientific Reports 7, 1–9. https://doi.org/10.1038/srep44289.

Lyashenko, V., Sotnik, S., Ma Babker, A., 2018. Features of packaging from polymers in pharmaceutics. Saudi Journal of Medical and Pharmaceutical Sciences, 4 (2), 166–174. https://doi.org/10.21276/sjmps.2018.4.2.2.

Mangolim, C.S., Silva, T.T.D., Fenelon, V.C., Koga, L.N., De Ferreira, S.B.S., Bruschi, M.L., Matioli, G., 2017. Description of recovery method used for curdlan produced by agrobacterium sp. IFO 13140 and its relation to the morphology and physicochemical and technological properties of the polysaccharide. PLoS One 12, 1–19. https://doi.org/10.1371/journal.pone.0171469.

Marsh, K., Bugusu, B., 2007. Food packaging - roles, materials, and environmental issues: scientific status summary. Journal of Food Science 72. https://doi.org/10.1111/j.1750-3841.2007.00301.x.

Neo, Y.P., Ray, S., Jin, J., Gizdavic-Nikolaidis, M., Nieuwoudt, M.K., Liu, D., Quek, S.Y., 2013. Encapsulation of food grade antioxidant in natural biopolymer by electrospinning technique: a physicochemical study based on zein-gallic acid system. Food Chemistry 136, 1013–1021. https://doi.org/10.1016/j.foodchem.2012.09.010.

Nuge, T., Tshai, K.Y., Lim, S.S, Nordin, N., Hoque, M.E., 2020. Characterization and optimization of the mechanical properties of electrospun gelatin nanofibrous scaffolds. World Journal of Engineering 17, 12–20. https://doi.org/10.1108/WJE-04-2019-0119.

Nuge, T., Tshai, K.Y., Lim, S.S., Nordin, N., Hoque, M.E., 2017. Preparation and characterization of CU-, FE-, AG-, ZN- and NI- doped gelatin nanofibers for possible applications in antibacterial nanomedicine. Journal of Engineering Science & Technology 12, 68–81.

Ortega-Toro, R., Collazo-Bigliardi, S., Talens, P., Chiralt, A., 2016. Influence of citric acid on the properties and stability of starch-polycaprolactone based films. Journal of Applied Polymer Science 133, 1–16. https://doi.org/10.1002/app.42220.

Paridah, M.T, Moradbak, A., Mohamed, A.Z., Owolabi, F. abdulwahab taiwo, Asniza, M., Abdul Khalid, S.H.P., 2016. We Are IntechOpen , the World ' S Leading Publisher of Open Access Books Built by Scientists , for Scientists TOP 1 %. Intech, vol. 13. https://doi.org/10.5772/57353.

Peelman, N., Ragaert, P., De Meulenaer, B., Adons, D., Peeters, R., Cardon, L., Van Impe, F., Devlieghere, F., 2013. Application of bioplastics for food packaging. Trends in Food Science & Technology 32, 128–141. https://doi.org/10.1016/j.tifs.2013.06.003.

Pellicer, E., Nikolic, D., Sort, J., Baró, M.D., Zivic, F., Grujovic, N., Grujic, R., Pelemis, S., 2017. Advances in applications of industrial biomaterials. Advances in Applications of Industrial Biomaterials 1–214. https://doi.org/10.1007/978-3-319-62767-0.

Ramos, M., Valdés, A., Beltrán, A., Garrigós, M., 2016. Gelatin-Based Films and Coatings for Food Packaging Applications, Coatings. Elsevier. https://doi.org/10.3390/coatings6040041.

Ramos, Ó.L., Pereira, R.N., Cerqueira, M.A., Martins, J.R., Teixeira, J.A., Malcata, F.X., Vicente, A.A., 2018. Bio-based nanocomposites for food packaging and their effect in food quality and safety. In: Food Packaging and

Preservation. Elsevier, pp. 271–306. https://doi.org/10.1016/B978-0-12-811516-9.00008-7.

Reddy, K.R., Rajgopal, K., Maheswari, C.U., Lakshmi Kantam, M., 2006. Chitosan hydrogel: a green and recyclable biopolymer catalyst for aldol and Knoevenagel reactions. New Journal of Chemistry 30, 1549–1552. https://doi.org/10.1039/b610355c.

Ross-Murphy, S.B., 1995. Structure–property relationships in food biopolymer gels and solutions. Journal of Rheology 39, 1451–1463. https://doi.org/10.1122/1.550610.

Sapper, M., Talens, P., Chiralt, A., 2019. Improving functional properties of cassava starch-based films by incorporating xanthan, gellan, or pullulan gums. International Journal of Polymer Sciences 2019, 1–8. https://doi.org/10.1155/2019/5367164.

Schmid, M., Müller, K., 2018. Whey Protein-Based Packaging Films and Coatings, Whey Proteins. Elsevier Inc. https://doi.org/10.1016/b978-0-12-812124-5.00012-6

Schmitt, C., Turgeon, S.L., 2011. Protein/polysaccharide complexes and coacervates in food systems. Advances in Colloid and Interface Science 167, 63–70. https://doi.org/10.1016/j.cis.2010.10.001.

Sharif, A., Mondal, S., Hoque, M.E., 2019. Polylactic acid (PLA)-based nanocomposites: processing and properties. In: Sanyang, M.L., Jawaid, M. (Eds.), Bio-Based Polymers and Nanocomposites : Preparation, Processing, Properties & Performance. Springer International Publishing, Cham, pp. 233–254. https://doi.org/10.1007/978-3-030-05825-8_11.

Shipra, T., K, M.G., K, D.P., 2008. Chitosan based antimicrobial films for food packaging applications. e-Polymers 8, 1–7. https://doi.org/10.1515/epoly.2008.8.1.1082.

Süfer, Ö., 2018. Poly (lactic acid) films in food packaging systems. Food Science and Nutrition Technology 2. https://doi.org/10.23880/fsnt-16000131.

Swain, S.N., Biswal, S.M., Nanda, P.K., Nayak, P.L., 2004. Biodegradable Soy-Based Plastics : Opportunities and Challenges 12.

Tang, X.Z., Kumar, P., Alavi, S., Sandeep, K.P., 2012. Recent advances in biopolymers and biopolymer-based nanocomposites for food packaging materials. Critical Reviews in Food Science and Nutrition 52, 426–442. https://doi.org/10.1080/10408398.2010.500508.

Tolstoguzov, V., 2003. Some thermodynamic considerations in food formulation. Food Hydrocolloids 17, 1–23. https://doi.org/10.1016/S0268-005X(01)00111-4.

Tolstoguzov, V.B., 2000. The importance of glassy biopolymer components in food. Nahrung/Food 44, 76–84. https://doi.org/10.1002/(sici)1521-3803(20000301)44:2<76::aid-food76>3.3.co;2-4.

Trinetta, V., Cutter, C.N., 2016. Pullulan: A Suitable Biopolymer for Antimicrobial Food Packaging Applications, Antimicrobial Food Packaging. Elsevier Inc. https://doi.org/10.1016/B978-0-12-800723-5.00030-9

Tripathi, S., Mehrotra, G.K., Dutta, P.K., 2008. Chitosan based antimicrobial films for food packaging applications. e-Polymers 093, 1–7. https://doi.org/10.1515/epoly.2008.8.1.1082.

Ullah, H., Santos, H.A., Khan, T., 2016. Applications of bacterial cellulose in food, cosmetics and drug delivery. Cellulose 23, 2291–2314. https://doi.org/10.1007/s10570-016-0986-y.

Vartiainen, J., Shen, Y., Kaljunen, T., Malm, T., Vähä-Nissi, M., Putkonen, M., Harlin, A., 2016. Bio-based multilayer barrier films by extrusion, dispersion coating and atomic layer deposition. Journal of Applied Polymer Science 133, 1–6. https://doi.org/10.1002/app.42260.

Index

Note: Page numbers followed by "f" indicate figures and "t" indicate tables.

Kolibri 360 GmbH
Ferdinand-Jühlke-Straße 7
99095 Erfurt, Deutschland
produktsicherheit@kolibri360.de